T0140178

Measuring and Understanding Complex Phenomena

Rainer Bruggemann • Lars Carlsen • Tugce Beycan
Christian Suter • Filomena Maggino
Editors

Measuring and Understanding Complex Phenomena

Indicators and their Analysis in Different Scientific Fields

 Springer

Editors
Rainer Bruggemann
Leibniz-Institute of Freshwater Ecology
and Inland Fisheries
Berlin, Germany

Lars Carlsen
Awareness Center
Trekroner, Roskilde amt, Denmark

Tugce Beycan
Institut de Sociologie
Université de Neuchâtel
Neuchâtel, Neuchatel, Switzerland

Christian Suter
Institut de Sociologie
Universite de Neuchatel
Neuchatel, Switzerland

Filomena Maggino
Dipartimento di Scienze Statistiche
Sapienza University of Rome
Roma, Italy

ISBN 978-3-030-59685-9   ISBN 978-3-030-59683-5   (eBook)
https://doi.org/10.1007/978-3-030-59683-5

This Springer imprint is published by the registered company Springer Nature Switzerland AG
The registered company address is: Gewerbestrasse 11, 6330 Cham, Switzerland

Preface

The international conference in Neuchatel, Switzerland, 2018, about partial order and its applications motivated to edit a new book representing the state of the art in the development and applications of partial order methodology. The very idea is to analyse multi-indicator systems. Part of the lectures presented in Neuchatel are now part of this book. However, due to many other actual contributions, this book is not a proceeding of the conference of 2018, but a monograph with a focus on indicators and their analysis.

Consequently, in this book (beside an introductory text) the reader will find

- Five chapters specifically concerned with indicators
- Six chapters where the methodological aspect of applied partial order is the main topic
- Three chapters with a sociological background
- Two chapters with an environmental background
- Finally, two chapters where software aspects are in the foreground

An introductory chapter may be helpful for interested scientists to understand how partial order in combination with multi-indicator systems can be applied. Furthermore, a brief overview about all 18 chapters is given.

For the future it is hoped that more scientists will be interested in the exciting field of applied partial order.

Schwandorf	Rainer Bruggemann
Roskilde	Lars Carlsen
Neuchatel	Tugce Beycan
Neuchatel	Christian Suter
Rome	Filomena Maggino

Indicators and Partial Orders – An Introduction

Role of Indicators

Our world will increasingly be more and more complex. Hence, evaluation of the state (in order to find decisions for management in the future) will be correspondingly difficult. In many cases deterministic mathematical models can be sufficiently sophisticated to support decisions. In the evaluation of chemicals, such as EUSES (Heidorn et. al. 1997) or the former E4CHEM (Bruggemann and Drescher-Kaden 2003) are suitable examples. Even agent-based models, cannot encompass all eventualities of our daily life. (Agent based modelling within a general context is described in Wikipedia, 2020; within geographical simulations in Castle and Crooks, 2006 and within an ecological context in Hüning et al. 2016.) Hence, one can find everywhere indicators, e.g., Fragile State Index (FSI) 2019, (Carlsen and Bruggemann 2013, 2014, 2017) or the Human Environment Interface Index (HEI), Environment Performance Index (EPI) (for both within the Partial order context, see (Bruggemann and Patil 2011), World happiness Index (Helliwell et al. 2019), Human development Index (Human Development Report 2019), Gender equality Index (Gender Equality Index 2019), Bruggemann and Carlsen 2020, Sustainable Cities Index (Sustainable Cities Index 2018), Sustainable Society Index (Europe Sustainable Development Report 2019), Food Sustainable Index (Barilla 2019) or indicator helping to measure the quality of life in cities (El Din et al. 2013), just to mention some typical indicators. The general problem is, how to quantify these indicators (examples are mentioned above). Often sub-indicators (we will call them "preliminary indicators") are defined which can be measured, or estimated by mathematical models or for which an ordinal scale is obvious. In the next step, this series of indicators typically is condensed to form a single quantity, sometimes called 'the index', or more precisely the composite indicator. In fact, this procedure, defining subsystems of indicators, leads to hierarchies of indicator systems, for example, that applied for the definition of the food index (Barilla 2019)).

The mathematical problem is how to carry out this condensation, or aggregation step, in the most sensible way possible. Bruggemann and Patil (2011) denoted the series of preliminary indicators a multi-indicator system (MIS). The information within a certain MIS is often important within a holistic point of view (see for instance Maggino and Zumbo 2012). The aggregation, independent of which method is applied, must be more or less considered as an averaging. Thus, it seems to be appropriate to evaluate the MIS as an interim aspect by mathematical methods, which are able to analyse multiple indicators with respect to the objective under which the MIS was constructed. The mathematical method of partial order theory is very helpful in this aspect, and, therefore, indicators and partial order are closely interrelated when an evaluation by ranking is wanted. Clearly, the partial order methodology is not the only possibility for studying an MIS (see, e.g., (Brans and Vincke 1985; Figueira et al. 2005; Colorni et al. 2001; Munda 2008; Munda and Nardo 2009; Roy 1972; Roy and Vanderpooten 1996; Maggino 2017)).

Here, however, the interplay of MIS and partial order is the main topic.

Partial Order Methodology

When one takes a closer look at the mathematics of partial ordering, it is closely, although not exclusively related to the regime of indicators. Partial ordering is a theory of binary relations and is as such especially well-suited for those indicators, which are ordinal in nature. The reason is that partial order is mathematically deeply intertwined with

- Graph theory
- Combinatorics
- Algebra

but not with numerical evaluation in the field of real numbers, see, for instance, (Trotter 1992). In the following, the three items are described in more detail.

Graph Theory

One of the most important visualization techniques of partial orders is the Hasse diagram. The Hasse diagram is a transitively reduced, acyclic digraph. This characterization may be enough for mathematicians, but not for scientists interested in applications. Thus, a few more details are given here.

Partial order is a binary relation among elements x_i and x_j of a set X which can be interpreted as 'better than', e.g., $x_i > x_j$. This relation obeys three axioms:

- Reflexivity, i.e., an element can be compared with itself.
- Antisymmetry, i.e., if an element x is 'better' than an element y, then y cannot be better than x, unless x and y are identical.

- Transitivity, i.e., if x is 'better' than y, and y is 'better' than z, then x is 'better' than z. A classical counterexample is the tournament. One may define 'better than' as team x beats team y. However, although team y may beat team z, it cannot be excluded that team z beats team x, which is a violation of the transitivity. On the other hand, when the order relation is associated with numerical, ordinal indicator values, the order relation between two elements is governed by the numerical relation between the indicator values. Thus, the elements can be ordered, i.e., fulfilling the axiom of transitivity.

If two elements of a set X have an order relation, one can define two vertices for the elements and connect them. Because the relation is oriented, the orientation for the two elements is indicated by an arrow and the relation can be described by our usual symbol '<'. When this recipe is performed for all elements of a set, a directed graph is obtained. When the order relation is based on only one single indicator, then a complete – linear – order is developed and each of the two elements of X are connected by an arrow. When there are three elements x, y, z and it is found $x < y$ and $y < x$, then transitivity demands that $x < z$. Hence, for most applications the arrow for $x < z$ can be omitted, as this relation follows due to the transitivity. The process of eliminating arrows is called a transitive reduction. Furthermore, a sequence of arrows such as $x_0 < x_1, x_1 < x_2, \ldots, x_{n-1} < x_n$, but $x_n < x_0$ is obviously not possible as it would be a violation of the transitivity as the transitive reduction would cause a cyclic graph. Eventually, the arrows can be replaced by simple lines, when the orientation is governed by the vertical position in the drawing plane. The resulting graph is called a Hasse diagram. Hasse diagrams or comparability graphs (graphs of the order relation, however without an orientation) can be analysed theoretically. Note that a sequence of lines which can be followed strictly upwards or downwards may be called an order theoretical connection, the set of objects within an order theoretical connection is called a chain.

An analysis can, e.g., investigate whether or not subsets of X dominate others, or whether subsets of X are strikingly not connected or only weakly connected with other parts of the graph. It is clear that 'weakly' needs a definition. Here it is used in the sense of 'only few connections'. As an example of a Hasse diagram, we can look at the development in Germany (2008–2015) in switching to more sustainable energy according to the UN Sustainable Development Goal No. 7, using three indicators (Table 1); for details see (Europe Sustainable Development Report 2019).

Instead of observing three line graphs for each indicator, the Hasse diagram shows at once some essential facts:

(1) All three indicators are not decreasing in their values for the time evolution: 2008-2009-2011-2012-2015. Other time series can be found, where the indicator values are simultaneously non-decreasing. One can see that for these special set of years, the pattern of indicator values is co-monotone with the time. Such subsets of objects, mutually comparable are called chains.
(2) 2010, 2011 and 2013 cannot be compared, because of a counter current development of indicator values. They are connected (in a general graph theoretical

Table 1 The three indicators

Indicator	Short	Description	Direction
sdg7_warm	sdg7_w	Population unable to keep homes adequately warm (%)	Low better
sdg7_eurenew	sdg7_e	Share of renewable energy in gross final energy consumption (%)	High better
sdg7_co2twh	sdg7_co2	CO_2 emissions from fuel combustion per electricity output (MtCO2/TWh)	Low better

context, but not order theoretically). These three objects are members of a so-called antichain.

(3) The group {2010, 2012} has no order theoretical connections with {2013}. The identification of the reasons in terms of indicator values is one main task in the applications of Hasse diagrams.

Combinatorics

Combinatorics comes mainly into play when directed graphs of the order relations are extended to form graphs with more connections maintaining the already given ones. This enrichment process can be continued until a complete order is obtained. However, when the Hasse diagram has elements of X that are not in an order relation, then the enrichment process delivers a set of complete orders, i.e., the set of linear extensions. However, the generation of linear extensions from a given Hasse diagram is computationally extremely difficult. Here combinatorics helps to find algorithms or even to find closed formulas. These, e.g., play an important role in an approximation, known as a local partial order model. An example would be the Hasse diagram (Fig. 1). It is possible to extend the graph to a linear order, where the sequence of years follows its natural order.

Algebra

It seems to be plausible to try to understand empirical partial orders as being composed of simpler graph structures. Any two partially ordered sets (posets) can be combined by following strict composition rules. These composition rules, such as addition, multiplication and disjoint union, have only little to do with the operations known for numbers. Nevertheless, this kind of composition is an important guideline to understand empirical posets. A remarkably richer algebra is obtained, when the order relations obeys additional requirements. The crucial concept is the uniqueness. Within an empirical poset, two elements of a set X can be in order relation to several others. However, when the additional requirement is uniqueness, then any

Fig. 1 Hasse diagram of
Germany, years 2008–2015.
Details in a publication,
submitted

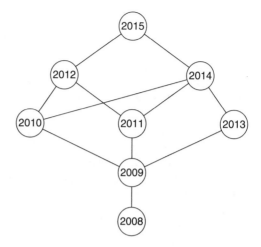

two elements are downwards and upwards, respectively, related to only one other
graph theoretical neighbored element. Such posets are called lattices and a special
realization is the formal concept analysis, deeply studied by the school of Wille
(Ganter and Wille 1986; Ganter 1987; Ganter and Wille 1996) and Kerber (2017).
The resulting lattices, formal concept lattices, are powerful tools in the analysis
of multi-indicator systems, especially when the indicators can only take discrete
numbers. The extension of the so-called formal concept analysis to indicators,
having continuous data in concept, bears additional theoretical difficulties; see, e.g.,
(Kerber 2017).

When data are metric data, the obvious question is, how to deal which such data
and what is the role of order relations compared to powerful statistical methods, such
as correlation or regression analyses, principal component and cluster analyses, just
to mention a few tools most often used in (multivariate) statistics.

When data are measured, then automatically data uncertainty comes into play.
By comparing partial order and (conventional) statistical tools it should initially
be made clear that partial order as a method to analyse data clearly belongs to
statistics. So why is a discussion needed? The reason is that multivariate statistics is
commonly associated with tools, which are already exemplified above. The aspect
of evaluation, especially evaluation in multi-indicator systems makes partial order
an important tool in this respect. Whereas applying conventional tools, a 'good' or
'bad' within a data set is not known, and partial ordering is specifically adapted
to that. Applied partial order methodology, together with the analysis of the graph
theoretical structure could be a relevant tool in decision making, operation research
and, to some degree optimization.

The role of uncertainty in data analysis by partial ordering goes back to papers
of Sørensen et al. (1998, 2000). Data, continuous in concept, cannot be considered
as ideally suitable items for partial ordering. There are two main reasons: (i) The
aforementioned role of uncertainty which often arises when data are measured. (ii)
The information due to distances is lost. Both aspects can be methodologically

handled within the framework of partial ordering, however at the price of its elegance. Furthermore, if the weights in linear sums of uncertain indicator values are not sharp, application of partial ordering, as we know it today, comes to its limits. Studies in this direction are for now just landmarks on a long way.

For readers interested in the mathematical aspect of partial order, we recommend (Maggino 2017; Trotter 1992; Neggers and Kim 1998; Schröder 2003; Davey and Priestley 2002).

This book, *Indicators and their Analysis in different Scientific Fields* reports recent developments in the field of partial order applications. Some chapters are based on presentations at the International Conference on Partial Orders in Applied Sciences in Neuchatel, October 2018. This conference series was initialized 1998 (cf. Table 2) and is a forum for the scientific community with special interests in the theory and application of indicators.

In the 18 chapters, a variety of new developments within the area of partial ordering can be found.

Indicators and Theoretical Developments

As mentioned above, indicators play an increasing role in characterizing complex systems and in decision problems. Indicators are necessary to understand system behaviour. Hence, several chapters focus on the various aspects of indicators, addressing subjects like scaling level, relevance and the role of the inherent characteristic of partial orders, i.e., the incomparability (See J. Wittmann, p. 3, F. Maggino et al., p. 17). Further chapters discuss the functionality of indicators, the workflow for building indicators, the structure of complex indicators and the sensitivity of indicator values, as well as assessment of inhomogeneous indicator-based typologies through the reverse clustering approach (See J. Owsinski et al., p. 31), using a typology of spatial units of Polish municipalities as an illustrative example.

Indicator values often are considered as continuous in concept, thus the evaluation and exploration is of some fuzzy character. This aspect is considered in two chapters (see pp. 83–101) where a strict generalization is given central importance. Evaluations using parameters can usually be considered as sets over lattices. These two chapters (See A. Kerber and R. Bruggemann, p. 83, and R. Bruggemann and Kerber, p. 91) are devoted to this approach, whereby the theoretical concept is exemplified in a study of heavy metals and sulphur pollution along the southern part of river Rhine.

Very often a strict linear order is wanted, which in the case of multiple indicators typically is obtained as a result of aggregation of indicators, e.g., leading to a weighted sum. Although attractive due to its simplicity, the disadvantages are, e.g., that potential conflicts expressed by the values of single indicators are suppressed. A chapter is devoted to the idea of combining the advantages of linearly weighted sums and partial order theory in order to relax the requirements for a strict linear

Table 2 List of international conferences concerning applied partial order theory

Year	Organisers, site of conference	Main ideas	Remarks	Reference
1998	Bruggemann, Simon, Grell, Berlin, Germany	Hasse diagrams and their use in different fields Basics of POT	Initialization Workshop in honour of E. Halfon, who was a guest scientist in the Leibniz Institute of Freshwater Ecology and Inland Fishery	1
1999	Sørensen, Carlsen, Roskilde, Denmark	Role of uncertainty		2
2000	Bruggemann, Pudenz, Lühr, Berlin, Germany	Analyses of larger sets Decision systems		3
2001	Voigt, Welzl, Iffeldof, Germany	Statistics meet Partial order		4
2002	Sørensen, Carlsen, Lerche, Roskilde, Denmark	Needs of decision makers Integrative approach	Decision to continue every two years	5
2004	Frank, Bruggemann, Bayreuth, Germany	Concept of Local Partial Order Bringing general chemistry		6
2006	Todeschini, Pavan, Verbania, Italy	Towards a tool for decision making	The first time, where there was a need for a historical overview about this kind of international conference	7
2008	Owsinski, Bruggemann, Warsaw, Poland	Probabilistic concepts Combinatorial and algebraic concepts Multicriteria ranking		8
2010	De Baets, De Meyer	Combinatorial aspects, new topics coming from social sciences		9
2012	Bruggemann, Wittmann, Berlin, Germany	Posetic coordinates	Decision to continue every three years	10

(continued)

Table 2 (continued)

Year	Organisers, site of conference	Main ideas	Remarks	Reference
2015	Fattore, Maggino, Florence, Italy	Ordinal data *vs.* metric data Social science		11
2018	Beycan, Suter, Neuchatel, Switzerland	Development and application of indicators	The activity around partial order is now a part of the more general topic of indicators	12

The table clearly shows the trend from partial order theory to the more general theory of indicators.

References of Table 2

1:**1998** Group Pragmatic Theoretical Ecology, ed. Proceedings of the Workshop on Order Theoretical Tools in Environmental Sciences, Berichte des IGB, Heft 6, Sonderheft I, 1998. IGB, Berlin.

2:**1999** Sørensen, PB, Carlsen, L, Mogensen, BB, Bruggemann, R, Luther, B, Pudenz, S, Simon, U, Halfon, E, Bittner, T, Voigt, K, Welzl, G and Rediske, F, eds. Order Theoretical Tools in Environmental Sciences - Proceedings of the Second Workshop , October 21st, 1999 in Roskilde, Denmark. National Environmental Research Institute, Roskilde.

3:**2000:** Pudenz, S, Bruggemann, R, Lühr, H-P, eds. Order theoretical tools in Environmental Science and Decision Systems. Proceedings of the third Workshop, November 6th-7th, 2000, in berlin, Germany, Leibniz Institute of freshwater Ecology and Inland Fisheries, Heft 14 , Sonderheft IV

4:**2001:** Voigt, K and Welzl, G, eds. Order Theoretical Tools in Environmental Sciences - Order Theory (Hasse diagram technique) Meets Multivariate Statistics-. Shaker - Verlag, Aachen.

5:**2002:** Sørensen, PB, Bruggemann, R, Lerche, DB, Voigt, K, Welzl, G, Simon, U, Abs, M, Erfmann, M, Carlsen, L, Gyldenkaerne, S, Thomsen, M, Fauser, P, Mogensen, BB, Pudenz, S and Kronvang, B, eds. Order Theory in Environmental Sciences - Integrative approaches, Proceedings of the 5th workshop held at NERI, 2002. NERI, Ministry of the Environment, Denmark, Roskilde.

6:**2004:** Bruggemann, R, Frank, H and Kerber, A., Proceedings of the Conference "Partial Orders in Environmental Sciences and Chemistry" (Bayreuth, 15-16 April 2004). MATCH Commun.Math.Comput.Chem. 54 (2005) 487-690.

7: **2006: No proceedings.** However, an impression may be obtained, visiting: Bruggemann, R and Carlsen, L, eds. Partial order in environmental sciences and chemistry, Springer, Berlin, 2006

8: **2008:** Owsinski, J. and Bruggemann, R, eds. Multicriteria Ordering and Ranking: Partial Orders, Ambiguities and Applied Issues. Systems Research Institute Polish Academy of Sciences, Warsaw.

9: **2010:** Bruggemann, R., Partial Order in Applied Sciences, Statistica Applicazioni, Special issue 2011

10: **2012:** Bruggemann, R, Carlsen, L and Wittmann, J, eds. Multi-indicator Systems and Modelling in Partial Order. Springer, New York, 2014

11: **2015:** Fattore, M, Bruggemann, R, Partial Order Concepts in Applied Sciences. Springer, Cham, Switzerland, 2017

12: **2018:** Bruggemann, R and Carlsen, L, eds. Indicators and Partial Orders, Springer, This book

order (See R. Bruggemann and L. Carlsen, p. 63). A further study along these lines is reported in a separate chapter (See R. Bruggemann et al., p. 63) focusing on the possible generation of a weak order from a partially ordered set without the need for subjectively defining parameters beyond the data matrix.

Indicators for Special Purposes

Two chapters (see N. Pankow et al., p. 105 and G. Al-Sharrah and H.M.S. Lababidi, p. 119) focus on the selection of indicators for specific purposes. One chapter focuses on the development, assessment for their applicability and relevance of indicators for sustainability assessment in the procurement of civil engineering services, whereas a second chapter reports on dependent indicators for environmental evaluation of desalination plants with a special focus on which types of correlations between environmental indicators may affect decision-making when it is done by ranking.

One chapter presents some efficient sampling designs based on partial order sets and (sampled) linear extensions as a more flexible process than other designs and is executable with acceptable initial sample size, the new design in general being more efficient than its rival designs (See B. Panahbehagh and R. Bruggemann, p. 135).

It is often seen that potentially harmful substances are actually in their own sense beneficial for their specific purposes, but, e.g., harmful to the environment. As an exemplary case, partial order methodology has been applied for the search for suitable alternatives to lead split shots (See L. Carlsen, p. 153).

A chapter with elements from both the environmental and social area puts forward the question: who is paying for our happiness? The well-defined index for happiness, the World Happiness Index, was used for ranking 157 countries based on 7 indicators, the result being compared to a similar ranking of the countries applying the Happy Planet Index focusing on the exploitation of our planet's resources (See L. Carlsen, p. 205).

Activated carbon is used for many purposes, e.g., for wastewater treatment as a strong sorbent. It has a long history and has been prepared from a variety of material using methods involving physical and/or chemical activation. One of the latest attempts has been based on *Miscanthus* straw. One chapter is devoted to a study that compares 21 different methods for obtaining activated carbon from various materials (See L. Carlsen and K. Abit, p. 165).

Organisms such as bacteria, fungi or algae have the ability to trap and immobilize Uran, U; however, bioremediation does not reduce widespread U contamination. One chapter is dedicated to investigating the ability to concentrate U in bio-organisms. Partial order methodology discloses which organisms are the optimal U trappers (See N.Y. Quintero, p. 181).

Indicators in Social Sciences

The area of social science is the subject of two chapters, where one chapter focuses on the main motivations (see M. Fattore and A. Arcagni, p. 219) for applying partial order theory in the statistical analysis of socio-economic data, whereas the second chapter demonstrates the use of partial order methodology to an analysis of subjective well-being data from a European harmonized official statistical survey based on indicators for life satisfaction, meaning of life and emotional status (See L.S. Alaimo and P. Conigliaro, p. 243).

Software

One of the most popular software packages for studying partial ordering is the PyHasse. The package contains today more than 100 specialized modules, many of which are developed for specific purposes. However, it has been argued that PyHasse constitutes as a tool for 'connoisseurs'. Hence, web-based versions of PyHasse were developed (See R. Bruggemann et al., p. 291). However, they include only a limited number of modules.

However, other approaches to ranking are available, e.g., the Deep Ranking Analysis by Power Eigenvectors (DRAPE), which is illustrated in a chapter by a study of the sustainability of 154 countries based on 21 human, environmental and economic well-being criteria (See C. Valsecchi and R. Todeschini, p. 267).

Schwandorf, Germany   Rainer Bruggemann
Roskilde, Denmarks   L. Carlsen

References

Agent based modelling. Wikipedia, https://en.wikipedia.org/wiki/Agent-based_model. Assessed 17.05.2020.

Barilla. (2019). *Food sustainability Index.* A study on global food sustainability. https// www.barillacfn.com/en/food_sustainability_index

Brans, J. P., & Vincke, P. H. (1985). A preference ranking organisation method (the PROMETHEE method for multiple criteria decision – Making). *Management Science, 31*, 647–656.

Bruggemann, R., & Carlsen, L. (2020). The UN Sustainable Development Goal No. 7, Sustainable energy. Ranking of EU countries in: Simulation in Umwelt- und Geowissenschaften, Workshop 2020, J. Wittmann, ed., ASIM-Mitteilung AM 173, 29–47.

Bruggemann, R., & Drescher-Kaden, U. (2003). *Einführung in die modellgestützte Bewertung von Umweltchemikalien – Datenabschätzung, Ausbreitung, Verhalten, Wirkung und Bewertung.* Berlin: Springer.

Bruggemann, R., & Patil, G. P. (2011). *Ranking and prioritization for multi-indicator systems - introduction to partial order applications.* New York: Springer.

Carlsen, L., & Bruggemann, R. (2013). An analysis of the 'failed states Index' by partial order methodology. *Journal of Social Structure, 14*(3), 1–32.

Carlsen, L., & Bruggemann, R. (2014). The "failed state Index" offers more than just a simple ranking. *Social Indicators Research, 115*, 525–530.

Carlsen, L., & Bruggemann, R. (2017). Fragile state Index trends and developments. A partial order data analysis. *Social Indicators Research, 133*, 1–14.

Castle, C. J. E., & Crooks, A. T. (2006). *Principles and concepts of agent-based modelling for developing geographical simulations* (CASA Working Paper Series, 110) (p. 60). London: University College London.

Colorni, A., Paruccini, M., & Roy, B. (2001). *A-MCD-A, aide multi Critere a la decision, multiple criteria decision aiding*. Ispra: JRC European Commission.

Davey, B. A., & Priestley, H. A. (2002). *Introduction to lattices and order* (2nd ed.). Cambridge: Cambridge University Press.

El Din, H. S., Shalaby, A., Elsayed Farouh, H., & Elariane, S. A. (2013). Principles of urban quality of life for a neighborhood. *HBRC Journal, 9*, 86–92.

The Europe Sustainable Development Report. (2019). *Sustainable Development Solutions Network and Institute for European Environmental Policy Paris and Brussels.* https://s3.amazonaws.com/sustainabledevelopment.report/2019/2019_europe_sustainable_development_report.pdf

Figueira, J., Greco, S., & Ehrgott, M. (2005). *Multiple criteria decision analysis, state of the art surveys*. Boston: Springer.

Fragile State Index. (2019). *The Fund for Peace*. Washington, DC, USA. https//fundforpeace.org/2019/04/10/fragile-states-index-2019/

Ganter, B. (1987). Algorithmen zur Formalen Begriffsanalyse. In B. Ganter, R. Wille, & K. E. Wolff (Eds.), *Beiträge zur Begriffsanalyse* (pp. 241–254). Mannheim: BI Wissenschaftsverlag.

Ganter, B., & Wille, R. (1986). Implikationen und Abhängigkeiten zwischen Merkmalen. In P. O. Degens, H.-J. Hermes, & O. Opitz (Eds.), *Die Klassifikation und ihr Umfeld* (pp. 171–185). Frankfurt: Indeks-Verlag.

Ganter, B., & Wille, R. (1996). *Formale Begriffsanalyse Mathematische Grundlagen*. Berlin: Springer.

Gender Equality Index. (2019). *Still far from the finish line*. European Institute for Gender Equality. https//eige.europa.eu/news/gender-equality-index-2019-still-far-finish-line

Heidorn, C. J. A., Hansen, B. G., Sokull-Kluettgen, B., & Vollmer, G. (1997). EUSES – A practical tool in risk assessment. In K. Alef, J. Brandt, H. Fiedler, W. Hauthal, O. Hutzinger, D. Mackay, M. Matthies, K. Morgan, L. Newland, H. Robitaille, M. Schlummer, G. Schüürmann, & K. Voigt (Eds.), *ECO-INFORMA '97 information and communication in environmental and health issues* (pp. 143–146). Bayreuth: ECO-INFORMA Press.

Helliwell, J., Layard, R., & Sachs, J (2019). *World Happiness Report 2019*. New York Sustainable Development Solutions Network. https//worldhappiness.report/ed/2019/

Human Development Report (2019). *Beyond income. Beyond averages. Beyond today. Inequalities in human development in the 21st century*. http//hdr.undp.org/sites/default/files/hdr2019.pdf

Hüning, C., Dalski, J., Adebahr, M., Lenfers, U., Thiel-Clemen, T., & Grundmann, L. (2016). Modeling & simulation as a service with the massive multi-agent system MARS. *In Simulation Series, 48*, 1–8.

Kerber, A. (2017). Evaluation, considered as problem Orientable mathematics over lattices. In R. B. Marco Fattore (Ed.), *Partial order concepts in applied sciences* (pp. 87–103). Cham: Springer.

Maggino, F. (2017). Developing indicators and managing the complexity. In F. Maggino (Ed.), *Complexity in society: From indicators construction to their synthesis* (Social Indicators Research Series, vol 70). New York: Springer.

Maggino, F., & Zumbo, B. D. (2012). Measuring the quality of life and the construction of social indicators. In K. C. Land (Ed.), *Handbook of social indicators and quality of life research* (pp. 201–238). Cham: Springer.

Munda, G. (2008). *Social multi-criteria evaluation for a sustainable economy* (operation, p. 227). Heidelberg/New York: Springer.

Munda, G., & Nardo, M. (2009). Noncompensatory/nonlinear composite indicators for ranking countries: A defensible setting. *Applied Economics, 41*, 1513–1523.

Neggers, J., & Kim, H. S. (1998). *Basic Posets*. Singapore: World Scientific Publishing Co.

Pollandt, S. (1997). *Fuzzy-Begriffe - Formale Begriffsanalyse unscharfer Daten*. Berlin: Springer.

Roy, B. (1972). Electre III: Un Algorithme de Classements fonde sur une representation floue des Preferences En Presence de Criteres Multiples. *Cahiers du Centre d'Etudes de Recherche Operationelle, 20*, 32–43.

Roy, B., & Vanderpooten, D. (1996). The European school of MCDA: Emergence, basic features and current works. *Journal of Multi-Criteria Decision Analysis, 5*, 22–38.

Schröder, B. S. W. (2003). *Ordered sets - an introduction*. Boston: Birkhäuser.

Sørensen, P. B., Mogensen, B. B., Gyldenkaerne, S., & Rasmussen, A. G. (1998). Pesticide leaching assessment method for ranking both single substances and scenarios of multiple substance use. *Chemosphere, 36*, 2251–2276.

Sørensen, P. B., Mogensen, B. B., Carlsen, L., & Thomsen, M. (2000). The influence of partial order ranking from input parameter uncertainty. Definition of a robustness parameter. *Chemosphere, 41*, 595–560.

Sustainable Cities Index. (2018). https//www.arcadis.com/en/united-states/our-perspectives/sustainable-cities-index-2018/united-states/

Sustainable Development Goal 7. Ensure access to affordable, reliable, sustainable and modern energy for all https://sustainabledevelopment.un.org/sdg7

Sustainable Development Goals, https://sustainabledevelopment.un.org/?menu=1300

Trotter, W. T. (1992). *Combinatorics and partially ordered sets. Dimension Theory*. Baltimore: The Johns Hopkins University Press.

Acknowledgement

The editors thank the reviewers.

Name	Affiliation
Ghanima Al-Sharrah	Chemical Engineering Department, College of Engineering & Petroleum, Kuwait University, P.O.Box 5969, Safat 13060, Kuwait
Rainer Bruggemann (retired)	Leibniz-Institute of Freshwater Ecology and Inland Fisheries, Berlin, Germany (private address: Oskar-Kösters-Str. 11, D-92421, Schwandorf, Germany)
Lars Carlsen	Awareness Center, Linkøpingvej 35, Dk-4000 Roskilde, Denmark
Paola Conigliaro	Italian National Institute of Statistics, Istat
Enrico Di Bella	Department of Economics and Business Studies, University of Genoa, Genoa, Italy
Marco Fattore	Department of Statistics and Quantitative Methods, University of Milan-Bicocca, Milan, Italy
Jan Owsinski	Systems Research Institute, Polish Academy of Sciences Newelska 6, 01-447 Warszawa, Poland
Bardia Panabehagh	Department of Mathematical Sciences and Computer, Kharazmi University, Tehran, Iran
Guillermo Restrepo	Max Planck Institute for Mathematics in the Sciences, Leipzig, Germany
Claudio Rocco (retired)	Universidad Central de Venezuela, Facultad de Ingeniería. Claudio Miguel Rocco, Via Calligherie 21, Faenza (RA) 48018 Italy
Christian Suter	Department of Sociology, University of Neuchatel, Neuchatel, Switzerland
Jochen Wittmann	HTW Berlin, University of Applied Sciences. Environmental Informatics, Berlin, Germany

Contents

Contributors

Kamilya Abit al-Farabi Kazakh National University, Almaty, Kazakhstan

Robert Ackermann Institute of Environmental Science & Technology, Technische Universität Berlin, Berlin, Germany

Leonardo Salvatore Alaimo Department of Social Sciences and Economics, Sapienza University of Rome, Rome, Italy
Italian National Institute of Statistics, Istat, Rome, Italy

Ghanima Al-Sharrah Chemical Engineering Department, College of Engineering & Petroleum, Kuwait University, Safat, Kuwait

Alberto Arcagni Department MEMOTEF, Sapienza University of Rome, Rome, Italy

Rainer Bruggemann Leibniz-Institute of Freshwater Ecology and Inland Fisheries, Berlin, Germany

Lars Carlsen Awareness Center, Roskilde, Denmark

Paola Conigliaro Italian National Institute of Statistics, Istat, Rome, Italy

Marco Fattore Department of Statistics and Quantitative Methods, University of Milan-Bicocca, Milan, Italy

Regina Gnirss Berliner Wasserbetriebe, Berlin, Germany

Janusz Kacprzyk Systems Research Institute, Polish Academy of Sciences, Warszawa, Poland

Adalbert Kerber Department Mathematics, University Bayreuth, Bayreuth, Germany

Peter Koppatz TH Wildau Technical University of Applied Sciences Wildau, Wildau, Germany

Haitham M. S. Lababidi Chemical Engineering Department, College of Engineering & Petroleum, Kuwait University, Safat, Kuwait

Filomena Maggino Department of Statistical Sciences, Sapienza University of Rome, Rome, Italy

Jan W. Owsiński Systems Research Institute, Polish Academy of Sciences, Warszawa, Poland

Bardia Panahbehagh Department of Mathematical Sciences and Computer, Kharazmi University, Tehran, Iran

Nora Pankow Institute of Environmental Science & Technology, Technische Universität Berlin, Berlin, Germany

Stergios Pirintsos Department of Biology, University of Crete, Heraklion, Greece

Valentin Pratz University of Heidelberg, Germany

Nancy Y. Quintero Secretaría de Educación Municipal, Colegio Santos Apóstoles, Cúcuta, Colombia
CHIMA, Mathematical Chemistry Group, Universidad de Pamplona, Pamplona, Spain
Corporación Colombiana del Saber Científico (Bogotá), SCIO, Bogotá, Colombia

Jaroslaw Stańczak Systems Research Institute, Polish Academy of Sciences, Warszawa, Poland

Roberto Todeschini Milano Chemometrics and QSAR Research Group, Department of Environmental Sciences, University of Milano-Bicocca, Milano, Italy

Cecile Valsecchi Milano Chemometrics and QSAR Research Group, Department of Environmental Sciences, University of Milano-Bicocca, Milano, Italy

Jan Waschnewski Berliner Wasserbetriebe, Berlin, Germany

Jochen Wittmann HTW Berlin, University of Applied Sciences. Environmental Informatics, Berlin, Germany

Slawomir Zadrożny Systems Research Institute, Polish Academy of Sciences, Warszawa, Poland

Part I
Indicators and Theoretical Developments

Some Basic Considerations on the Design and the Interpretation of Indicators in the Context of Modelling and Simulation

Jochen Wittmann

1 Indicators: In General, in Mathematics, in Modelling and Simulation

Indicators are necessary and widely used means to understand system behavior. An overview on the work concerning multi-indicator systems with focus on a ranking of the indicator quantities gives (Bruggemann et al. 2014).

This paper does not focus on a ranking of different indication aspects a system provides, but on aggregation these aspects to a single compressed value.

The definition of "indicator" bases on the fact, that the system (or model) quantity of interest is difficult to observe or completely hidden within the system. This kind of definition can be found e.g. in the field of economic sciences as "Measurable variable used as a representation of an associated (but non-measured or non-measurable) factor or quantity." (Businessdictionary 2017). The same reference gives the representative example for an indicator with the "consumer price index (CPI) [that] serves as an indicator of general cost of living which consists of many factors some of which are not included in computing CPI." (Businessdictionary 2017).

Beside economics, there is a wide range of other domains using indicators intensively: Biology knows indicator plants or organisms that are representatives for special types of ecosystems (see e.g. Haseloff 1982), but also indicators in the sense of summarizing measures for the state of the environment such as the index of biodiversity for example as a measure for the intactness of an ecosystem (Campbell and Reece 2003).

J. Wittmann (✉)
Environmental Informatics, HTW Berlin, University of Applied Sciences, Berlin, Germany
e-mail: Jochen.Wittmann@HTW-Berlin.de

© The Author(s), under exclusive license to Springer Nature Switzerland AG 2021
R. Bruggemann et al. (eds.), *Measuring and Understanding Complex Phenomena*,
https://doi.org/10.1007/978-3-030-59683-5_1

Medicine as well knows indicators, e.g. the vital signs as a measure for the state of a patient especially in intensive medicine (for example the wide range of patient monitoring systems (Elliott and Coventry 2012). However, another interpretation gains in importance with respect on sending alarms if the situation becomes instable or dangerous. The Early-Warning-Score (HealthcareInstitute 2017; Helios-Kliniken 2017) provides an indicator for the over-all state of the patient and combines a list of vital signs to a single value. Thus, the indicator excerpts the information of the n vital signs and combines them to a single, highly aggregated measure.

So far, we know an indicator as an aggregating measure for at least partially hidden or inaccessible system quantities. In the context of system analysis, modelling, and simulation, however, an indicator is required quite in the sense of medical applications as a tool to sign whether the systems situation is normal, critical, or catastrophic. The intention is to aggregate the "control panel" of the system (or model) under observation with its lots of parameters (levels, tachometers, diagrams ...) to one single value. The expression range of a traffic sign with its colors green, yellow, and red is the desired level of aggregation for the system manager.

At the end of this short introduction stands the observation, that indicators in modelling context loose the function of making hidden quantities visible and measurable because the model description is man-made and virtual and, therefore, transparent and accessible on every level. What remains is the aggregating and/or ranking function of indicators, which should be discussed more in detail in the following sections.

2   Functionality of Indicators

2.1   Typical Application Types for Indicators

Before we deal with the structure of indicators, a distinction should be made at the application level as to which functional tasks are to be solved by indicators or indicator systems. In the course of this paper it will be worked out that an exact specification of the expected function of an indicator is the decisive key for an effective and efficient use. Therefore, at this point, an (incomplete) list of possible fields of application for indicators.

2.1.1   Warnings

A relatively simple requirement is to interpret the indicator or the current indicator value as an indication of whether the current system status is within the normal range or is cause for concern. In this case, exceeding a previously set limit will result in a warning about the current system status.

This use is based on two basic ideas: firstly, the fact that the indicator value serves as an indicator for a more or less complex and less transparent system state, and secondly, that several influencing variables can be combined in such an indicator value, which, as an aggregated value, provide indications of the current behavior of the overall system.

Typical examples are warnings regarding the condition of complex industrial plants or in intensive care medicine to summarize the values from various vital parameters.

2.1.2   Decisions Between Alternatives

While in the first case the scale of the indicator together with an absolute threshold value comes to the fore and requires special design considerations, the second field of application requires an indicator design that evaluates different decision alternatives and thus allows a comparison of these alternatives. Typical examples are the classic advantage and disadvantage lists for previously given decision scenarios and a decision for the overall problem derived from the individual arguments collected in these lists. In this case, the focus is not so much on the state of the system itself, but much more on a relative evaluation, a ranking that relates the different scenarios of the decision problem to one another.

2.1.3   Optimization

A much more complex use of indicators is found in the solution of system optimization tasks. In this case it is assumed that the behavior of a system can be influenced by setting parameters (the so-called manipulated variables), whereupon the value of the target variable changes. Through targeted changes of the manipulated variables, an optimum of the target function value is to be achieved iteratively during optimization.

The description of the optimization procedure clearly shows the use of indicators: The indicator fulfils the function of the so-called target function and thus summarizes the system state achieved by setting the manipulated variables on a single scale. On this scale, the mechanism of the optimization algorithm then takes effect and iteratively minimizes or maximizes the target function or indicator value.

2.1.4   Modelling Real Systems

Similar to the use for triggering warnings, several indicators can be recorded and observed simultaneously and their dynamic change in indicator values can be interpreted as an image of the underlying real system. Once again, the example from intensive care medicine is the most vivid: the measured vital parameters are not combined into one indicator variable, but their value progression is visualized as if in a control station. The expert observer interprets the dynamics of the different indicator values as an image of the real system and its dynamics and draws conclusions about the future behavior of the system. The measured values are therefore not interpreted in the actual sense as indicators, but as current values of system variables that determine dynamics.

A significant difference to system modelling must be noted at this point: The dynamic courses of the indicators are exclusively visualized and must be interpreted by the observer himself. Relationships between the individual variables are not explicitly specified in the sense of a model, but can only be assumed by attentive observation of the functional processes. However, the internal structure of the observed system always remains hidden. Findings about the structure as well as predictions about future system behavior remain pure hypotheses in the mind of the observers; they cannot be derived from the set of indicators. Here lies a substantial difference to the structural models (glass box), as they are set up in the system modelling for example by systems of differential equations.

2.2 Structural Alternatives for Indicators

Main confusing fact defining and using indicators seems the definition of the functionality of the indicator. Not only the reachability of a value seems to be of importance but also the aggregating and valuating character of an indicator. With the applications of Sect. 1 in mind and together with the differentiation of the application types from Sect. 2.1, there is a differentiation concerning the functions of an indicator easily possible that leads to the following three levels of functionality:

2.2.1 Level 1: Observation and Transmission

The intention is to observe a certain system or model quantity. If this quantity is not measurable directly, transmission becomes necessary. Transmission means taking the indicator value instead of the value of the hidden or less accessible model quantity.

A very simple example should illustrate the distinction between the different functional levels when using indicators: a box of muesli is given together with the question of how high the proportion of fruits and cereals is. The box is opaque, so that a direct answer to the question is not possible. An indirect measure must be found. In this case, the different specific gravity of the proportions is used to calculate the quotient between the weight of the box and its volume. This value serves as an "indicator" for the ratio of grain to fruit in the interior. The indicator value thus provides information about a system quantity that is inaccessible to the black box (Fig. 1).

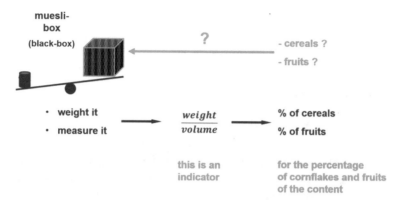

Fig. 1 Example muesli box I

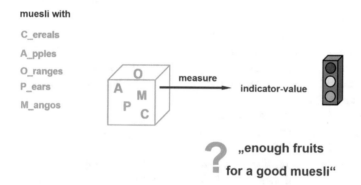

Fig. 2 Example muesli box II

2.2.2 Level 2: Judging

In the second step, the user is interested in a rating beyond the mere value of the indicator variable. In addition to the scale of the indicator, a decision must therefore be made as to whether the measured value is "good" or "bad" or how the indicator values should be ranked between alternatives when making a decision.

Let's extend the example of the muesli box by distinguishing the fruit content in apples, oranges, pears and mangoes and ask the evaluative question: Does the muesli contain enough fruit to make it taste good? It is obvious that a distinction between "good" and "not good" is necessary depending on a threshold value of the indicator.

Thus, judging means introducing a classification for the values of the indicator quantity and thus introducing classes of interpretation as well (Fig. 2).

2.2.3 Level 3: Aggregation

Aggregating is the usage mainly applied: not only one but several aspects of a system are

(a) measured (see level 1)
(b) classified (see level 2),
(c) weighted to each other, and
(d) functionally combined to one resulting single value.

As can already be seen from the example in 2.2.2, the task of judging is in most cases not a one-dimensional problem but a multidimensional one. In the context of indicators and system optimisation, it is better to speak of a multi-criteria problem. Different aspects should be considered when assessing the state of the system measured by the indicator. As in Level 1 and Level 2, a scale must be introduced for each of these individual aspects and an evaluation or definition of threshold values must be carried out.

At this point the problem arises that the evaluation of sub-criteria is contradictory and therefore no simple and unambiguous decision can be made. Ultimately, the Hasse diagrams, which are the subject of many contributions in this volume, represent an alternative solution to this decision problem by defining a partial order for the subcriteria.

The second fundamental alternative is to combine the individual criteria into an aggregated value. This can be done by arbitrary mathematical operations. Usually, the values of the subcriteria are added, but multiplication, exponentiation and any other connections are also conceivable. In order to compensate for imbalances with regard to the dimensions of the criteria but also to realize an application-specific weighting of the subcriteria, the values are usually weighted before they are subjected to the aggregation function. The aim of this aggregation is always to determine a one-dimensional indicator value with only one scale, on the basis of which a clear ranking or a clear decision can then be made.

In the muesli example, the quantity available for each type of fruit must be determined, a weighting factor must be assigned, and the individual values determined in this way must be aggregated (for example, by forming totals) to determine the final indicator value for the "quality" of the muesli. Obviously, the problem of the weighting of the individual aspects ("Can 2 slices of mango compensate for the lack of 20 pieces of apple?") and the decision for the aggregation operation (sum formation? product? ...) come to light. The advantage of this alternative, however, is that there is a single indicator value at the end and no incomparability has to be discussed, as occurs with the use of partial orders.

2.2.4 Hierarchy of the Levels

The hierarchy of the levels is obvious: level 1 describes the access to a quantity under observation, level 2 deals with the range of the values of the quantity observed, and at last, level 3 broadens the functionality by permitting n inputs of level-2-type mathematically combined to an aggregated level-3-indicator. Of special interest is the structure of the mathematical mapping calculating from n values one.

Two degrees of freedom offers this mapping to the user: First, the possibility to give the measured parameter values an additional weight before composition. Second, the kind of functional composition of the n input parameters itself.

These two degrees of freedom influence the design of a hierarchically aggregated indicator essentially. After the following, more procedural section concerning the workflow for building indicators, Sect. 4 will focus on the design of a complex, hierarchically structured indicator and will discuss weights and composition in some more detail.

2.3 Fitting Structural Alternatives to the Application Types

Before we dedicate ourselves to the workflow with the design of an indicator in the 3rd section, a short comparison between the levels just explained and the typical application fields from the previous section should be made at this point.

The selection of suitable criteria is always connected with the specification of a scale (level 1) and in the vast majority of cases additional classes are formed on this scale which correspond to level 2 (judging). Thus the application fields "Warnings" and "Decisions between alternatives" can be treated. For system optimization, it is necessary that the indicator value be designed in such a way that a new value assignment for the manipulated variables is constructively possible from the current value. In addition to judging, the indicator must also constructively allow the calculation of feedback on the input variables of the system. In the case of optimization, it is sufficient to consider this system as a black box under observation. This changes, if the claim of the investigation lies in the modelling of the real system. Then it is not sufficient to observe and visualize the current values of system variables as indicator values; rather, in the sense of a glass box, knowledge about the static and dynamic relationships between the observed variables is necessary. Consequently, a pure indicator system cannot replace a real model of a system.

3 The Workflow for Building Indicators

If the focus lies on how to get an indicator, it will be essential to bring the corresponding workflow to mind and reflect its steps in detail. The Fig. 3 shows the actions in green and the resulting objects in blue colour.

Step 1: scope and borders

The first step is the decision, which model quantities among the complete set (given by the system or the model) are of interest for the indicator objective. Thus,

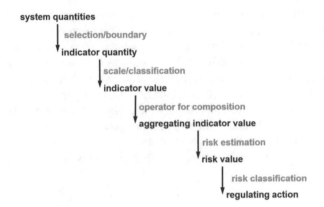

Fig. 3 The workflow of indicator design

the system's parameter set is reduced to a subset containing the relevant information. In modelling and simulation theory, this step is analogous to defining the borders of the system under consideration within the detailed "real world" (Schmidt 1985).

Step 2: scale

The second step is to define an appropriate scale for the values of the selected indicator quantity. In general, there a two cases to distinguish: First, a metric or nominal scale defines the indicators scale: In this case, the scale gives no hints for a ranking and may be supposed as "objective" so far. Second, the indicator originally comes with an ordinal scale or the designer of the indicator defines discrete classes for the measured indicator values: In this case, by classification, a first valuating influence is given and thus certain "subjectivity" is brought into the indicator development process. This corresponds to functionality level 2 from the previous section.

Step 3: aggregation method

The third step deals with the aggregation functionality. If there are different indicator quantities selected, they must be combined to an aggregated single value, now. Two degrees of freedom have to be determined:

Which weight has a certain value in comparison to the other selected indicator values? Moreover, which operations shall be used for aggregation? The first question is often discussed and masks the expressiveness of the second one. A wide choice of operators is possible: addition, multiplication, potentiation, integration, minimum/maximum, and any other mathematical operation possible. The choice of the aggregation function is one of the most neglected decisions along the indicator design workflow although it offers a wide range of opportunities in modelling the resulting aggregated value. In this step lies great potential for expressing the relationships between the single indicator values. Much more can be achieved already in this step by modelling these relations individually and accurately according to the original intention and objective of the indicator.

Step 4: risk measure

The next step in the workflow is the determination of a scale for the expected risk in consequence of a certain indicator value. Now, the value of the indicator has to be valued not only according to its own value-range but according to the risk, the indicator value implies for the system. It is of importance to introduce this as a properly separated step in the workflow because it contains a new level of valuating and thus demands for an explicit design decision, too.

Step 5: risk threshold

Normally, the last step transforms the risk-value to a classified scale and thus gives a simplified status of the system. Often this classification is combined with setting one or more thresholds for appropriate reactions if the threshold is crossed. Here we finally find for example the traffic light symbology as one possible occurrence at the end of the indicator design workflow.

4 The Structure of Complex Indicators with Sub-Indicators and Weightings

In consequence of Sect. 2 "functionality" and Sect. 3 "design-workflow", the general structure of an indicator can be depicted as in Fig. 4. At the bottom, the system (or model) with all its quantities available serves as a source for the indicator apparatus. In addition to the explanation so far, not only a modular, but also a hierarchical structure to build an aggregated indicator is allowed. Thus, in general, a modular hierarchical indicator design combines the originally measured values of the system to the combined and interpreted final indicator value on top. The guiding visualization for an indicator value is the picture with the indicator as the top of an iceberg representing the complete structure not seen under sea-level respective in the "black-box" with the complex system under observation in it.

However, what users really get with an indicator is a complex model for interpreting system state. Therefore, users have to know even the invisible parts of the iceberg and have to handle indicator design in complete analogy to model design and model construction as explained in the following section.

5 Complex Indicators as a Distinguished Valuation Model

Following the argumentation of the preceding subsections, indicators in modelling and simulation are much more complex than simple indicators for inaccessible sys-

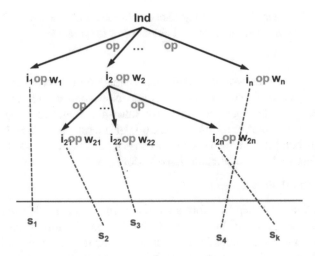

Fig. 4 The general structure of an indicator (s system quantity, i indicator value, w weight, op aggregation/weighting operation)

tem variables. In fact, indicators can have a complex internal structure, represented by

– a set of different sub-indicators
– the hierarchical structure of these sub-indicators
– a weighting system
– a freely definable aggregation function.

The conclusion of the deliberations is the fact, that indicators themselves are constructs that model a valuating system within a more comprehensive system structure. If we transfer the terms and definitions from modelling and simulation to the process of building complex indicators, the specification of the indicator seems to be very analogue to the building of a dynamic model. In fact, indicators can be considered as a "valuation model" that complements the original "dynamic model" with its defining parts as declaration of model quantities and the dynamic description.

For indicator purposes, the dynamic model has to provide a set of model quantities MQ, thus the indicator part for the model has to include.

A. the set of model quantities
   MQ.
B. the set of indicator quantities with
   $IQ = \{iq_1, \ldots, iq_n \,|iq_i \varepsilon MQ|\}$.
C. the set of weights
   $W = \{w_1, \ldots, w_n \,|w_i\ \varepsilon \Re|\}$.
D. The aggregated indicator value AIV as result of the aggregation function AF
   $AIV = AF(iq_1, \ldots, iq_n, w_1, \ldots, w_n)$.

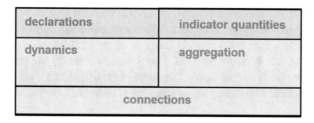

Fig. 5 The general structure of a model component with integrated valuating model by hierarchical indicator

This is the necessary and sufficient indicator structure for an indicator based on a simple model. However, to transfer the situation to models based on a modular-hierarchical model specification, the situation grows more complex. With respect to Sect. 4 that introduced a hierarchically structured indicator-tree the specification for the aggregating indicator has to be defined more precisely.

Structural, the situation is as follows: The hierarchical model consists of model components with three defining parts (without loss of generality, we follow the model description language of (Zeigler 1990)):

- declarations,
- dynamic description,
- specification of the connections.

The indicator part of a model component consists of:

- indicator quantities (A–C in the list above)
- aggregation description (D in the list above)
- specification of the connections (see Fig. 2)

In result, the model component has to be extended according to Fig. 5.

Some special remarks are necessary concerning the interpretation of the connections-part of this extended model component specification: As in model hierarchy, the component has to specify to which other indicator-calculation the value of the aggregated indicator function AIV has to be connected.

Structural, this connection is completely analogues to the connections of model quantities in the model hierarchy. Therefore, the connection part for the indicator is common with the connection part of the standard-modelling part of the component. Doing so, the indicator-hierarchy according to Fig. 2 can be built by already existing concepts for building and maintaining model-hierarchies.

Beside this structural identity, it has to be emphasized that the indicator hierarchy might completely differ from the model hierarchy.

Following precisely the steps for model design (Wittmann 2016) in building indicators, it becomes obvious that the indicator model has to be submitted to an indicator validation (according to model validation), and an intensive sensitivity analysis.

All these steps of the design workflow have to be done with respect to the initial intention for introducing the indicator. This argumentation follows in full analogy the argumentation that good models have to be as detailed as necessary and as abstract as possible to allow efficient model experiments on the one hand and to meet the targets on the other hand. The same objective oriented design and usage should be direct the design and the interpretation of indicators.

6   Test and Validation of Indicator Models

Even if the preceding strict distinction between the (dynamic) model of the real system and the indicator and evaluation model based on it may be too theoretical for some practical examples, it emphasizes that even for the simple use cases, at each step of the argumentation and interpretation it must be differentiated whether this step refers to an abstract representation of reality or whether it deals with evaluation in the sense of an indicator model. The system model can do completely without indicators and be executable, the indicator model as introduced in the previous section, on the other hand, cannot work without the system model.

In addition to this differentiation in design, the differentiation between the system model and the indicator model has a further advantage: firstly, it makes it obvious that the indicator model must also be validated, and secondly, it constructively determines the way in which such validation is to be carried out. Validation can only take place according to the rules that apply to the validation of the system model: There must be sufficient agreement with regard to the requirements between the results derived from reality and the results generated with the aid of the indicator model.

This throws us back to the fields of application and functionalities for indicators classified at the beginning and hopefully also makes clear why this classification work was carried out in advance. The validity of an indicator (system) can only be validated in relation to the specific requirements to be specified for each individual application. An abstract "better" or "worse" for one or the other indicator approach, detached from the objective in the specific application, cannot exist.

Constructively and in addition to the work steps for the creation of an indicator, the user has the task of specifying

(a) the purpose of the indicator
(b) the accuracy with which the indicator is intended to fulfil this purpose
(c) the examples to be used to test the correspondence between the results of the indicator and reality.

In the field of software engineering, one speaks of the requirements for a system (a), the degree to which these requirements must be fulfilled (b) and the test cases with which the fulfilment of the requirements can/must be proven.

Strictly speaking, the task to develop an indicator begins much earlier than normally assumed with an explicit specification of the validity criteria for the planned indicator and it ends much later than assumed with testing the indicator against the previously established validity criteria.

7 Some Summary Remarks

By classifying indicators according to fields of application and internal structure, the article leads to the insight that an indicator is an independent evaluation model that is independent of the actual system model. Due to its model character, however, like all models it is subject to quality control by careful validation with the usual consideration of validity, accuracy and sensitivity (e.g. with regard to weighting factors in the aggregation of composite indicators).

In contrast to multi-criteria methods based on partial orders, aggregated indicator systems lead to the one-dimensional decision aids popular with the user (e.g. in the form of a traffic light that reflects the system state), which very quickly convey the current system state, but in the case of a system warning require knowledge about the structure of the aggregation rules in order to determine the partial indicator responsible for the error display. Partial orders, on the other hand, do not make a final decision for incomparable alternatives. Instead, they highlight them and leave the final decision to the user. In the event of a warning, however, the user also has knowledge of the structures that lead to the warning and is usually better able to detect its cause.

References

Bruggemann, R., Carlsen, L., & Wittmann, J. (2014). *Multi-indicator systems and modelling in partial order*. New York: Springer.

Businessdictionary. (2017, December 07). Von http://www.businessdictionary.com/definition/indicator.html. Abgerufen.

Campbell, N. A., & Reece, B. J. (2003). *Biologie*. Heidelberg, Berlin: Spektrum Verlag.

Elliott, M., & Coventry, A. (2012). Critical care: The eight vital signs of patient monitoring. *British Journal of Nursing, 21*(10), 621–625.

Haseloff, H.-P. (1982). Bioindikatoren und Bioindikation. *Biologie in unserer Zeit.* , S. 12, Nr. 1, 20–26.

HealthcareInstitute. (2017, June 25). *Early-warning-score*. Von http://www.ihi.org/resources/Pages/ImprovementStories/EarlyWarningSystemsScorecardsThatSaveLives.aspx. Abgerufen.

Helios-Kliniken. (2017, December 07). *Early-warning-score*. Von http://www.heliosaktuell.de/nachrichten/ein-sticker-hilft-dabei-komplikationen-vorauszusagen. Abgerufen.

Schmidt, B. (1985). *Systemanalyse und Modellaufbau*. Berlin: Springer.

Wittmann, J. (2016). Komplexität beim Modellieren und Simulieren: Eine Analyse und ein Plädoyer für schlanke Modelle. In *Wiedemann, T.: Tagungsband ASIM 23. Symposium Simulationstechnik 2016 Dresden* (S. 99–106). Wien: Argesim.

Zeigler, B. (1990). *Object-oriented simulation with hierarchical, modular models*. London: Academic.

Indicators in the Framework of Partial Order

Filomena Maggino, Rainer Bruggemann, and Leonardo Salvatore Alaimo

1 Introduction: Indicators and Management of Complexity

The topic of indicators has been and continues to be considered a "niche field" in the methodological scientific debate. However, during the last decades, this issue has been discussed in any conference, workshop and seminar on measuring socio-economic dimensions. This is not a specific issue of the natural and social sciences. Indicators are used and constructed everywhere and their functions in contemporary societies are widespread. We can observe their increasing importance also in the media and in the public debate. Governments and international organizations use them to compare and rank countries on some particular topics, like quality of life, wellbeing or sustainability.

Although the term indicator is often used as a synonym for *index*, in statistics it represents a more recent term indicating indirect measures of phenomena not directly measurable. In this perspective, an indicator is not simple crude statistical information but represents a measure organically connected to a conceptual model aimed at describing different aspects of reality (Maggino 2017a). In brief, an indicator is what relates concepts to reality through observation. Indicators should

F. Maggino (✉)
Dipartimento di Scienze Statistiche, Sapienza University of Rome, Rome, Italy
e-mail: filomena.maggino@uniroma1.it

R. Bruggemann
Leibniz-Institute of Freshwater Ecology and Inland Fisheries,
Berlin, Germany

L. S. Alaimo
Sapienza University of Rome, Rome, Italy

Italian National Institute of Statistics – Istat, Rome, Italy
e-mail: leonardo.alaimo@istat.it

© The Author(s), under exclusive license to Springer Nature Switzerland AG 2021
R. Bruggemann et al. (eds.), *Measuring and Understanding Complex Phenomena*,
https://doi.org/10.1007/978-3-030-59683-5_2

be developed and managed so that they represent different aspects of the reality. They picture the reality in an interpretable way, allow meaningful stories to be told and support evaluations and decisions.

Therefore, any discussion of indicators must start from two fundamental questions: what are indicators? and why are they so important?

In order to fully understand the importance of indicators, complexity needs to be addressed. In his "millennium" interview on January 23, 2000 (San Jose Mercury News), Stephen Hawking said: *I think this century will be the century of complexity* (Gorban and Yablonsky 2013). We can encounter this concept in different fields (e.g., physics, chemistry, biology, engineering, software, social sciences). It is sometimes abused and used interchangeably with other ones (e.g., large, complicated), nevertheless having different meanings. It has no precise meaning and no unique definition (Erdi 2008). This notion does not belong to a particular theory or discipline, but rather to a "discourse about science". According to Morin (1984), we cannot approach the study of complexity through a preliminary definition: there is no such thing as "one" complexity, but "different" complexities.

"Complex" is often associated with the concept of "system" and the topic of "complex systems" is a subject of great scientific debate and interest. While a simple system has a small number of components with defined roles and clear rules, a complex one contains many elements, which are interdependent and interact non-linearly. *A system isn't just any old collection of things. A system is an interconnected set of elements that is coherently organized in a way that achieves something* (Meadows 2009, 11). This definition takes up and updates the idea expressed by Aristotle in *The Politics*: the whole is something over and above its parts, and not just the sum of them all. The analysis and understanding of complex systems require approaches allowing more concise views. The guiding concept is synthesis. Generally speaking, synthesizing responds to a need for concreteness in the relation with things. It is justified by the fact that knowledge of complex phenomena involves some form of *reductio ad unum* (Sacconaghi 2017). The correct way of understanding those phenomena is to conceive them as a whole, adopting a synthetic approach.

Getting in contact with reality always involves some process of synthesis, more or less conscious, consisting in the reduction of a multiple in units. This *reduction* could be a risk. Any synthesis should be a *stylization* and not an over-simplification of reality.

Indicators play a key role in describing, understanding and controlling complex systems. An indicator is, therefore, a tool for understanding reality. It is not necessarily a number. It can be an object, a map, an image. It is what allows us to grasp the complexity and guide us in understanding it. There is a large amount of literature on the use of metaphoric images for the representation of phenomena, especially for complex ones (Lima 2013; Tufte 2015). In Fig. 1, we can see an example of the representation of multidimensional poverty in Italy (2008) and its dynamics (Lima 2013). This infographic shows how poverty "red thread" has various weight, that depends on the different criterion and perspective that the Italian National Institute of Statistics – Istat used to photograph the society. As

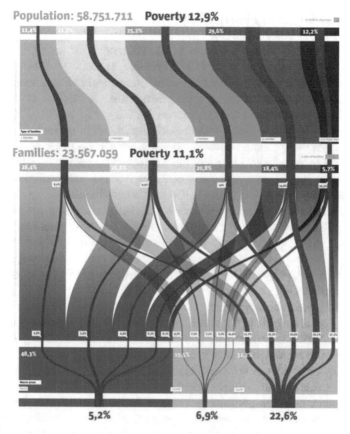

Fig. 1 The Italian Poverty Red Thread 2008. This map of the poverty line in Italy, realized by Mario Porpora, is organized according to family typologies (number of family members), and further categorized by location (the north, center, or south of Italy). (Source: Lima 2013)

clearly shown, percentage values numerically alike have a different absolute value. In addition, the information is immediately understandable. It is clear that poverty is spread differently in different macro-areas of the country: 5.2% in the North; 6.9% in the Centre and 32.2% in the South.

According to Porter (2001), the "soft" power of numbers and indicators is characteristic of our time. If we hope to use indicators and other measures to make the world navigable in simpler terms, let us be careful what we wish for. It is essential that what we are going to build is an authentic representation of the reality, preserving the systemic characteristics of the phenomena defined by elements and their relationships. In this perspective, each indicator measures and represents a distinct constituent of the defined phenomenon and all of them do not represent a pure and simple collection of indicators but are part of a complex system, a multi-indicators system, in which. In other words, only a complex instrument (a multi-indicators system) allows a full and correct understanding of complexity (see e.g., Bruggemann and Patil 2011).

Meadows (2009) defines a system as "an interconnected set of elements that is coherently organized in a way that achieves something" (Meadows 2009, 11). This definition identifies the three main components of a system: *elements*, *interconnections* and *functions*. A system is not just a collection of things; they must be interconnected and have a purpose, i.e. they must be aimed at achieving an objective. The purpose of a system is often difficult to understand. "The best way to deduce the system's purpose is to watch for a while to see how the system behaves" (Meadows 2009, 14). From this Meadows' statement, it can be deduced that a system has its own behaviour, different from its parts and that, like any behaviour, it can change over time. Each system is based on a *stock*, i.e. the elements that constitute it in a given time. These stocks change over time due to the effect of *flows*. "Flows are filling and draining, births and deaths, purchases and sales, growth and decay, deposits and withdrawals, successes and failures" (Meadows 2009, 18). Meadows highlights the dynamism of the systems, their adaptation over time. One cannot understand them without understanding their dynamics of stocks and flows. Obviously, the change can concern both the system as such and one or even all of its essential components. Change can also be traumatic and unexpected. Most of systems are able to withstand the impact of drastic changes thanks to one of their fundamental characteristics, *resilience*. "It is both the ability to adapt to change by evolving and the ability to resist it by restoring its initial state. Resilience presupposes change: it is not static being, but becoming" (Alaimo 2020, 21). A system is, therefore, an organic, global and organized entity, made up of many different parts, aimed at performing a certain function. If one removes a part of it, its nature and function are modified; the parts must have a specific architecture and their interaction makes the system behave differently from its parts. Systems evolve over time and most of them are resilient to change.

Simple systems are characterized by few elements and few relationships between them; they can be analyzed analytically. Complex systems, on the contrary, are made up of many elements and many relations of different types; they can be analyzed only in a synthetic way. In a complex system, elements and connections, besides being numerous, are various and different.

A particular type of complex system is the *Complex Adaptive System* (CAS). They add to the other characteristics typical of complex systems the ability to adapt. CASs are able to adapt to the world around them by processing information and building models capable of assessing whether or not adaptation is useful. The elements of the system have the main purpose of adapting and, in order to achieve this purpose, they constantly look for new ways of doing things and learning, thus giving rise to real dynamic systems. These systems challenge our ability to understand and predict. "It is evident that the main characteristics of complex adaptive systems are typical of social organizations and phenomena. Each of them is made up of a network of elements, which interact both with one another and with the environment. They are multidimensional and their different elements or dimensions are linked together in a non-linear way. They evolve over time, modifying both

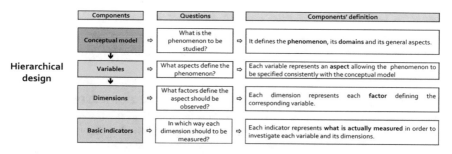

Fig. 2 The components of the hierarchical design

their dimensions and the links between them. The measurement and analysis of social organizations and phenomena requires the definition of systems of indicators capable of capturing their different aspects. As can be easily understood, these systems are dynamic, since they have to adapt to the changes in the measured phenomena. In simple terms, they are CASs and can be monitored and measured through systems of indicators that are CASs themselves" (Alaimo 2020, 26–27).

Developing indicators starts from a need of knowledge. But in most cases the only way that can be followed is generating indicators, thus projecting a system into a collection of indicators. Indicators should be developed, through a hierarchical design, requiring the definition of the following components, shown in Fig. 2:

1. **Conceptual model**

In social sciences, all measurement processes start with the definition of the concept to be measured (Lazarsfeld 1958). This operation is a *process of abstraction*, a complex stage that allows us the identification and the definition of:

- the model aimed at data construction,
- the spatial and temporal ambit of observation,
- the aggregation levels (among indicators and/or among observation units),
- the models allowing interpretation and evaluation.

2. **Latent variables and their dimensions**

Each variable represents an aspect to be observed and reflects the nature of the considered phenomenon consistently with the conceptual model. The identification of the latent variable is founded on theoretical assumptions (requiring also an analysis of the literacy review) also about its *dimensionality*. According to its level of complexity, the variable can be described by one or more factors, called *dimensions*. Thus, we can observe *uni-dimensional* (when the definition of the considered variable assumes a unique underlying dimension) or *multidimensional* (when the definition of the considered variable assumes different underlying factors) variables.

This identification will guide the selection of the indicators. The correspondence between the defined dimensionality and the selected indicators has to be demonstrated empirically by testing the selected *model of measurement*. The specification of the latter refers to the relationship between constructs and indicators. In literature, we distinguish two different conceptual approaches: *reflective* and *formative*.[1]

3. Basic indicators

In the majority of the cases the defined variable can be measured only indirectly through observable elements which are called *indicators* of the reference variable. Each basic indicator represents what can be actually measured in order to investigate the corresponding variable. In other words, the indicator is what relates concepts to reality.

We must specify that in all phases of hierarchical design is involved *subjectivity*. Measurement is not an arbitrary process, but necessarily involves subjectivity. There will always be the influence of the subject's point of view: in the definition of phenomena; in the definition of the hypotheses on reality; in the selection of indicators; in the choice of statistical tools. Subjectivity represents one of the dimensions inevitably involved in defining concepts, making measurement a *complex exercise*.

The proper and accurate application of the hierarchical design allows defining a complex structure in which each indicator measures and represents a distinct component in the description of the phenomenon. Different types of indicators can be present within a system, contributing to its complexity.

Within a system, indicators may show different characteristics related to (i) the perspective through which the indicators are reporting the phenomenon to be observed, (ii) the level of observation (e.g. micro/macro, internal/external), (iii) the nature of the observed characteristics (e.g., objective/subjective, qualitative/quantitative), (iv) the level of dis/aggregation, (v) the communication context in which the indicators are used, (vi) the interpretation attributed to the indicators in statistical analyses, (vii) the criteria of their adoption, and (ix) their quality.[2]

Indicators are also classified according to the type of data they contain. We can have continuous (metrics), discrete (counts, we can perform operations of a linear space can be performed, if meaningful, such as weightings, weighted sum as utility function), ordinal (ratings/ranks, for which at least comparisons are possible) and nominal data (descriptive, but restricted applicability, for instance stratifications).

One important question we have to address is how many indicators there should be within a system. There is no single answer. Choosing too few of them carries

---

[1]For the main characteristics of these two different approaches, see Maggino (2017a). The literature about the difference between these two tipes of models is rich. As shown by Alaimo and Maggino (2020), the state of the theory on formative models has been in intense discussion for some years. Several authoritative scholars (for instance, Edwards 2011; Aguirre-Urreta et al. 2016) have questioned the validity of this method and published appeals to no longer host its applications in scientific journals. The debate seems to be far from being resolved. We would like to point out that the choice between the two types of model does not depend directly on the researcher, but exclusively on the nature and direction of relationships between constructs and measures.

[2]For more information, see Maggino 2017a.

the risk of not considering fundamental aspects of the phenomenon. In general, multiple indicators make it possible to measure conceptual dimensions with greater precision (multiple measurements make it possible to compensate for random errors), accuracy and discriminant capacity. But, using too many of them can lead to errors and disturbances in the measurement (information noise, like redundancy). The selection must always be guided by the conceptual model, the basis of each system.

Dealing with a multi-indicators system also raises the question of finding tools and methods that allow us to analyse how they are related to each other. The increasing dissemination and use of indicators in recent years have highlighted the weakness of traditional approaches for their treatment. The multi-indicators systems require approaches allowing more concise views able to summarizing the complexity. In this case, using traditional statistical techniques (in particular, those of dimensional reduction: principal component analysis, factor analysis, etc.) is not functional. The guiding concept, as previously written, crossing all possible strategies is *synthesis* (Maggino 2017b). It may concern two different aspects of the system (Maggino 2009), the units (which aims at aggregating the individuals value of one indicator observed at micro level; this synthesis should allow the created macro units to be compared – social groups, age groups, geographic areas – with reference to the indicators of interest) or the basic indicators (which aims at aggregating the values referring to several indicators for each unit, micro or macro).

Focusing on the latter, synthesis can be faced through two different approaches, aggregative-compensative and non-aggregative. The aggregative-compensative approach[3] is the mainstream method to the synthesis. In fact, the term "aggregation" is often use as a synonym of synthesis is and, implicitly or not, it is generally taken for granted that "evaluation implies aggregation". However, many critical issues affect this approach. First of all, the treatment of ordinal data. To be aggregated and processed in an effective way, we must consider them as "numbers"; thus, they must be scaled to numerical values. Unfortunately, this often turns out to be inconsistent with the nature of phenomena and produces results that may be largely arbitrary, poorly meaningful and hardly interpretable.

Another question regards the relationships among indicators. According to Fattore and Maggino (2014), it is clear that many data systems available to social scientists often comprise weakly interdependent attributes; this situation is a major obstacle to effective synthesis through aggregative procedures. The composite indicators approach results inappropriate in these cases, because all its procedures are aggregative and (even partially) compensative. Another problem is linked to the interpretation of results. The values assumed by composite indexes often tend to be representative of situations profoundly different from each other, as a result of

---

[3]For a review of the aggregative-compensative approach, please see Maggino 2017b, Mazziotta and Pareto 2017.

different values in the elementary indicators, or similar situations between them.[4] This can lead to misleading conclusions. One possible solution to the weakness of the aggregative-compensative methods can be to synthesize not necessarily through aggregation. This need has led the research to focus on developing alternative methods, the non-aggregative approaches. They respect the ordinal nature of the data and the process and trends of phenomena (not always linear but more frequently monotonic) and avoid any aggregation among indicators. One of the most useful references in this perspective is the Partial Order Theory (see for instance Davey and Priestley 1990). Non-aggregative approaches are focused not on dimensions but on profiles, which are combinations of ordinal scores, describing the «status» of an individual. Profiles can be mathematically described and analyzed through tools referring to that theory, in particular Partially Ordered Set (POSET). Through these tools, information can be extracted directly from the relational structure of the data, obtaining robust results, not based on binding hypotheses. This approach gives an effective representation of data and their structure.[5] The application of POSET methodologies lead to conclusions much more meaningful, robust and consistent than those based upon traditional statistical tools. Moreover, focusing on the profiles allows having a "synthesis" always representative of the effective combinations among basic indicators. This avoids flattening the differences between different combinations in a single numerical value and misinterpreting the results.

2   Overview About Concepts in Partial Order Theory

2.1   Two Basic Approaches

Partial order theory is a mathematical discipline which combines elements of Graph theory and Combinatorics. In the broadest sense is Graph theory that branch of Discrete Mathematics which studies relations. Within the context of the analysis of indicators the relations are defined on the basis of profiles. Depending on the type of data of the indicators two main lines of analysis approaches can be defined:

1. Ordinal data: The corresponding indicators may have finite and discrete values of different degrees. Let $Q(k)$ the set of values of the kth indicator, then $\prod Q(k)$, $k = 1, \ldots, m$, m the number of indicators, is the set of all possible value's combinations, i.e. of all possible profiles. The analysis of this set as described

---

[4]For instance, in two papers on sustainable development and regional differences in Italy, Alaimo and Maggino show how similar values in composite indicators assumed by different regions can represent similar or even completely different combinations in basic indicators (Alaimo and Maggino 2018, 2020).

[5]In particular, the computations performed to assign numerical scores to the statistical units involve only the ordinal features of data, avoiding any scaling procedure or any other transformation of the kind.

by Fattore 2016, Fattore et al. 2011 allows powerful conclusions and should be called the value-based approach, because the *realization* of a profile by objects does not play a leading role.

2. Metric data: The set Q(k) is not finite and $\prod Q(k)$, k=,1,...,m, is an infinite set. Although a powerful approach exists to analyse this infinite set by methods described by Kerber 2017, Kerber and Bruggemann (this book, and 2015), Bruggemann and Kerber 2018, in practical applications the *set of objects* is considered, together with their profiles. This approach is focusing on objects, which have a profile, corresponding to the set of indicators and should therefore be called the object-based approach. As in biology, chemistry and physics measured data are the basis for research, the object-based approach is usually applied. See for instance Carlsen and Bruggemann 2017, Carlsen 2018.

2.2 Interplay: Indicators and Objects

The interplay of indicators and the objects and thus the role of graph theory, is more easily seen in the object-based-approach, which will be described in more details in the following.

In the field of indicators applied to objects the graph -theoretical relations can be

- Between indicators
- Between objects (being described and quantified by indicators) and
- Between objects and indicators.

In many cases indicators are applied in the framework of ranking studies, hence the relations studied by graph theory are *directed* relations; and the graphs *directed graphs*. Ranking has to do with *ordering*, hence the directed relations should obey axioms, namely those of partial order theory: They have to be reflexive (an object must be comparable with itself), antisymmetric, if an object x is better than object y then the reverse can only be true if the two objects are identical (often in partial order theory this axiom is relaxed by replacing 'identical' with 'equivalent') and finally transitive, if object x is better than object y, and object y is better then object z, then this implies that object x is better than object z.

The study of directed graph based on order relations leads to a certain type of directed graphs, namely to *acyclic triangle free directed* graphs, often called Hasse diagrams (or simply line diagrams). Hasse diagrams are an appropriate visualization of partially ordered sets (but not the only ones). Although the axioms of order theory sound plausible, they impose a very important additional requirement on indicators (if applied in ranking studies): Any single indicator implies an order among objects. A consequence is that even if indicators induce an order among objects they may be counter current. A drastic example is provided by environmental chemistry: A chemical may be considered as hazardous, if its tendency to accumulate in biota is high. This accumulation tendency can be measured or estimated Log KOW. Another typical aspect is its toxicity. Toxicity is usually measured as that concentration (of the chemical) where p% of test organisms show an adverse effect (LC). In contrast to accumulation a large value of LC implies a *lower* toxicity. Hence both indicators together will be meaningless, a *re-orientation* is necessary to find a co-monotone behaviour in both aspects. When a multi-indicator system is conceptualized in order to support decisions in a complex system then the above requirements will select out many candidates.

2.3   *Indicator Systems*

A final system of indicators leading to a Hasse diagram (or a poset if the visualization is not in the focus of the study) can be seen as a system, where the parts (the single indicators) are combined to a graph which indeed allows more insight into the indicator system and the objects described by the indicators, then an ensemble of single indicators. The Hasse diagram can be in two extremal states, both visualized by extremely simple Hasse diagrams:

 (i) No relation among the objects -AC (antichain)
(ii) All objects are related – CC (complete chain)

The first case (AC) may be a result of information noise or by an incorrect orientation, the second case (CC) leads to a ranking, i.e. to an ordering of all objects under all the indicators applied. In reality the Hasse diagram is in a state between the two extremal cases and in the mathematically oriented literature there are attempts to measure the complexity of Hasse diagrams (Luther et al. 2000; Restrepo 2014). The degree, measuring the state of a Hasse diagram (or of the partial order it is representing) with respect to AC or CC is of eminent importance and is certainly not available if the indicator-system are seen only as an ensemble of many single indicators.

The fact that a Hasse diagram is somewhere between the two states leads immediately to the second mathematical component, namely combinatorics. The question is, as to how far a ranking can be found without an aggregation of the indicators, for example by weighted sums. The conceptual idea is very simple: Can we find an order preserving map, by which a poset is mapped into an order, where all objects are mutually comparable. For example: Three indicators leading to a Hasse diagram of type (AC) imply that there are six such mappings! When a Hasse diagram belongs to type (CC) then the requirement of "order preserving" implies that only and only one mapping can be found. Generally, however one obtains a number of order preserving mappings between 1 and 2^m, with m being the number of indicators. Once such a set of mappings is obtained, statistical measures can be applied to characterize this set of mappings. In the literature often just mean values are used, to describe in the average the position each object has in the image of each mapping (Rkav: average rank). Although numerical devices are known (Bubley and Dyer 1999) and also a mathematically extremely elegant approach (De Loof et al.

2006), the task to derive from the Hasse diagrams *easily understandable* estimation methods for Rkav is still an important object of research in the future (see also Bruggemann and Annoni 2014; Bruggemann and Carlsen 2011, 2014; Bruggemann et al. 2004). It should be noted that in the field of research of indicator systems the pure rendering of results by an effective but hardly understandable method cannot be satisfying, because finally it is of main importance to understand, i.e. to trace back the reason, why certain objects under certain indicator systems get a certain position in the ranking!

The extremal case (AC) throws another light on the idea that an indicator system is more than its parts: The Hasse diagram is of no direct help as there is no order relation among the objects. Nevertheless, the indicators themselves have a different influence why order relations among objects are broken, thus inducing new relations, namely among indicators (Bruggemann and Voigt 2011, 2012; Bruggemann and Carlsen 2014).

Even if the Hasse diagram is somewhere in the middle between AC and CC, algebraic concepts (taken from universal algebra, Davey and Priestley 1990) such as congruence, linear sums and separated sums, are providing powerful tools to analyse the indicator system. Congruence, for instance provides a methodological framework, as to how far a family of subsets of objects can be considered as a CC-system, i.e. not the objects themselves can be ranked, but appropriate selected subsets of objects (Carlsen and Bruggemann, Soc.Ind. Res., Febr., 2019 submitted).

References

Aguirre-Urreta, M. I., Rönkkö, M., & Marakas, G. M. (2016). Omission of causal indicators: Consequences and implications for measurement. *Measurement: Interdisciplinary Research and Perspectives, 14*(3), 75–97.

Alaimo, L. S. (2020). *Complexity of social phenomena: Measurements, analysis, representations and synthesis.* Unpublished Doctoral Dissertation, University of Rome "La Sapienza", Rome, Italy.

Alaimo, L. S., & Maggino, F. (2018). Sviluppo sostenibile e differenze regionali. In E. di Bella, F. Maggino, & M. Trapani (Eds.), *AIQUAV 2018. V Convegno dell'Associazione Italiana per gli Studi sulla Qualità della Vita. Libro dei Contributi Brevi* (pp. 199–206). Genova: Genova University Press.

Alaimo, L. S., & Maggino, F. (2020). Sustainable development goals indicators at territorial level: Conceptual and methodological issues—The Italian perspective. *Social Indicators Research, 147*, 383–419. https://doi.org/10.1007/s11205-019-02162-4.

Bruggemann, R., & Annoni, P. (2014). Average heights in partially ordered sets. *MATCH Communications in Mathematical and in Computer Chemistry, 71*, 101–126.

Bruggemann, R., & Carlsen, L. (2011). An improved estimation of averaged ranks of partially orders. *MATCH Communications in Mathematical and in Computer Chemistry, 65*, 383–414.

Bruggemann, R., & Carlsen, L. (2014). Incomparable-what now? *MATCH Communications in Mathematical and in Computer Chemistry, 71*, 699–714.

Bruggemann, R., & Kerber, A. (2018). Fuzzy logic and partial order; first attempts with the new PyHasse-program L_eval. *MATCH Communications in Mathematical and in Computer Chemistry, 80*, 745–768.

Bruggemann, R., & Patil, G. P. (2011). *Ranking and prioritization for multi-indicator systems – Introduction to partial order applications*. New York: Springer.

Bruggemann, R., & Voigt, K. (2011). A new tool to analyze partially ordered sets – Application: Ranking of polychlorinated biphenyls and alkanes/alkenes in river main, Germany. *MATCH Communications in Mathematical and in Computer Chemistry, 66*, 231–251.

Bruggemann, R., & Voigt, K. (2012). Antichains in partial order, example: Pollution in a German region by Lead, Cadmium, Zinc and Sulfur in the herb layer. *MATCH Communications in Mathematical and in Computer Chemistry, 67*, 731–744.

Bruggemann, R., Sørensen, P. B., Lerche, D., & Carlsen, L. (2004). Estimation of averaged ranks by a local partial order model. *Journal of Chemical Information and Computer Sciences, 44*, 618–625.

Bubley, R., & Dyer, M. (1999). Faster random generation of linear extensions. *Discrete Mathematics, 201*, 81–88.

Carlsen, L. (2018). Happiness as a sustainability factor. The world happiness index: A posetic-based data analysis. *Sustainability Science, 13*, 549–571.

Carlsen, L., & Bruggemann, R. (2017). Fragile state index: Trends and developments, a partial order data analysis. *Social Indicators Research, 133*, 1–14.

Davey, B. A., & Priestley, H. A. (1990). *Introduction to lattices and order*. Cambridge: Cambridge University Press.

De Loof, K., De Meyer, K. H., & De Baets, B. (2006). Exploiting the lattice of ideals representation of a poset. *Fundamenta Informaticae, 71*, 309–321.

Edwards, J. R. (2011). The fallacy of formative measurement. *Organizational Research Methods, 14*(2), 370–388.

Erdi, P. (2008). *Complexity explained*. Berlin, Heidelberg: Springer. https://doi.org/10.1007/978-3-540-35778-0.

Fattore, M. (2016). Partially ordered sets and the measurement of multidimensional ordinal deprivation. *Social Indicators Research, 128*, 835–858.

Fattore, M., & Maggino, F. (2014). Partial orders in socio-economics. A practical challenge for poset theorists or a cultural challenge for social scientists? In R. Bruggemann, L. Carlsen, & J. Wittmann (Eds.), *Multi-indicator systems and modelling in partial order* (pp. 197–214). New York: Springer.

Fattore, M., Maggino, F., & Greselin, F. (2011). Socio-economic evaluation with ordinal variables: Integrating and poset approaches. *Statistica & Applicazioni, Special Issue, 2011*, 31–42.

Gorban, A. N., & Yablonsky, G. S. (2013). Grasping complexity. *Computers & Mathematics with Applications, 65*(10), 1421–1426.

Kerber, A. (2017). Evaluation, considered as problem orientable mathematics over lattices. In R. Bruggemann & M. Fattore (Eds.), *Partial order concepts in applied sciences* (pp. 87–103). Cham: Springer.

Kerber, A., & Bruggemann, R. (2015). Problem driven evaluation of chemical compounds and its exploration. *MATCH Communications in Mathematical and in Computer Chemistry, 73*, 577–613.

Lazarsfeld, P. F. (1958). Evidence and inference in social research. *Daedalus, 87*(4), 99–130.

Lima, M. (2013). *Visual complexity: Mapping patterns of information*. New York: Princeton Architectural Press.

Luther, B., Bruggemann, R., & Pudenz, S. (2000). An approach to combine cluster analysis with order theoretical tools in problems of environmental pollution. *Match, 42*, 119–143.

Maggino, F. (2009). *The state of the art in indicators construction in the perspective of a comprehensive approach in measuring Well-being of societies*. Firenze: Firenze University Press, Archivio E-Prints.

Maggino, F. (2017a). Developing indicators and managing the complexity. In F. Maggino (Ed.), *Complexity in society: From indicators construction to their synthesis* (pp. 87–114). Cham: Springer.

Maggino, F. (2017b). Dealing with synthesis in a system of indicators. In F. Maggino (Ed.), *Complexity in society: From indicators construction to their synthesis* (pp. 115–137). Cham: Springer.

Mazziotta, M., & Pareto, A. (2017). Synthesis of indicators: The composite indicators approach. In F. Maggino (Ed.), *Complexity in society: From indicators construction to their synthesis* (pp. 161–191). Cham: Springer.

Meadows, D. H. (2009). *Thinking in systems: A primer*. London: Earthscan.

Morin, E. (1984). Le vie della complessità. In G. Bocchi & M. Cerruti (Eds.), *La sfida della complessità* (pp. 49–60). Milano: Feltrinelli.

Porter, T. M. (2001). *Trust in numbers: The pursuit of objectivity in science and public life*. Princeton: Princeton University Press.

Restrepo, G. (2014). Quantifying complexity of partially ordered sets. In R. Bruggemann, L. Carlsen, & J. Wittmann (Eds.), *Multi-indicator systems and modelling in partial order* (pp. 85–103). New York: Springer.

Sacconaghi, R. (2017). Building Knowledge. *Between Measure and Meaning: A Phenomenological Approach*. In Complexity in Society: From Indicators Construction to their Synthesis, ed. F. Maggino, 51–68. Cham: Springer.

Tufte, E. R. (2015). *The visual display of quantitative information*. Cheshire: Graphics Press.

Assessing Inhomogeneous Indicator-Related Typologies Through the Reverse Clustering Approach

Jan W. Owsiński, Jarosław Stańczak, Sławomir Zadrożny, and Janusz Kacprzyk

1 Introduction

The paper addresses the following pragmatic problem: We are given a typology of spatial units (here: Polish municipalities, close to 2500 in number), elaborated for definite planning purposes (see Śleszyński and Komornicki 2016). The typology resulted from a complex procedure, involving a number of indicators. Moreover, the set of indicators used was not uniform across all (types of) municipalities, for the procedure had a "branching" character, implying different subsets of features for particular types. At the same time, the number of types had to be kept "reasonable" for pragmatic purposes. This gives rise to several questions, not only on the "validity" of the typology, but also its "meaning", and, last but not least, "intuitive appeal", so important from the policy making standpoint.

In view of these questions a study was performed, aimed at (1) providing a comparative material for the typology elaborated, (2) basing this comparative material on a uniform set of data (variables, indicators), (3) identifying the effects of inhomogeneity of the original criteria and use of incommensurable variables, hard to express on a par with the others.

This exercise was based on the "reverse clustering" approach, developed by the authors (Owsiński et al. 2017a, b; 2021). This approach consists in attempting to recreate a given partition, P_A, of a set of n objects, on the basis of a set of data on these objects, X, composed of vectors x_i, $x_i = [x_{i1}, \ldots, x_{im}]$. We wish to obtain a partition P_B of the analysed set of objects that is as close to P_A as possible (e.g. in terms of the Rand index), by applying an optimisation procedure, described in the references mentioned.

J. W. Owsiński (✉) · J. Stańczak · S. Zadrożny · J. Kacprzyk
Systems Research Institute, Polish Academy of Sciences, Warszawa, Poland
e-mail: owsinski@ibspan.waw.pl; Jaroslaw.Stanczak@ibspan.waw.pl; zadrozny@ibspan.waw.pl; Janusz.Kacprzyk@ibspan.waw.pl

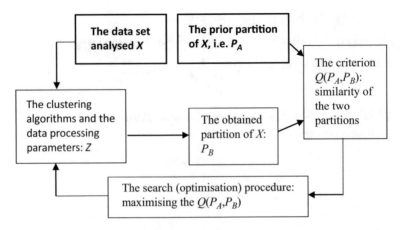

Fig. 1 Schematic view of the reverse clustering procedure

The optimisation procedure is applied to the vector Z, describing the selected clustering algorithm that yields the partition P_B. This vector is composed of: (i) the choice of the clustering algorithm, (ii) the choice of its essential parameter(s) – e.g. the number of clusters, or some threshold distance etc., (iii) the weights, or choice, of variables, (iv) the distance definitions used (e.g. as expressed through Minkowski exponent).

The working of the entire procedure of reverse clustering is schematically shown in Fig. 1.

The preliminary results from the concrete study, considered here, were reported in Owsiński et al. (2018). Now, besides presenting an ampler view of the results, we focus on the broader implications for the use of a similar approach in other settings, where the "composite indicator" context applies, while either continuous variables are used along with the "strongly" discrete ones to categorise objects, or the procedure applied involves branchings, so that, in effect, a single-axis-indicator might not render appropriately the resulting categories.

2 The Study with Its Narrow and Broader Motivations

We present here a study, in which the data on all of Polish municipalities are analysed in the presence of a definite typology of these municipalities, elaborated for a concrete (spatial planning) purpose by the specialists from the Institute of Geography and Spatial Organization of the Polish Academy of Sciences (Śleszyński and Komornicki 2016). We apply in this context the reverse clustering approach, meaning that we try to find the parameters of the broadly conceived clustering procedure, which, when applied to the data on the municipalities, yield a possibly similar typology to the one we are given at the outset, in this case – provided by the geographers.

The study had, therefore, three essential motivations:

1. Yet another check on the capacities and effectiveness of the reverse clustering approach (Owsiński et al. 2017a, b; 2021), for yet another set of data and for a different kind of substantive prerequisites;
2. Analysis of a typology, given by a complex and "branching" process, so as to derive conclusions on the "deviations" of this typology from the one obtained on the basis of a coherent, unified data set on the same subject, this analysis leading, hopefully, to some broader conclusions; and
3. The substantive analysis of the given data set, for comparison with the original typology, and, perhaps, some tangible substantive conclusions.

The broader meaning of this exercise (implied in the motivation 2 above) results from the following image of the situation:

There is a set of data, concerning a collection of entities, describing certain features of these entities. We wish to categorise these entities into a relatively small number of categories (much smaller than the number of entities). (We abstract here from the question whether we deal with an entire "population" or a "sample". In the latter case we assume the "sample" is "representative".) In general, it would often be convenient, if the categories formed a linear order (for the reason of categorization in many cases refers somehow to the "composite indicator" context), although it may happen that they do not. Namely, in many contexts, even if we are aware of the essential multidimensionality of the subject matter, we deal with some sort of "general axis", corresponding to the potential or hypothetical "composite indicator", this axis representing the "magnitude / intensity / graveness of the phenomenon", to which the indicator is supposed to refer. In this particular case we deal with the "urban-rural-peripheral" axis, and the potential divergences are associated with some special phenomena and corresponding groups of units, e.g. urban areas featuring different patterns of development (or, indeed, decay), or peripheral rural areas, where the share of settling urbanites plays an important role (see also Fig. 3 further on).

We assume that we do dispose of a certain categorization of the entities in question, this categorization coming out of a special procedure, which involves, say, categorical variables, branchings, various data sets in various branches of the proce-dure, etc., like in the examples of Fig. 2, showing two typical procedures, related to social care / unemployment benefit registration and relevant data production, which can hardly be translated into a unified data set and easily processed as such.

Having the data set and its partition, we wish to recreate the partition for this data set as faithfully as possible, using clustering. We shall not be using the "decision variables" of the procedure (or, if used, they will be treated like other variables), and the data will be the same for all objects. In the here analysed case we used a different set of variables, as our assumption was to use only the publicly available data. The substantive sense of the data remains, though, except for one or two original variables, not accessible to us, very much the same.

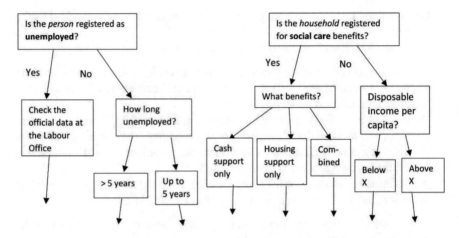

Fig. 2 Two examples of the procedures, leading to the kind of prior categorization of interest here

From the analysis we wish to get an image more "naturally" related to the data and to compare it with the original categorization, in terms of: (a) existence of potential "twists" or "artifacts" in the original partition, resulting from the application of "decision variables", "thresholds", etc.; (b) the number and the nature of categories (clusters) obtained in a more flexible environment; and (c) detecting the potential outliers, which might, again, be overlooked in the original partition. In the case there is a definite need of the categories to form an ordering (along the hypothetical axis of a "composite indicator"), the exercise might also serve to (d) confirm or put to doubt the possibility of actual formation of such an order in a more "natural" manner. This kind of situation is schematically depicted in Fig. 3. We would like to indicate, in the context of this illustration, that two quite typical issues arise, in connection with a potential partition of a data set that in its general shape is distributed along the already mentioned "main indicator axis", namely: **1.** Frequent arbitrary manner of cutting into pieces the "cloud" of data points, stretching along this main axis; **2.** In the cases of "branchings" out of this main axis the question of their relation to the main axis (which ones lie along the main axis, and which one diverge from it?) and the existence of an actual separation from the main axis.

An example of such a situation may be also provided by the case of poverty measurement, and the categories, established in this context. In the measurement the leading variable seems to be income per capita in the household, social and other benefits included, while other variables, such as number of persons in the household, ages of household members, their education, health conditions, housing situation, etc., being usually either highly or at least significantly correlated. There may, however, be yet another variable, or group of variables, that are less correlated, but for some definite reason (e.g. crosschecking) included in the measurement. Say: driving license? obesity? political attitude? arms possession?

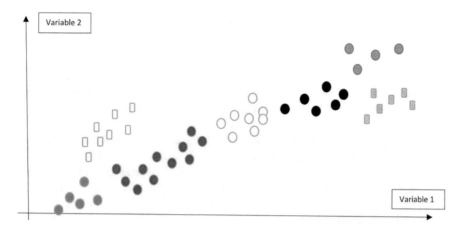

Fig. 3 An image of categorization, for which the reverse clustering would reveal divergences from a "natural" partition. Rounded shapes indicate objects, belonging to (five) clusters "along the primary indicator axis", while the rectangular ones – belonging to (two) "divergent clusters"

3   On the Reverse Clustering

As outlined in the Introduction, the reverse clustering technique aims at identifying the partition P_B of a certain set of n entities into clusters $(A_q, q = 1, \ldots, p)$, which, for the given set X of (descriptions of) these entities, $X = \{x_i = \{x_{i1}, \ldots, x_{im}\}\}_{i = 1, \ldots, n}$, and a given a priori partition P_A of these entities, is the closest to P_A. In the search for the best P_B we apply a natural measure of similarity, or distance, between the partitions, namely the Rand index (Rand 1971) *or some* of its variations (see, e.g., Hubert and Arabie 1985). The Rand index in its original form counts, for two partitions, the following four numbers: of pairs of objects which fall into the same cluster in both partitions (a), those that are in different clusters in both partitions (b), and that are in the same cluster in one partition and in different clusters in the other $(c$ and $d)$. The original Rand index is simply the quotient: $(a + b)/(a + b + c + d)$, with $a + b + c + d = \frac{1}{2}n(n - 1)$, of course. The search is performed with the evolutionary algorithm of own design (Stańczak 2003), featuring two-level selection: of individuals and of the operations.

Although it could be argued that the very problem of the reverse clustering is equivalent to some kind of "supervised classifier choice", it is, in general, not. Namely, (1) the aim is not to provide the basis for classifying individual incoming new entities, one after another, but rather to map entire new samples (of cardinality perhaps even much bigger than n) into the obtained partition P_B; (2) the new partition P_B needs not be composed of the same number of clusters as P_A; (3) in particular, the approach may lead to the identification of outliers, not entering any of the essential clusters, forming the partition.

In addition, let us emphasise that the generic problem statement encompasses, in fact, a whole variety of the potential situations, having quite different interpretations, and therefore also the statuses of the obtained solutions. A broader account on this is provided in Owsiński et al. (2018, 2021), and we shall only sketch here the respective image. We can speak, namely, of two "axes" in this image:

(I) the degree of "certainty" of P_A, meaning that this partition may be either a "solid" one, e.g. in our case, the classification of the municipalities according to provinces (a province constituting a cluster), or just a hypothesis (an expert opinion);

(II) the degree of association of the partition P_A with the data set X (the known or assumed foundation of P_A on the actual data set X) – in the above case of provinces and municipalities there is no such association, while an expert would most probably base her/his opinion on the data from X, when, for instance, designing a functional typology, like in the present case, although not necessarily in an exact manner.

In this perspective, Table 1 presents the relevant examples of situations potentially encountered.

As already indicated, the vector Z of the clustering procedure parameters sought, is composed of the choice of the algorithm itself, the crucial parameter(s) of the algorithm, the weights of variables, and the scaling of the distance measure (Minkowski exponent). The algorithms accounted for are k-means and similar (Steinhaus 1956; Lloyd 1957), general hierarchical aggregation (parameterised with the Lance-Williams formula, see Lance and Williams 1966, 1967), and DBSCAN (Ester et al. 1996), as a representative of the local density-based algorithms. Thus, quite a broad range of algorithms is covered, with, indeed, a very significant scope of search, regarding the variables composing Z.

Given the composition of the optimised Z one could include in it, for instance, an explicit feature selection procedure or another operation, oriented at the shaping of the description space, as, say a preprocessing stage. We preferred, though, to encapsulate the entire procedure in one optimisation task, aiming integrally at getting possibly close to P_A, including all the parameters we thought would be important.

4   The Case Studied

The particular case here studied (described also preliminarily in Owsiński et al. 2018) concerns the typology of close to 2500 Polish municipalities, elaborated for definite planning purposes by a team from the Institute of Geography and Spatial Organization of the Polish Academy of Sciences (see Śleszyński and Komornicki 2016). As already indicated, the typological procedure was quite complex, with the use of a variety of variables and criteria, and including branching decisions. For our purposes here suffice to quote the "headings" of the typology elaborated, as given in Table 2.

Table 1 A rough classification of the potential situations, treated with the reverse clustering approach: examples of relevant cases

Examples of respective situations:		Degree of certainty of the reference partition P_A		
		Low (or none)	Medium	High
Degree of association of P_A with X	Low (or none)	A random partition of the set of entities	Expert opinion, not based on the data set	Formal partition, based on a criterion from outside of X^a
	Medium	Partition based on one or few of variables in X	Partition based on a part of variables in X^b	Formal partition, based on some of variables from X
	High	Partition based on an arbitrary breakdown of variables from X	Partition based on an arbitrarily selected clustering algorithm	Partition based on a well verified clustering study

[a]Case of classification P_A of municipalities by provinces, when P_B is based on other socio-economic data

[b]The case, treated in Owsiński et al. (2017b), where the partition P_A was based on two out of four variables, while P_B was based on all four variables

Table 2 Functional typology of Polish municipalities

Functional types	Number of units		Population		Area		Population density
	No.	%	in '000	%	'000 km²	%	Persons / sq. km
1. Urban functional cores of provincial capitals	33	1.3	9557	24.8	4.72	1.5	2025
2. Outer zones of urban functional areas of provincial capitals	266	10.7	4625	12.0	27.87	8.9	166
3. Cores of the urban areas of subregional centres	55	2.2	4446	11.6	3.39	1.1	1312
4. Outer zones of urban areas of subregional centres	201	8.1	2409	6.3	21.38	6.8	113
5. Multifunctional urban centres	147	5.9	3938	10.2	10.39	3.3	379
6. Communes having pronounced transport function	138	5.6	1448	3.8	20.06	6.4	72
7. Communes having pronounced non-agricultural functions	222	9.0	1840	4.8	33.75	10.8	55
8. Communes with intensive farming function	411	16.6	2665	6.9	55.59	17.8	48
9. Communes with moderate farming function	749	30.2	5688	14.8	93.83	30.0	61
10. Communes featuring extensive development	257	10.4	1878	4.9	41.59	13.3	45
Totals for Poland	**2479**	**100**	**38,495**	**100**	**312.59**	**100**	**123**

Source: Śleszyński and Komornicki (2016)

Table 3 Ordering of categories according to population densities, persons per sq.km

Categories	1	3	5	2	4	6	9	7	8	10
Population density	2025	1312	379	166	113	72	61	55	48	45

It is obvious from Table 2 that the main axis of the distinctions introduced is – quite naturally – the urban-rural one, meaning the degree of urbanisation, here mostly reflected through population density. As we order the categories from Table 2 conform to population density, we obtain an image as in Table 3.

Thus, there is a clear axis, along which the categories are situated, with, perhaps, definite divergences, related to some of the less densely populated municipality categories (e.g. the positions of categories 7 and 9), very much like in the schematic Fig. 2. This implies even a possibility of devising some kind of aggregate indicator, e.g. along the lines of Owsiński (2017), but such an attempt was not the aim of the study. On the other hand, there is the question of the number of categories, which might have been deliberately minimised in the prior functional typology, with the effect of designing, actually, categories that are not representing the functional types in a similar (balanced) manner. This is best visible on two examples: categories 6 and 7, one referring to a relatively narrowly specialised communes, and the other – in reality – composed of several sub-categories (e.g. municipalities with mining activities, but also with extensive tourist activities!).

In this context, the study intended to possibly accurately recreate the typology shortly characterised above in order to identify potential divergences and their sources, possibly with substantive underpinning. However, in view of the complications of the original procedure, the decision was taken of using a unified set of variables, describing the municipalities, selected so as to possibly faithfully render the general diversity of these units, the result of this selection being shown in Table 4.

Actually, two sets of variables were used in the experiments: the entire set of 21 variables, as shown in Table 4, and the set of 18 variables, from no. 4 till the end. In the latter case, only relative variables are used, while in the former case, the two first variables are very important absolute drivers of differentiation. It must be added, though, that in all calculations the values of variables are unitarised. The data, used by us, were by 1–2 years more recent than those, which constituted the basis for the prior functional typology, but, in view both of the inertia of respective processes and the nature of the variables, this is of no importance for the content of this study.

5 The Outline of Results

An exemplary image of one of the results obtained is provided in Fig. 4. This particular result, composed of 10 clusters (the 11th one, signalled in the map legend, does not appear on it), was obtained with the k-means algorithm. Table 5 presents

Table 4 The choice of variables used in the reverse clustering study

1. Population number	12. Average farm acreage indicator
2. Overbuilt area	13. Registered employment indicator
3. Share of transport related areas	14. Registered businesses per 1000 inhabitants
4. Population density	15. Average business employment indicator
5. Share of agricultural land	16. Share of manufacturing and construction businesses
6. Share of overbuilt areas	17. Pupils per 1000 inhabitants
7. Share of forest areas	18. Students of over-primary schools per 1000 inhabitants
8. Share of population over 60 years of age	19. Own revenues of municipality per inhabitant
9. Share of population below 20 years of age	20. Share of revenues from personal income tax in own communal revenues
10. Birthrate for the last 3 years	21. Share of social care expenses in total communal budget
11. Migration balance for the last 3 years	

Table 5 Comparison (contingency/confusion table) of the original partition and an obtained one, illustrated in Fig. 4. Correspondence between clusters in two partitions was established on the basis of the biggest numbers of objects falling into the same cluster

Clusters in the prior partition	Clusters obtained from the reverse clustering:										Error	Relative error
	1	2	3	4	5	6	7	8	9	10		
1	**16**	0	14	0	2	0	0	0	0	1	17	0.52
2	0	**88**	13	84	26	36	9	2	7	0	177	0.67
3	3	0	**45**	0	7	0	0	0	0	0	10	0.18
4	0	9	3	**76**	9	50	24	4	26	0	125	0.62
5	0	0	5	8	**126**	1	0	0	2	0	16	0.11
6	0	0	0	13	18	**34**	15	33	24	0	103	0.75
7	0	6	0	16	19	11	**98**	27	45	0	124	0.56
8	0	0	0	6	4	56	0	**384**	46	0	112	0.23
9	0	1	0	37	20	97	34	146	**330**	0	335	0.50
10	0	0	0	9	11	11	109	34	88	0	262	1
Error	3	16	35	173	116	262	191	246	238	0	1280	0.517

the contingency (or confusion) table, comparing the original partition from Table 2 with the one illustrated in Fig. 4, definitely implying that the closeness to the original partition is by no means even close to satisfactory, at least in quantitative terms (roughly half of units being "misclassified"). At the same time, though, this result is very much telling in qualitative terms. Let us enumerate at least some of the essential points with this respect.

Thus, first, although the number of clusters obtained is the same as in the original partition, it can be said that one of the original clusters "disappeared" (no. 10), and one "appeared" (the single-item cluster of the capital city of Warsaw). The latter

Fig. 4 Map of Poland with the image of clusters of municipalities, obtained in one of the calculation runs (k-means), with 10 clusters

effect is quite understandable, for even though Warsaw is a relatively small city for the capital of the country of 38 million inhabitants, it has close to two million inhabitants, while there exists a group of cities with 500–700,000 inhabitants, followed by another group of 200–400,000. Here the deviation from the Zipf's law is obvious.

Secondly, the "errors" occur mainly between the clusters of similar character. Thus, e.g. for the distinctly urban clusters 1, 3, 5 and the new metropolitan cluster 10, when taken together, the error is at less than 5%.

Third, in connection with the above, there are two kinds of the initial clusters most affected by the errors: (1) the relatively poorly defined clusters (suburban areas, initial clusters 2 and 4), supposedly actually separated out of quite continuous fragments of the data set, and (2) the clusters, defined not conform to the overwhelming logic of the urban-rural axis (initial clusters 6 and 7, composed of units, performing very specific functions, not really matched by the remaining prior types, e.g. through other similar distinctions).

6   Some Comments on the Relation to Indicator Dimension

Let us start with Table 6, in which the obtained weights of variables are shown for the previously illustrated solution from the reverse clustering. The weights, which are also subject to the optimisation procedure, add up to 1. It can be said that they determine the contribution of individual variables to the solution obtained – the best partition, i.e. the closest to the prior one, that the procedure could find. Note that the two first variables account together for more than 70% of weight of all the 21 variables.

This is, indeed, a very powerful indication that the urban-rural axis plays here the truly dominating role. If we add to this other variables, whose weights exceed 1% (registered businesses, registered employment, migration balance, students, . . .), the significance of this main axis even increases.

When, however, we use the limited set of variables (18 variables, all relative ones, without the first three), the image is different: there is no such strong "pulling force" along the urban-rural axis, and the four leading variables (employment, businesses, own revenues of the municipality, students) account together for 57% of weight. The shape of the obtained clusters is also different, but in general outline similar to the one here presented, with one important exception that Warsaw belongs to her

Table 6 Variable weights obtained in the calculation, illustrated in Fig. 4 and Table 4

Weight	Variable
0.3802	Population
0.3278	Overbuilt area
0.026	Share of transport-related areas
0	*Population density*
0.0186	Share of agricultural land
0.0017	Share of overbuilt areas
0.0043	Share of forest areas
0.0008	*Share of population over 60 years of age*
0.0026	Share of population below 20 years of age
0.0013	Birthrate for last 3 years
0.0397	Migration balance for last 3 years
0.0109	Average farm acreage indicator
0.0441	Registered employment indicator
0.0566	Registered businesses per 1000 inhabitants
0.0064	Employment-based average business scale indicator
0.0119	Share of businesses from manufacturing and construction
0.0009	*Number of pupils per 1000 inhabitants*
0.0338	Number of students of over-primary schools per 1000 inhabitants
0.0098	Own revenues of municipality per inhabitant
0.0227	Share of revenues from personal income tax in own communal revenues
0	*Share of social care expenses in total communal budget*

original category of large cities, while the tenth – new – cluster is again a singleton, but this time an extremely rich provincial commune, where large-scale mining and industrial development is located. The original clusters are, however, even less well reconstructed.

7   Final Observations

The approach here applied allowed for an easy reconstruction of the main axis of partitioning, even if not all of the clusters obtained lie (exactly) along this axis. It is easy to point out these of the original categories of municipalities, which – confirming the quite obvious intuitive supposition – are not in conformity with this axis. This is, largely, what we intended to obtain, when trying to check the capacity of the approach to handle similar typologies, in which some of the types may not necessarily be situated roughly along the "leading indicator", whether explicit or implicit. At the same time, we got the confirmation as to the primary hypothesis of existence of the distinct subgroups, even if in a sense distributed along the "leading indicator" axis. This motivates us to apply the approach to other, similar cases, in which the interpretations, in terms of relation to the "main axis" or the "leading indicator" may not be as simple as in this particular one.

On the substantive side only few conclusions, which appear to be truly justified against the background of the results obtained, namely: (1) it is necessary to consider separately Warsaw and its zone of influence; (2) more generally, treatment of the suburban zones / zones of influence of bigger and smaller agglomerations ought to be reconsidered in the direction of appropriate adjustment / identification of the thresholds / criteria accounted for (a similar observation applies, anyway, also to the rural farming communities); (3) regarding the "special types" of rural municipalities, they should be better justified in terms of their more general socio-economic characteristics.

References

Ester, M., Kriegel, H.-P., Sander, J., & Xu, X.-w. (1996). A density-based algorithm for discovering clusters in large spatial databases with noise. In E. Simondis, J. Han, & U. M. Fayyad (Eds.), *Proceedings of the Second International Conference on Knowledge Discovery and Data Mining (KDD-96)* (pp. 226–231). Menlo Park: AAAI Press.

Hubert, L., & Arabie, P. (1985). Comparing partitions. *Journal of Classification, 2*(1), 193–218. https://doi.org/10.1007/BF01908075.

Lance, G. N., & Williams, W. T. (1966). A generalized sorting strategy for computer classifications. *Nature, 212*, 218.

Lance, G. N., & Williams, W. T. (1967). A general theory of classification sorting strategies. 1. Hierarchical systems. *Computer Journal, 9*, 373–380.

Lloyd, S. P. (1957). Least squares quantization in PCM. Bell Telephone Labs Memorandum, Murray Hill, NJ; reprinted in *IEEE Trans. Information Theory*, IT-28 (1982), 2, 129–137.

Owsiński, J. W. (2017). Is there any 'Law of Requisite Variety' in construction of indices for complex systems? *Social Indicators Research, 136*(3), 1125–1137. https://doi.org/10.1007/s11205-016-1545-5.

Owsiński, J. W., Kacprzyk, J., Opara, K., Stańczak, J., & Zadrożny, S. (2017a). Using a reverse engineering type paradigm in clustering: An evolutionary programming based approach. In V. Torra, A. Dalbom, & Y. Narukawa (Eds.), *Fuzzy sets, rough sets, multisets and clustering. Dedicated to Prof. Sadaaki Miyamoto* (Studies in computational intelligence 671). Cham: Springer.

Owsiński, J. W., Opara, K., Stańczak, J., Kacprzyk, J., & Zadrożny, S. (2017b). Reverse clustering. An outline for a concept and its use. *Toxicological & Environmental Chemistry, 99*, 1078. https://doi.org/10.1080/02772248.2017.1333614.

Owsiński, J. W., Stańczak, J., & Zadrożny, S. (2018). Designing the municipality typology for planning purposes: The use of reverse clustering and evolutionary algorithms. In P. Daniele & L. Scrimali (Eds.), *New trends in emerging complex real life problems. ODS, Taormina, Italy, September 10–13, 2018* (AIRO Springer Series, vol. 1). Cham. ISBN/EAN 9783030004736: Springer. https://doi.org/10.1007/978-3-030-00473-6_33.

Owsiński, J. W., Stańczak, J., Zadrożny, S. & Kacprzyk, J. (2021). *Reverse Clustering: Formulation, Interpretation and Case Studies*. Springer Verlag (forthcoming).

Rand, W. M. (1971). Objective criteria for the evaluation of clustering methods. *Journal of the American Statistical Association, 66*(336), 846–850. https://doi.org/10.2307/2284239.

Śleszyński, P., & Komornicki, T. (2016). Functional classification of Poland's communes (gminas) for the needs of the monitoring of spatial planning (in Polish with English summary). *Przegląd Geograficzny, 88*, 469–488.

Stańczak, J. (2003). Biologically inspired methods for control of evolutionary algorithms. *Control and Cybernetics, 32*(2), 411–433.

Steinhaus, H. (1956). Sur la division des corps matériels en parties. *Bulletin de l'Academie Polonaise des Sciences, IV*(C1.III), 801–804.

Uncertainty in Weights for Composite Indicators Generated by Weighted Sums

Rainer Bruggemann and Lars Carlsen

1 Introduction

Checking the literature with respect to decision support systems (for an overview see for example Colorni et al. 2001; Figueira et al. 2005; Munda 2008; Munda and Nardo 2008), the concepts of partial order theory seem to play a minor role or are at best interim results (cf. ELECTRE family (Roy 1972, 1990) or PROMETHEE (in step I) (Brans and Vincke 1985)). There are two arguments why partial order theory does not play the role which it could have: (1) The appearance of incomparability, i.e., the fact that a ranking, i.e., a total order cannot be obtained, because of non-resolved conflicts in data and (2) the inability to include stakeholders' knowledge.

Originally, the fact that data beyond the data matrix are not needed was considered by the authors of this paper as an advantage, because partial order theory based purely on the data matrix leads to "data driven results" and is therefore free of any subjectivisms, beside those which already were included in the data matrix. All parameters used in ELECTRE (Roy 1990) or PROMETHEE (Brans and Vincke 1985) or other decision support systems beyond the data matrix are considered as subjective and are often difficult to be obtained. Nevertheless, the stakeholder's qualitative knowledge should be included within the framework of partial order theory.

One method in the area of decision support systems is the weighted sum of (normalized) indicator values for each object. Objects are the issue of interest for which a decision is to be found. The advantage of the weighted – sum – approach

R. Bruggemann (✉)
Leibniz-Institute of Freshwater Ecology and Inland Fisheries, Berlin, Germany
e-mail: brg_home@web.de

L. Carlsen
Awareness Center, Roskilde, Denmark
e-mail: LC@AwarenessCenter.dk

© The Author(s), under exclusive license to Springer Nature Switzerland AG 2021
R. Bruggemann et al. (eds.), *Measuring and Understanding Complex Phenomena*,
https://doi.org/10.1007/978-3-030-59683-5_4

45

is its simplicity, i.e., its high potential for transparency if results are to be discussed in, e.g., public meetings. This paper acknowlegdes that modeling or stakeholders' knowledge by weights is a useful step in decision making, however – due to the often vague nature of knowledge about weights – the uncertainty of finding weights must be integrated in the analyses. This idea is already explained in several publications (Bruggemann et al. 2008b, 2012, 2013, Bruggemann and Carlsen 2017, 2018). The number of incomparabilities, U (see below), depending on the measure of uncertainty degree, with respect to the numerical value of the weights plays a central role in this connection. The measure of uncertainty is called s and will be explained in detail below. Between U and s a very simple linear relation U(s) can be derived (cf. Bruggemann et al. 2008b).

In this paper the origin of slight deviations from the predicted results of U(s) is further investigated. To understand these deviations the most simple indicator system is studied, namely consisting of two indicators only. The paper is organized as follows. (1) where basic assumptions and equations are introduced, (2) discussing the concept of "crucial weights", (3) describing examples, initially some fictitious examples, followed by an example taken from the field of sociology, and (4) concluding with a critical discussion.

2 Materials and Method

2.1 Basic Concepts of Partial Order

Let X be a set of objects, labeled by x(i) (i = 1,. . . ,n). Objects could be but not limited to

- chemical compounds
- nations, characterized by for example child well-being indicators
- strategies, characterized by performance indicators
- geographical units, characterized for example by pollution, or (as in another contribution for this book described), by poverty indicators

To define an order relation among them, the relation "≤" has to obey the following order axioms:

- reflexivity: the object can be compared with itself
- antisymmetry: if x ≤ y and y ≤ x ⇒ x = y
- transitivity: if x ≤ y and y ≤ z ⇒ x ≤ z

There are many possibilities to find a realization of an order relation. A special realization of order relations is given by Eqs. 1, 2 and 3:

$$x\,(i) \rightarrow (x\,(i,\,1)\,,x\,(i,\,3)\,,\ldots.x\,(i,\,m))\,\Big)$$

where $x(i, j)$ is the value of the ith object and the jth indicator $(j = 1, .., m)$ (1)

$$x(i1) \leq x(i2) : \iff (x(i1, 1), \ldots, x(i1, m)) \leq \left(x(i2, 1), \ldots x(i2, m)\right)$$
(2)

Equation 2 needs clarification, as it is not yet clear under which conditions one tuple (that of $x(i1)$ is to be considered less or equal to that of $x(i2)$. The way how Eq. 2 can be given a meaning, opens the door to many variant.

By Eq. 3

$$(x(i1, 1), \ldots, x(i1, m)) \leq (x(i2, 1), \ldots x(i2, m)) :$$
$$\iff x(i1, j) \leq x(i2, j) \text{ for all } j = 1, .., m$$
(3)

a partial order is defined, which is close to a statistical interpretation of the data matrix (dm). The reason is that now the properties of the entries of the dm, i.e., of $x(i,j)$ are decisive whether or not an order relation can be established. Two objects, following Eq. 3 are called "comparable", otherwise "incomparable".

The immediate relation to the data and the corresponding indicators has three consequences:

1. Any order relation $x \leq y$ is a direct reflection of the data values of x and y. This is in contrast to many decision support systems, where an order relation cannot easily traced back to the original data, i.e., to the dm.
2. The partial order methodology, based on Eq. (3) is applicable wherever a data matrix is available and where a ranking aim can be defined.
3. It may be necessary to express the ranking aim by a set of indicators

Since the set of indicators $\{q_1, \ldots, q_m\}$ is of main importance for all partial order results based on Eqs. (1, 2 and 3), this set is called the information basis, or - focusing on the role of indicators – a multi-indicator system (MIS) (cf. Bruggemann and Patil 2011). In the literature the method, based on Eq. 3 together with appropriate supporting software, is often denoted Hasse diagram technique (HDT) (Bruggemann and Halfon 2000; Bruggemann et al. 2001, 2008a; Patil and Taillie 2004; Simon et al. 2006; Helm 2006; Bruggemann and Voigt 2011, 2012; Carlsen and Bruggemann 2011, 2014; Newlin and Patil 2010; Annoni et al. 2014; Sørensen et al. 2006) with reference to the German mathematician Helmut Hasse introducing these diagrams (Hasse 1967). Sets X, equipped with a partial order (and here in this paper by the Eqs. (1), (2) and (3)) is called partially ordered sets and is conveniently denoted as posets.

Equation 3 is obviously an extremely hard one, as it demands that

1. all properties considered must follow Eq. 3. Thus, even if m-1 indicators (columns of dm) obey Eq. 3, a single exception would obviously break the order and

2. even minor numerical differences (possibly being considered as scientifically irrelevant) are evaluated and lead to comparabilities/incomparabilities, although the objects should better be considered as equivalent.

These two aspects have led to many activities, such as fuzzy concepts that were introduced (Wieland and Bruggemann 2013; Bruggemann et al. 2011). The role of incomparabilities were analyzed in details (cf. e.g. Bartel and Mucha 2014; Bruggemann and Carlsen 2014a, b, 2015, 2017). Two objects x,y mutually incomparable are denoted as x ∥ y. When the orientation $x \leq y$ or $x \geq y$ is of minor interest then the mere fact of comparability is denoted by $x \perp y$.

2.2 Modelling a Decision Support System

As already mentioned, for the sake of public acceptance, the model for decision support should be simple and the weighted sum of indicator values of an object seems to be the best starting point. The synthetic indicator, also known as composite indicator CI of an object x will be calculated, according to Eq. 4.

$$CI(x) = \Sigma \ (g(j) * x(i,j)) \ i = 1, \ldots, n; j \Big) \ 1, \ldots, m \qquad (4)$$

The entries of a data matrix $x(i,j)$ must be metric in order to combine them by multiplication with a scalar $g(j)$ and subsequent additions. Furthermore, the entries $x(i,j)$ and the weights $g(j)$ should conveniently be normalized, i.e., being elements of the range [0,1]. The boundary condition for the weights $g(j)$ is:

$$\Sigma \ g(j) = 1 \ j =, 1, \ldots, m \qquad (5)$$

and

$$0 \leq g(j) \leq 1 \qquad (6)$$

The crucial point is that (as already mentioned) most often a sharp value for the weights $g(j)$ cannot be given. Therefore the theoretical concept (already published and discussed in more detail in Bruggemann and Carlsen 2017) can be described by the following six items:

1. $g(j)$ is taken from an interval $[g(j)_{min}, g(j)_{max}]$, for the sake of simplicity it is written: $g(j)_{min} = gjmin$ and $g(j)_{max} = gjmax$.
2. $s(j):=$ gjmax − gjmin is called the degree of uncertainty with respect to the weights.
3. it is assumed that $s(j) = s$ for all j.
4. the evolution concept
5. identification of the number of incomparabilities, U, as a leading quantity and
6. in refinement of (5), the concept of Us and other quantities derived from U.

The concepts (4)–(6) are explained in more details below:

4. *Evolution:*

Let g(j) be a value taken from [0,1] then the selection of a series of the pairs (gjmin, gjmax) induces a set of posets. Let for example consider asystem with m = 2 and select as weights g(1) = 0.5, g(2) = 0.5. If s = 0 then there is a sharp knowledge and no uncertainty with respect to g(1) and g(2). Consequently, there is one and only one CI, and U, i.e., the number of incomparabilities equals 0. Now let gjmin = 0.45 and gjmax = 0.55., then the resulting s equals 0.1 and weights g(1), g(2) may be selected within this range. There is obviously a slight uncertainty, which, however, may not lead to incomparabilities. Let now be gjmin = 0.3 and gjmax = 0.7, then s = 0.4 and there is considerable uncertainty concerning the selection of weights. With other words: Around a fixed tuple g defined by (g(1), g(2), . . . ,g(m)) a (mathematical) environment, env(g,s), is of interest, where the starting tuple g and its uncertainty s is to be specified. Consequently to each environment belongs a set of composite indicators, which can, but most no be co-monotonic. Increasing s from 0 (sharp knowledge about the weights) until s = 1 (no knowledge at all) and regarding Eqs. (4, 5 and 6) leads to a series of environments as follows:

$$\text{env}\,(g, 0) \subseteq \text{env}\,(g, 0.1) \subseteq \ldots .\text{env}\,(g, 1) \tag{7a}$$

$$\text{env}\left(g', 0\right) \subseteq \text{env}\left(g', 0.1\right) \subseteq \ldots .\text{env}\left(g', 1\right) \tag{7b}$$

. . . .

$$\text{env}\left(g'', 0\right) \subseteq \text{env}\left(g'', 0.1\right) \subseteq \ldots .\text{env}\left(g'', 1\right) \tag{7c}$$

Equations (7a, 7b, . . . ,7c) result from the fact that g also can be varied (symbolized by g, g', g//). The Scheme (7a-. . . 7c) makes clear that it will be difficult, to check all resulting posets, due to the set of CI, caused by a certain environment. Therefore a controlling quantity is needed to observe the development of posets in a general manner:

5. *Incomparability as controlling quantity:*

The number of incomparabilities U(g,s) for each env(g,s) is introduced, measuring the number of pairs x ∥ y of each poset induced by the tuple g and s. In fact, by Eq. 4 the order relations among the objects are no more a matter of the indicators but on the values of the set of composite indicators {CI} possible within the actually used environment env(g,s). To stress this, we will sometimes write x ≤ {CI} y or x ∥CI y . Furthermore, the values of the composite indicator will be denoted CI(g,x) (or CI(g) to stress the role of g.

6. Us and other quantities derived from U:

The poset, based on the data matrix alone, with n objects and m indicators will have a number of incomparabilities which is the "total incomparability" and which is called U0. It is then clear that

$$U(g, 0) = 0 \tag{8a}$$

and

$$U(g, 1) = U0 \tag{8b}$$

In former publications (Bruggemann and Carlsen 2017) it is shown that the Eq. (9) describes sufficiently the general situation for a certain g and s varying from 0 to 1:

$$Us = s * U0 \tag{9}$$

In the following the incomparabilities are called

- *Udirectly,* if the determination of U is done, by checking each poset in env(g,s)
- *Uan* if U can be determined by means of other quantities, such as the crucial weights (see below)
- *Us* from Eq. (9)
- *U0* the number of incomparabilities of the poset under m indicators, i.e. without any additional information beyond the data matrix.

In Fig. 1 the evolution of U on the basis of Eq. (9) and Udirectly, i.e. directly determined from all the posets resulting from different values of s is shown. To be clear: each environment env(g,s) allows a set of weights, which in turn allows different CI, and these CI, evaluated similar to Eqs. 1, 2 and 3 (replace x(I,j) by CI(x(i),k), k being a label for the resulting CI and check x $\|_{CI}$ y).

Hereto, a fictitious data matrix with 14 objects and m = 3 indicators was applied.

Fig. 1 Evolution of U, based on Eq. (9), and of the number of incomparabilities directly determined from all posets, resulting from env(g,s), "Udirectly" is shown. Data are shown in Table 1

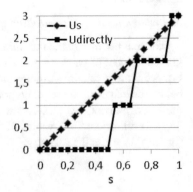

Fig. 2 The Hasse diagram resulting from Table 1. The total number of incomparabilities, U0 = 3

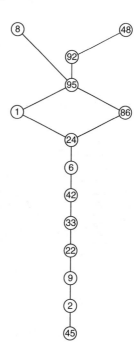

Table 1 Fictitious data matrix

SEQ	q(1)	q(2)	q(3)
1	4,356,708,827	4,110,873,864	3,891,820,298
2	0,693,147,181	0	0
6	4,189,654,742	3,850,147,602	363,758,616
8	5,017,279,837	4,976,733,742	4,905,274,778
9	2,890,371,758	2,772,588,722	1,386,294,361
22	3,044,522,438	2,833,213,344	2,564,949,357
24	4,204,692,619	3,850,147,602	3,663,561,646
33	3,218,875,825	3,044,522,438	2,833,213,344
42	336,729,583	336,729,583	3,218,875,825
45	0	0	0
48	5,003,946,306	5,164,785,974	5,030,437,921
86	4,234,106,505	4,248,495,242	3,828,641,396
92	49,698,133	5,081,404,365	4,955,827,058
95	4,418,840,608	4,465,908,119	4,043,051,268

The Hasse diagram is shown in Fig. 2:

Checking Fig. 1 the deviations from Eq. 9 seem to be not too large Thus, Eq. 9 can still be used as a general guide for the evolution of posets due to increasing values of s (given a certain weight tuple g). Nevertheless, it is of interest, to understand the deviation, which we are considering as a "fine-structure" of the evolution. It is clear that the low number of incomparabilities (U0 = 3) implies at maximum 3 jumps in the curve for Udirectly. So far, the number of "jumps" in Fig. 1 is obvious. However, where will the jumps appear?

2.3   Understanding the Fine-Structure

It is completely clear that the ordinal structure of orderings within partial order theory react on the evolution due to s only in discrete steps. The reason is that the partial order is changing, when a transition from $x \perp y$ to $x\|y$ appears. Nevertheless it is of interest at which values of weights, within a series of env(g,s) such a transition and hence an enhancement of U is to be expected. In order to analyze the fine-structure, the simplest system is assumed, where still incomparabilities can appear, i.e., the m = 2 system. Starting from low s-values there are only so few CI's possible that almost for all object pairs (x, y) it will be found $x \perp y$. Increasing s, some CI's may exist such that a comparability relation is transferred to an incomparability relation. This means that the extension of the range for possible weight values can be formulated as follows:

s small : CI1 (x) ≤ CI1 (y) , CI2 (x) ≤ CI2 (y)   →   s enlarged : CI1 (x) ≤ CI1 (y) but CI2 (x) ≤ CI2 (y)

⇓   ⇓

s small g ∈ env (g, s) , such that $x \leq_{\{CI\}} y$ s enlarged : there are weights so that

$$CI (g, x) \leq CI (g, y) \text{ and}$$

$$CI\left(g', x\right) \leq CI\left(g', y\right)$$

Hence, within an m = 2-system, there must be a value gc1 (gc1, (gc2 = 1- gc1) such that CI((gc1, 1-gc1),x) = CI((gc1,1-gc1),y) and if g1 < gc1, then $x \leq_{CI} y$ whereas for g1 ≥ gc1, then $x\|_{CI}y$. These decisive weights (within the m = 2-system) are called crucial weights. Crucial weights can be calculated in closed form, see Eqs. 10a and 10b and for more details Bruggemann et al. 2008b.

$$gc1 = (x (2, 2) - x (1, 2)) / [(x (2, 2) - x (1, 2)) - (x (2, 1) - x (1, 1))] \quad (10a)$$

$$gc2 = (x (2, 1) - x (1, 1)) / [(x (2, 1) - x (1, 1)) - (x (2, 2) - x (1, 2))] \quad (10b)$$

If an environment env(g, s) encompasses one of the possible crucial weights, then there is a chance that composite indicators are generated which are countercurrent (i.e. not co-monotonic) to others and which therefore generate incomparabilities. Therefore the crucial weights in the g-space are the location, where the number of incomparabilities will increase.

As an object set may have more than two elements, many more possible crucial weights can be calculated, according to different object pairs. As the weights have to follow Eqs. 5 and 6, the relevant gc-values must also be in the range [0,1].

With the focus on gc1 (the other value in m-2-systems is just 1-gc1) a distribution of gc-values is possible and instead of a statistical oriented notation we write simply $0 \leq gc1(1) \leq gc1(2) \leq \ldots \leq 1$.

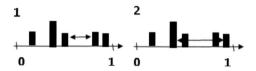

Fig. 3 The same set of crucial weights. In (1) the uncertainty degree is so low that no crucial weight is in the corresponding env(g,s)(1). In (2) there is a larger s assumed and the resulting env(g,s) encompasses two crucial weights (see text)

The visualization of the ordered set of crucial weights is called a spectrum of crucial weights. The situation may be characterized by Fig. 3:

In Fig. 3, the bar of the second crucial weight (counted from the left) is higher than the others. This means that this crucial weight is more often realized than the other crucial weights. Thus, there are more than 1 object pairs, leading to the same value of gc1. If an environment includes a crucial weight, then the number of incomparabilities is increasing according to the number of realizations for that specific gc-value.

Let $h(gc1(k))$ be the number of realizations for $gc1(k)$, then the considerations above lead to:

$$U(s) = \sum h\,(gc1(k))$$
$$gc1(k) \in env\,(g*, s)$$
$$\tag{11}$$

Depending on the start – value g, increasing s will imply different environments and the summation depends on how many crucial weights (together with their number of realization) fall into the actual environment.

2.4 Consequences for Decision Making

In the handbook of the OECD, Nardo (2008) recommends that weights should be selected in the range around (1/m), m being the number of indicators. Therefore, in the case of the system with m = 2 the distribution of the gc-values around 0.5 is of interest. If the distribution of the gc-values has its maximum around 0.5 then slight changes of the weights by increasing the environment env((0.5,0.5), s) may pass many gc-positions and the number of incomparabilities may strongly increase. In that case the selection of weights to calculate composite indicators need much more care than in the other extreme scenario, where the gc-distribution may have its maximum near 0 or near 1. Then the enlargement of environments env((0.5,0.5), s) by s will for non-extreme values of s not pass many gc-positions and hence the uncertainties about the weights (near 0.5) do not have much influence. It should be clear that the analysis of uncertainty and their influence on the incomparabilities need two arrangements:

1. the selection of g, the starting weight tuple and
2. the manner how s is increasing to model the uncertainty about the values of the weights.

   (a) In an m = 2-system one could for example start with g1 = 0, then the intervals of increasing [gmin, gmax] have always the same lower boundary and increasing s influences only gmax.
   (b) In an m = 2-system where we start with a g, with g1 = 0.5 increasing values of s affect both boundaries of [gmin, gmax].

3 Results

3.1 Three Fictitious Systems of Crucial Weights

A distribution of crucial weights may be derived from a MIS. Nevertheless, it is also possible to suppose a certain gc-distribution as an archetype. The latter strategy is followed within this subsection.

(a) h(gc(1j,k)) as function of k in the following form:

$$h\left(gc(1), k\right) = 1 - k \quad k = 0, .., 1$$

(once again: without referring to a specific, empirical MIS).

h(gc(1),k) = 1 − k means that the number of realizations of gc s: ts linear decreasing (just by construction).l.

Beside a normalization constant the behavior of h(gc(k)) as a function of k can be described by h = f(k), where k is assumed to vary continuously. Then based on Eq. 11 the following expression holds:

$$U(k) = \int_0^k f\left(k^{'}\right) dk^{'} \tag{12}$$

With f(k′) = 1 − k′, Eq. 12 leads to.

U(k) = N*k*(2 − k), where N is a scaling factor and k' is the integration variable of an integral with the limit 0 and k.

Beside the factor N the resulting function for U is as shown in Fig. 4:

(b) A maximum for the gc-distribution is assumed at gc = 0.5 and h(gc(1),k) is supposed to decrease symmetrically like a parabola. This situation is a more realistic with respect to decision support. Here the maximum of gc-realizations is thought of as being in the range around g1 = 0.5. A model for that is

$$h\left(gc(1), k = 4 * k * (1 - k)\,, \quad k = 0, .., 1.\right.$$

Fig. 4 The dotted line with rectangles is U as given by Eq. 9, whereas the solid line with the rhombic marker follows Eq. 11. Ordinate: the values of incomparabilities, following from different models, abscissa: s

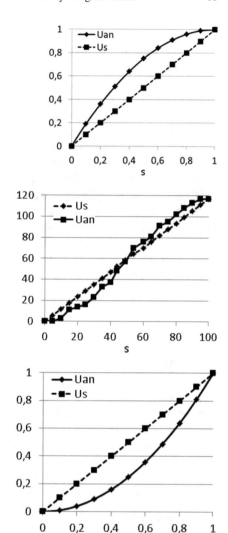

Fig. 5 U will follow a sigmoid function (solid line with cubic markers) based on Eq. 12, whereas U calculated by Eq. (9) is the dotted line with the rhombic markers. Abscissa: s-values

Fig. 6 The dotted line with cubic markers: U following Eq. 9, the solid line with rhombic markers follows Eq. 12, abscissa: s-values

The integral of such a h(k) – function leads to a behavior of U as sown in Fig. 5.

(c) Finally the third type of gc-distribution is modeled by

$$h\,(gc(1), k) = k,\ k = 0, \ldots, 1.$$

The resulting U-function, Uan, as well as Us, are shown in Fig. 6. Applying Eq. 12 and normalization to 1 leads to:

$$Uan = k^2,$$

The character of the resulting graph can easily be deduced by checking dh/dk: The slope at $k = 0$ is 0, whereas the slope at $k = 1$ equals 2.

In all these three situations the behavior of U is calculated assuming an evolution of uncertainty, corresponding to case 2a, i.e., while s is increasing and thus extending the integral [gjmin, gjmax], the starting value of $g(1)$ is always 0. It should be noted that this is not a realistic assumption, following Nardo (2008) recommending extreme weights (i.e. g1 near 0 or near 1) are to be avoided. However, it is a simple consideration to see that the evolution of uncertainties around $g1 = 0.5$ is similar to that of Fig. 5, which is the outcome of a linearly decreasing distribution of gc-values. Thus, env((0.5,0.5),s) sees many realizations for gc-values in the neighborhood of $g1 = 0.5$, then increasing s and thus symmetrically increasing the environments the number of realizations is diminishing. At $g1 = 0.5$ and $s = 0$ there is exact knowledge of the weight, there is just one composite indicator and $U = 0$. Extending s there are many gc-values with high realizations in the neighborhood of $g1 = 0.5$ therefore U by Eq. 11 will strongly increase and then reach the final value $U = U0$ asymptotically with a reduced slope.

3.2 Real World Example

3.2.1 Preliminaries

The following example is based on a report of UNICEF (UNICEF, Innocenti Research Centre, Report card 7, 2007). A comprehensive assessment of well-being of "children and young people in 21 nations of the industrial world" is given (quoted from the report). The following part is an abridged version taken from Bruggemann and Patil (2011).

The study provided 40 different indicators, which are aggregated through several interim steps into 6 main indicators as follows together with their identifiers:

1. Material well-being, wb
2. Health and safety, hs
3. Educational well-being, ed.
4. Family and peer relationships, fa
5. Behaviors and risks, br
6. Subjective well-being, sub

The sixth indicator, sub, needs some explanation: It is the attempt to "reflect children's own views and voices – for example, the surveys of reported family affluence, experience of bullying, or the frequency of communications with parents." (quoted from the report).

From the data matrix UNICEF defines a composite indicator with equal weights for all 6 indicators, i.e.:

$$CI(x) = \sum (1/6) * R_i(x), \quad \text{weight vector} = (1/6, 1/6, .., 1/6) \tag{13}$$

where, $R_i(x)$ is the rank by the ith indicator of nation x.

From CI, Eq. 13 the following ranking is deduced (from the worst to the best):

(UK, US, HU, AU, PT, FR, CZ, PL, GR, CA, DE, BE, IRE, IT, NO, SU, ES, FI, DK, SW, NE)

Any decision maker may see that "his" nation is "good" with respect to some indicator, even if his nation got a bad overall position in the ranking. Thus, Italy is good in the FAMILY indicator, "fa". Naturally the question arises, is not "fa" more important than the others, say "ed", education, and give "fa" a higher weight? UNICEF, however, used the same weight for each indicator. Hence, for any indicator, there is the same trade-off compensation: good points may compensate bad points. However, if, trade-off compensation is allowed, then questioning the uniform weight of any indicator in the index is indeed justified. Thus, Italy would get a better overall position if the weight for "fa" would get a higher value. However, such procedure would make decision makers, feeling responsible for their own nation unhappy. Poland for example is good in education, and would therefore like to see this indicator given a higher weight. Hence, it may be a good idea to keep the six indicators separated, but simultaneously analyzed rather than composited.

In Fig. 7, the Hasse diagram is shown based on the entries of the data matrix derived by UNICEF. We inverted them, so that the "good" nations are on the top of the Hasse diagram. The information base, IB, is {wb, hs, ed., fa, br, sub}. In order to include USA, the missing value in the indicator "sub" was given the mean value taken from the 20 nations.

Figure 8 shows that there are many conflicts among the different nations: Each nation has obviously some positive aspects with respect to child well-being, whereas there is no region that has not a bad value in at least one indicator compared with other nations. The longest chain contains three nations. Nations within a chain have incomparabilities with all nations outside the chain. This is a situation, which is good for an analysis of the reasons, why child well – being has some deficits, however for decision making the situation is not that comfortable. Because the analysis is based on two indicators only, the following subsection studies the child well-being by two indicators (namely fa (family) and ed. (education)) and intervals for the weights.

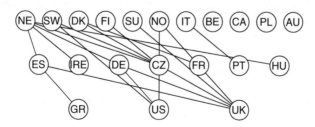

Fig. 7 Hasse diagram of 21 nations, the multi-indicator system: {wb, hs, fa, ed., br, sub}

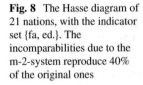

Fig. 8 The Hasse diagram of
21 nations, with the indicator
set {fa, ed.}. The
incomparabilities due to the
m-2-system reproduce 40%
of the original ones

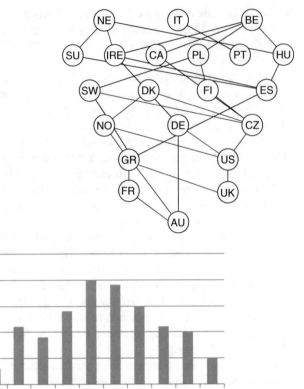

Fig. 9 gc-distribution of the poset of 21 nations and the indicators ed. and fa. The units at the
ordinate are arbitrary. The abscissa: gc-values

3.2.2 Adaptation to an m = 2-System

Once again the number of incomparabilities is used as a leading quantity. The
idea behind this is that any reduction in the number of indicators will reduce the
incomparabilities. However, the incomparabilities themselves are the reason, why
weights are introduced, because otherwise a complete order is already obtained
within the reduced set of indicators. Thus, the procedure is to select those two
indicators, which preserve the highest number of the incomparabilities of the
original indicator set. It turns out by use of the program weight2evolution_vs6_2
of PyHasse that with ed. and fa the number of incomparabilities, Fig. 8.

In Fig. 9 the distribution of the gc-values for the m = 2-system is shown. The
distribution is evidently similar to the second type, shown in Sect. 3.1 (item b).
Hence U based on Eq. 12 should follow a sigmoid function.

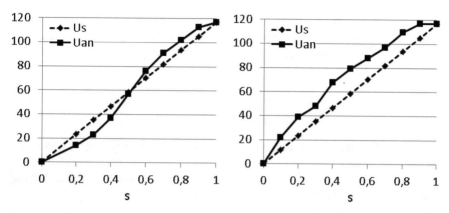

Fig. 10 The development of U, within the same evolution from s = 0 to s = 1, however with different starting points for the weights.Abscissa: s starting with 0 and ending at 1.Abscissa: describes the s-evolution

Indeed the values of U directly taken from all the posets possible in the series of env(**g**,s) as well of such, following Eq. 12 are coinciding and show the expected curve (Fig. 10).

It remains to check the behavior of the starting point g. Starting with g1 = 0.1 the left curve, starting with g1 = 0.5 the right curve in Fig. 10 results.

4 Discussion

Generally Eq. 9 appears as a good starting point to check, how many incomparabilities are to be expected, when the uncertainty in the numerical values of the weights is described by the parameter s and s is evolving from 0 (sharp knowledge) to 1 (every weight between 0 and 1) is possible. The actual study goes beyond Eq. 9 and analyzes why deviations from the straight line, represented by Eq. 9, are possible and how they differ from the straight line. In order to get an idea the most simple system, still allowing incomparabilities was studied: The system with m = 2. It becomes clear that the crucial influence is given by those weights, for which the composite indicator values of any two incomparable objects become nearly equal values. Furthermore, it becomes clear that the larger the number of U0 the more dense is the distribution of gc-values along the weight axis, as more evenly is the distribution of jumps and it can be expected that then the deviations from the U = s*U0 – line becomes less striking.

Clearly most of multi-indicator-systems will have by far more than 2 indicators. Can the study above provide us with an advise? No and yes! No, because an explicit formula for crucial weights in case of m > 2 leads to m-2-dimensional manifolds, and even if they can be algebraically determined, it is rather difficult to derive information about crucial weights in general. Yes, because following the device of Nardo (2008) the weights should not too much deviate from 1/m, m being the number of indicators. In the case of the weighted sum, Eq. 4, the condition for crucial weigths, namely CI(x) = CI(y), with x,y two objects of the object set, can be relaxed as follows.

$$\delta' (i1, i2) := \sum_{j=1,...,m} g(j) * (x (i1, j) - x (i2, j)) \qquad (14)$$

In Eq. 14 i1 and i2 are labelling different objects of the object set, j is labelling different indicators.

If the weights are taken the same value (g(j) = 1/m for all j) then the sum of differences of the matrix entries alone is decisive:

$$\delta (i1, i2) := \sum_{j=1,...,m} (x (i1, j) - x (i2, j)) \qquad (15)$$

Scanning the whole object set and calculating the δ-values for all pairs, where x(i1) ‖ x(i2) a distibution for δ is obtained. If then the distribution has high values in the range g(j) = 1/m then the number of incomparabilities will relatively high and thence the decision situation difficult. The acceptance of Eq. 15 implies that the values of the data matrix x are considered as sharp. When noise may perturb the entries of the data matrix x, then a probability scheme is to be developped whose result leads to an expectation value for δ(i1,i2). However, this as well as the following three points are future tasks.

Three further points are worth to be mentioned here:

1. Up to now the uncertainty degree s was the same for each weight. However in reality the conceptual uncerteinty for each weight may be different. This aspects needs still a lot of investigations.
2. The determination of a composite indicator due to Eq. 4 needs that the indicators fulfill certain scaling levels. At least they must be metric in nature, in order to let multiplication with a scalar (the weights) and summation a mathematically meaningful. In the case of the child well-being, it is clear that a) a normalization of ranks, as well as the subsequent algorithmic comination is at least mathematically questionable. Nevertheless the example is important enough, to consider the indicators as if they are metric quantities.
3. It is completely clear that the weighted sum, with its high potential for compensation (Munda 2008) has advantages because of its transparency, but there are other high sophisticated decision support systems (DSS). However an analysis as in this and the former papers will be extremely difficult for those other DSS, not only because of their more involved mathematical structure, but also, because usually more parameters influence the final result, i.e. the final ranking.

Hence there is still much of work possible, to clarify the role of weights and other parameters in other DSS.

References

Annoni, P., Bruggemann, R., & Carlsen, L. (2014). A multidimensional view on poverty in the European Union by partial order theory. *Journal of Applied Statistics, 42*, 535–554.

Bartel, H.-G., & Mucha, H.-J. (2014). Measures of incomparabilities and of inequality and their application. In R. Bruggemann, L. Carlsen, & J. Wittmann (Eds.), *Multi-indicator systems and modelling in partial order* (pp. 47–67). New York: Springer.

Brans, J. P., & Vincke, P. H. (1985). A preference ranking organisation method (the PROMETHEE method for multiple criteria decision – Making). *Management Science, 31*, 647–656.

Bruggemann, R., & Carlsen, L. (2014a). Incomparable-what now? *MATCH Communications in Mathematical and in Computer Chemistry, 71*, 699–714.

Bruggemann, R., & Carlsen, L. (2014b). Incomparable: What now II? Absorption of incomparabilities by a cluster method. *Quality & Quantity, 49*, 1633–1645.

Bruggemann, R., & Carlsen, L. (2015). Incomparable – What now III. Incomparabilities, elucidated by a simple version of ELECTRE III and a fuzzy partial order approach. *MATCH Communications in Mathematical and in Computer Chemistry, 73*, 277–302.

Bruggemann, R., & Carlsen, L. (2017). Incomparable: What now, IV. Incomparabilities: A modelling challenge. In M. Fattore & R. Bruggemann (Eds.), *Partial order concepts in applied sciences* (pp. 35–47). Cham: Springer.

Bruggemann, R., & Carlsen, L. (2018). Partial order and inclusion of stakeholder's knowledge. *MATCH Communications in Mathematical and in Computer Chemistry, 80*, 769–791.

Bruggemann, R., & Halfon, E. (2000). Introduction to the general principles of partial order ranking theory. In P. B. Sørensen, L. Carlsen, B. B. Mogensen, R. Bruggemann, B. Luther, S. Pudenz, U. Simon, E. Halfon, T. Bittner, K. Voigt, G. Welzl, & F. Rediske (Eds.), *Order theoretical tools in environmental sciences – Proceedings of the Second Workshop, October 21st, 1999 in Roskilde, Denmark* (pp. 7–43). Roskilde: National Environmental Research Institute.

Bruggemann, R., & Patil, G. P. (2011). *Ranking and prioritization for multi-indicator systems – Introduction to partial order applications*. New York: Springer.

Bruggemann, R., & Voigt, K. (2011). A new tool to analyze partially ordered sets. Application: Ranking of polychlorinated biphenyls and alkanes/alkenes in river Main, Germany. *MATCH Communications in Mathematical and in Computer Chemistry, 66*, 231–251.

Bruggemann, R., & Voigt, K. (2012). Antichains in partial order, example: Pollution in a German region by lead, cadmium, zinc and sulfur in the herb layer. *MATCH Communications in Mathematical and in Computer Chemistry, 67*, 731–744.

Bruggemann, R., Halfon, E., Welzl, G., Voigt, K., & Steinberg, C. (2001). Applying the concept of partially ordered sets on the ranking of near-shore sediments by a battery of tests. *Journal of Chemical Information and Computer Sciences, 41*, 918–925.

Bruggemann, R., Voigt, K., & Pudenz, S. (2008a). Partial ordering and Hasse diagrams: Applications in chemistry and software. In M. Pavan & R. Todeschini (Eds.), *Data handling in science and technology* (Vol. 27, pp. 73–95). Amsterdam: Elsevier B.V.

Bruggemann, R., Voigt, K., Restrepo, G., & Simon, U. (2008b). The concept of stability fields and hot spots in ranking of environmental chemicals. *Environmental Modelling & Software, 23*, 1000–1012.

Bruggemann, R., Kerber, A., & Restrepo, G. (2011). Ranking objects using fuzzy orders, with an application to refrigerants. *MATCH Communications in Mathematical and in Computer Chemistry, 66*, 581–603.

Bruggemann, R., Restrepo, G., & Voigt, K. (2012). Weighting intervals and ranking, exemplified by leaching potential of pesticides. *MATCH Communications in Mathematical and in Computer Chemistry, 69*, 413–432.

Bruggemann, R., Restrepo, G., Voigt, K., & Annoni, P. (2013). Weighting intervals and ranking. Exemplified by leaching potential of pesticides. *MATCH Communications in Mathematical and in Computer Chemistry, 69*, 413–432.

Carlsen, L., & Bruggemann, R..(2011). Decision support tools in environmental forensics. In Organizing Committee of the "Experts Workshop" on Environmental Forensics (Ed.), *Environmental forensics* (pp. 239–247). #, Tbilisi, Georgia.

Carlsen, L., & Bruggemann, R. (2014). Indicator analyses: What is important – And for what? In R. Bruggemann, L. Carlsen, & J. Wittmann (Eds.), *Multi-indicator systems and modelling in partial order* (pp. 359–387). New York: Springer.

Colorni, A., Paruccini, M., & Roy, B. (2001). *A-MCD-A, Aide Multi Critere a la Decision, Multiple Criteria Decision Aiding*. Ispra: JRC European Commission.

Figueira, J., Greco, S., & Ehrgott, M. (2005). *Multiple criteria decision analysis, state of the art surveys*. Boston: Springer.

Hasse, H. (1967). *Vorlesungen über Klassenkörpertheorie*. Marburg: Physica-Verlag.

Helm, D. (2006). Evaluation of biomonitoring data. In R. Bruggemann & L. Carlsen (Eds.), *Partial order in environmental sciences and chemistry* (pp. 285–307). Berlin: Springer.

Munda, G. (2008). *Social multi-criteria evaluation for a sustainable economy*. Berlin: Springer.

Munda, G., & Nardo, M. (2008). Noncompensatory/nonlinear composite indicators for ranking countries: A defensible setting. *Applied Economics, 41,* 1–11.

Nardo, M. (2008). *Handbook on constructing composite indicators -methodology and user guide*. Ispra: OECD.

Newlin, J., & Patil, G. P. (2010). Application of partial order to stream channel assessment at bridge infrastructure for mitigation management. *Environmental and Ecological Statistics, 17,* 437–454.

Patil, G. P., & Taillie, C. (2004). Multiple indicators, partially ordered sets, and linear extensions: Multi-criterion ranking and prioritization. *Environmental and Ecological Statistics, 11,* 199–228.

Roy, B. (1972). Electre III: Un Algorithme de Classements fonde sur une representation floue des Preferences En Presence de Criteres Multiples. *Cahiers du Centre d'Etudes de Recherche Operationelle, 20,* 32–43.

Roy, B. (1990). The outranking approach and the foundations of the ELECTRE methods. In C. A. Bana e Costa (Ed.), *Readings in multiple criteria decision aid* (pp. 155–183). Berlin: Springer.

Simon, U., Bruggemann, R., Pudenz, S., & Behrendt, H. (2006). Aspects of decision support in water management: Data based evaluation compared with expectations. In R. Bruggemann & L. Carlsen (Eds.), *Partial order in environmental sciences and chemistry* (pp. 221–236). Berlin: Springer.

Sørensen, P. B., Lerche, D., & Thomsen, M. (2006). Developing decision support based on field data and partial order theory. In R. Bruggemann & L. Carlsen (Eds.), *Partial order in environmental sciences and chemistry* (pp. 260–283). Berlin: Springer.

UNICEF, Innocenti Research Centre, Report card 7, 2007.

Wieland, R., & Bruggemann, R. (2013). Hasse diagram technique and Monte Carlo simulations. *MATCH Communications in Mathematical and in Computer Chemistry, 70,* 45–59.

A Study to Generate a Weak Order from a Partially Ordered Set, Taken Biomonitoring Measurements

Rainer Bruggemann, Lars Carlsen, Bardia Panahbehagh,
and Stergios Pirintsos

1 Introduction

Several scientific disciplines are involved in the study of chemical processes that occur in water, air, terrestrial and living environments, and the effects of human activity on them. Environmental chemistry, as one of these, is not only the discipline handling substances in difficult targets, such as sludge (Jin et al. 2017), trees (Ferretti et al. 2002), or lichens (Pirintsos et al. 2006), but also having the task to support decisions in environmental systems (Pirintsos and Loppi 2008).

Due to the complexity of environmental systems a series of well-defined indicators is constructed, representing the knowledge about the system and thus supporting decisions for an appropriate management (Buonocore et al. 2018; Grönlund 2019). Hence, the start for a management and a decision based on a ranking of chemicals (our example) is the analysis of multi-indicator systems (MIS). An example of MIS is the output of several lichen biomonitoring studies concerning metal pollution in the atmospheric environment. To be more specific, the indicators are related to

R. Bruggemann (✉)
Leibniz-Institute of Freshwater Ecology and Inland Fisheries, Berlin, Germany
e-mail: brg_home@web.de

L. Carlsen
Awareness Center, Roskilde, Denmark
e-mail: LC@AwarenessCenter.dk

B. Panahbehagh
Department of Mathematical Sciences and Computer, Kharazmi University, Tehran, Iran
e-mail: panahbehagh@khu.ac.ir

S. Pirintsos
Department of Biology, University of Crete, Heraklion, Greece
e-mail: pirintsos@biology.uoc.gr

© The Author(s), under exclusive license to Springer Nature Switzerland AG 2021
R. Bruggemann et al. (eds.), *Measuring and Understanding Complex Phenomena*,
https://doi.org/10.1007/978-3-030-59683-5_5

63

locations, their values are derived following a procedure described by Nimis and Bargagli (1999).

Lichens are perennial, slow-growing organisms, highly dependent on the atmosphere for nutrients. The lack of a waxy cuticle and stomata allows many contaminants, which are deposited on lichens by precipitation, fog and dew, dry sedimentation and gaseous absorption, to be absorbed over the whole lichen thallus surface, indicating levels of these contaminants in the surrounding environment (Loppi et al. 1999). By biomonitoring at specifically selected sites, for example near roads (Frati et al. 2006) or more pristine areas (Loppi and Pirintsos 2003), information is obtained about the transport and origins of pollution.

In Pirintsos et al. (2014) 11 metals/metalloids are investigated in 20 sites of an urban and industrial area based on the lichen biomonitoring data set of Demiray et al. (2012), where *Xanthoria parietina* lichen specimen have been used as a biomonitoring organism. The evaluation of the corresponding data matrix is based on the conception that the Hasse diagram technique (see below) can further be expanded and improved in the direction of (i) cumulative risk, (ii) the up-to-date formal presentation and (iii) the interpretation of results in biomonitoring studies of metal atmospheric pollution.

The analysis of biomonitoring results can be crudely characterized by two aspects: (a) attempts to support a decision, based on order relations and (b) attempts to present small scale spatial variations within a geostatistical approach. Here our focus is on the order theoretical aspects.

As the metals and metalloids are measured at m different sites the concentrations found in lichens of each site define, after transformations as recommended by Nimis and Bargagli (1999) an indicator. Hence the MIS contains m indicators, the values describing the pollution due to a single metal or metalloid.

The question arises how to derive a decision when confronted with m indicators. Here we show first a Hasse diagram, which is a visualization of the partial order, induced by the set of indicators (cf. Bruggemann and Patil 2011), then we discuss, as to how far a single ranking (a weak order (see below)) can be obtained without the need of a subjective weighting scheme of the indicators in order to aggregate them by a weighted sum. As an exact solution of the problem how to get a weak order is hardly computationally tractable, we investigate a new calculation method.

2 Material and Methods

2.1 Data Set

Eleven Metals and metalloids, i.e., Hg, Al, As, Cd, Cu, Fe, Mn, Ni, Pb, V and Zn for which their pollution has been monitored are included in the study (Pirintsos et al. 2014). For a management it is of importance which of these metals or metalloids (in the following we call them simply metals, although this is not a chemically correct

Table 1 data of 11 metals and their scores in lichens in 10 sites

metals										
Al	As	Cd	Cu	Fe	Hg	Mn	Ni	Pb	V	Zn
Sites										
st1	st2	st3	st4	st5	st6	st7	st8	st9	st10	
data matrix										
	st1	st2	st3	st4	st5	st6	st7	st8	st9	st10
Al	4	5	7	5	5	4	4	4	3	5
As	2	5	7	6	7	5	5	4	4	4
Cd	5	7	7	7	4	3	4	2	2	3
Cu	5	5	7	7	4	3	5	3	3	4
Fe	7	7	7	7	7	7	7	7	6	7
Hg	4	3	5	3	3	3	4	2	2	2
Mn	7	7	7	7	6	6	7	4	5	6
Ni	7	7	7	7	6	5	7	4	4	7
Pb	7	7	7	7	5	4	6	4	7	4
V	7	7	7	7	7	6	7	7	4	7
Zn	7	7	7	7	7	7	7	6	7	7

term) is highly concentrated. As sites we select only 10 (of 20) because of reasons which become clear in following sections. The concentrations are transformed into a scale of integers from 1 to 7, following the suggestion of Nimis and Bargagli (1999). 1 indicates a high naturality, whereas 7 express a high deviation from the natural state. The data are shown in Table 1, in the Results-section. An entry of the data matrix, dm(I,j) is associated with jth site and the ith metalloid. Details can be found in Pirintsos et al. (2014).

2.2 Basic Concepts of Partial Order

Let X be a finite set of n objects, labeled by $x(i)$ ($i = 1, \ldots, n$). Objects could be but not limited to

- chemical compounds (here: metals/metalloids)
- nations, characterized by for example child well-being indicators
- strategies, characterized by performance indicators
- geographical units, characterized for example by pollution, or (within a socio-economic context) by poverty indicators

Here, indeed the elements of the real example are "chemical elements", namely n = 11 metals.

To define an order relation among them, the relation "≤" has to obey the following order axioms:

- reflexivity: the object can be compared with itself
- antisymmetry: if $x \le y$ and $y \le x \Rightarrow x = y$
- transitivity: if $x \le y$ and $y \le z \Rightarrow x \le z$

A special realization of order relations is given by Eqs. (1, 2, and 3):

$$x(i) = (x\,(i,\,1)\,,\,x\,(i,\,2)\,,\,\ldots,\,x\,(i,\,m)) \tag{1}$$

The quantity $x(i, j)$ is the value of the ith object $(i = 1, \ldots, n)$ (here the ith metal), the jth indicator $(j = 1, .., m)$ (here the jth site) and m the number of indicators used (here m = 10).

Equation 1 describes a mapping $X \rightarrow IR^m$, wherein X is the set of objects (the metals) and IR^m is the set of tuples of real numbers with m components. Note that the tuples are also denoted as data profiles.

According to $m = 10$ sites, we will have a system of 10 indicators.

$$x(i1) \le x(i2) : \iff (x\,(i1,\,1)\,,\,\ldots,\,x\,(i1,\,m)) \le \left(x\,(i2,\,1)\,,\,\ldots,\,x\,(\underline{i2},\,m)\right) \tag{2}$$

Equation 2 needs clarification, as it is not yet clear under which conditions one tuple (that of x(i1) is to be considered less or equal to that of x(i2). The way how Eq. 2 can be given a meaning, opens the door to many variants. By Eq. 3

$$(x\,(i1,\,1)\,,\,\ldots,\,x\,(i1,\,m)) \le (x\,(i2,\,1)\,,\,\ldots.x\,(i2,\,m)) : \iff x\,(i1,\,j) \le x\,(i2,\,j)$$
$$\text{for all } j = 1, .., m \tag{3}$$

a special partial order is defined. Two objects, following Eq. 3 are called "comparable", otherwise "incomparable".

The immediate relation to the data and the corresponding indicators has two consequences:

1. Any order relation $x \le y$ is a direct reflection of the data values of x and y. This is in contrast to many decision support systems, where an order relation cannot easily be traced back to the original data, i.e., to the data matrix.
2. The partial order methodology, based on Eq. 3 is applicable wherever a data matrix is available and where a ranking aim can be defined.

In the literature the method, based on Eq. 3 together with appropriate supporting software, is often denoted Hasse diagram technique (HDT) (Galassi et al. 1996; Grisoni et al. 2015; Halfon and Reggiani 1986; Bruggemann et al. 2001, 2008; Patil and Taillie 2004; Klein and Ivanciuc 2006; Simon et al. 2004, 2006; Helm 2003; Bruggemann and Voigt 2008, 2011, 2012; Carlsen and Bruggemann 2011, 2014a, b; Carlsen 2008a, b, 2013, 2018; Newlin and Patil 2010; Annoni et al. 2014; Sørensen et al. 1998, 2000; Pavan and Todeschini 2004; Pudenz and Heininger 2006; Quintero et al. 2018; Restrepo and Bruggemann 2008; Restrepo et al. 2008a, b; Voigt et al. 2004a, b) with reference to the German mathematician Helmut Hasse (1967). Sets X, equipped with a partial order (and thus in this paper by the Eqs. 1, 2, and 3) are called partially ordered sets and are conveniently denoted as posets and indicated by $(X, \leq)$.

Two objects x, y mutually incomparable are denoted as $x \parallel y$. Within a poset (X, $\leq$) the number of incomparable pairs $x \parallel y$ is called U. When the orientation $x \leq y$ or $x \geq y$ is of minor interest then the mere fact of comparability is denoted by $x \perp y$.

Note that partial order methodology can also be applied, by evaluation of the space of all possible data profiles, when the indicators are discrete (cf. e.g. Fattore and Maggino 2014, as well as Maggino et al. this book).

2.3 Hasse Diagram

The construction of a Hasse diagram, starting from a set of partial order relations (as an outcome of Eq. 3) is frequently explained in the literature (see e.g. Bruggemann and Halfon 1997). For the sake of reader's convenience, some words about Hasse diagrams may nevertheless useful here: The basis is the order relation $x < y$. Usually the object x will be drawn below object y; both are vertices of a graph and presented by small circles, with the label of the object in the centre. In case $x < y$ a line is connecting x with y, called an edge, if the vertices are in a cover relation, i.e if there is no object z for which is valid: $x < z < y$. The orientation of the order relation is just obtained from the vertical position. When two objects are not connected by a system of oriented edges the two objects are incomparable.

By this construction a Hasse diagram allows a two-fold interpretation:

1. Upwards: The numerical values of the objects are nondecreasing along a system of edges. This "vertical" oriented analysis allows a ranking of objects of subsets of X, so-called chains.
2. In contrast to (1) there is also a "horizontal" evaluation. This evaluation has its focus on not connected objects. Following the construction principles of a Hasse diagram, the objects of in the same vertical position are mutually incomparably. A set of mutually incomparable objects is called an antichain.

2.4 Weak Order

When the general policy of decision is to find not only the optimal option but also alternatives, then ranking is a good starting point, since suboptimal objects can be easily identified if the optimal object is not suitable (e.g., due to political or economic reasons). The task is how to get a ranking, which is at least a weak order, if ties are accepted. Whereas a complete (i.e., total or linear) order is a set of objects, in which all elements $x, y \in X$ are mutually comparable with $x \neq y$, a weak order does not require the condition $x \neq y$, i.e., it accepts equivalent elements (or in terms of statistics: it accepts ties).

A Hasse diagram allows identifying rankings for subsets of X, without any subjectivity beyond the data matrix. It is clear that the task to get a weak order should be parameter free too. Hence, the typical procedure to aggregate the values of the m indicators into a composite indicator by a numerical procedure, where weights for each indicator and other parameters are required, is to be avoided. An important device, how to get a weak order without the need of finding additional parameters, such as weights for the indicators, out of a partially ordered set is found in the paper of Winkler (1982). The crucial term is the average height, denoted as *Hav*.

2.5   *Average Height*

Any poset can be represented by a set of linear order, whose elements are called linear extensions (Davey and Priestley 1990; Trotter 1992). A linear extension is a linear order, respecting all order relations within a poset. For example the set $X = \{a, b, c, d\}$ may have the following order relations:

$$a < b, a < c, a < d. \tag{4}$$

Obviously, $b \parallel c$ and $c \parallel d$, i.e. $U = 2$. Then the set of linear extensions is:

$$\{(a, b, c, d), (a, c, b, d), (a, c, d, b)\}.$$

Within the above set $\{(a, b, c, d), (a, c, b, d), (a, c, d, b)\}$ a linear extension is for example (a, b, c, d), others are (a, c, b, d) and (a, c, d, b). Each single linear extension indicates a complete ordered set, for example (a, b, c, d) denotes: $a < b, a < c, a < d, b < c, b < d, c < d$). All order relations of Eq. 4 are reproduced. The fact that the poset in Eq. 4 includes some incomparabilities leads to the necessity to consider the 3 linear extensions simultaneously. Within each linear extension any object x has a height that is the number of objects $\leq x$. For example, in the linear extension (a, b, c, d) object a has the height 1, b the height 2, whereas in the linear extension (a, c, d, b) object b has the height 4.

The idea of Winkler (1982) is to calculate the average of all heights of all objects, denoted as Hav(x). Let L(k) be the kth linear extension and h(L(k),x) the height of x in L(k), then, after Winkler (1982)

$$Hav(x) := \left(\sum h\left(L(k), x\right) \right) / LT \quad (k = 1, \ldots, LT) \tag{5}$$

where LT is the number of linear extensions derived from a specific poset.

Equation 5 could be a good starting point, when the set of linear extensions is small. Taking into mind that the number of linear extensions for an object set with n objects can be up to n! the problem to generate linear extensions and store them into a memory is computationally hard (see e.g. Atkinson and Chang 1986).

The above mentioned difficulty leads to several variants:

- There is still an exact method available. It is based on the fact that the storage of some sets derived from the poset needs less memory than the storage of the linear extensions. From a methodological, mathematical point of view this method transforms the original poset into a lattice and the quantities of interest can be directly derived from this lattice (De Loof et al. 2006). However, the lattice-method is only working, when $U*n$ (U: number of incomparabilities in a set of n objects) is not too large, for details see Bruggemann and Carlsen (2011).
- Some approximations seem to have found more applications, for instance the method of Bubley and Dyer (1999), which suggests a "good" sampling of linear extensions.
- Another one has a graph – theoretical background and considers the local environment around each object within a poset. There are two variants: (1) the LPOM0 (local partial order model 0) Bruggemann et al. 2004) and (2) an extended model (LPOMext) (Bruggemann and Carlsen 2011). Although the extended variant is thought of as delivering better results than LPOM0, it turned out (Rocco and Tarantola 2014) that the more simple method (LPOM0) may be in some cases a better approximation than the extended one.

2.6 Idea for an Alternative for the Hav-Calculation

Often partial order can be considered as being composed from simpler posets, here for example, the concept of linear sum is of specific interest. It is defined as follows: Let X_1, X_2 be disjoint subsets of X with

$$X = X_1 \oplus X_2 \tag{6a}$$

$$x \in X_1, y \in X_2 \text{ implies } x > y \text{ for every } x, y \tag{6b}$$

Equation (6b) can be formulated as follows: If two sets can be found where for an element of the first set, x, and for any element of the second set, y, is valid: $x > y$; the relations among the first, and the second set, resp., are not of interest.

We may speak of "X_1 is fully dominating X_2". Equation 6b does not imply that within X_1 or X_2 the elements are mutually comparable.

Let $Hav(x, X)$ denote the average height of x, considering the set X and the settings of Eqs. 6a and 6b. Then:

$$Hav\,(x, X) = |\,X_2\,| + Hav\,(x, X_1) \tag{7}$$

where $|\ldots|$ denotes the cardinality of the set. Eq. 7 is a simple conclusion found from Eq. 6b:

$$H\,(L(k), x\ in\ X) = |\,X_2\,| + H\,(L(k), x\ in\ X_1)\,.$$

Thus, a calculation method can be thought of, which can be formulated as follows:

$$x \in X_1 : Hav\,(x, X) = Hav\,(x, X_1) + |\,X_2\,| \tag{8}$$

$$y \in X_2 : Hav\,(y, X) = Hav\,(y, X_2)\,, \tag{9}$$

supposed that Eq. 6b is exactly fulfilled.

The concept of $X_1, X_2 \subset X$ with $X_1 \cap X_2 = \varnothing$ was already studied by (Restrepo and Bruggemann 2008) and lead to two quantities, the dominance of X_1 over X_2 and the separability of X_1 and X_2 (Eqs. 10 and 11).

$$\underline{\mathrm{Dom}}\,(X_1, X_2) := |\,\{(x, y)\ \text{with}\ x \in X_1, y \in X_2\ \text{and}\ x > y\}\,| \,/\,(|X_1| * |X_2|) \tag{10}$$

$$Sep\,(X_1, X_2) := |\,\left\{(x, y)\ \text{with}\ x \in X_1, y \in X_2\ \text{and}\ x \,\|\, y\right\}\,| \,/\,(|X_1| * |X_2|) \tag{11}$$

By a set of subsets the quantities, defined in Eqs. 10 and 11 can be conveniently denoted as matrices, dominance (Dom) and separability (Sep) matrices.

Equation (6b) demands that $Dom(X_1, X_2) = 1$ and $Sep(X_1, X_2) = 0$.

If $\mathrm{Dom}(X_1, X_2) = 0$, then X_1, X_2 are completely separated subsets, meaning that then $x \in X_1$, $y \in X_2$ implies $x \,\|\, y$. In that case it is easily seen that we find:

$$\text{Hav}(x, X_1) < \text{Hav}(x, X) < | X_2 | + \text{Hav}(x, X_1) \tag{12a}$$

$$\text{Hav}(y, X_2) < \text{Hav}(y, X) < | X_1 | + \text{Hav}(y, X_2) \tag{12b}$$

because as an extremal case the subposet based on X_1 can once be completely below the subposet $(X_2, <)$ or completely above $(X_2, <)$. Hence: When $Dom(X_1, X_2)$ < 0.5 then the role of the separability matrix is overwhelming (because the sum of Dom- and Sep-matrices is bounded, due to the finite number of comparabilities and incomparabilities and the Eqs. 8 and 9 fail. Therefore it is needed that the poset, to be considered, has more comparabilities than incomparabilities. This is the reason, why instead of 20 sites (the real example) only the 10 first sites were selected.

Summarizing: From a methodological point of view, we want to check, as to how far a deviation of $Dom(X_1, X_2)$ from 1 can lead to acceptable results.

3 Results

3.1 Randomly Generated Datasets

In order to test as to how far deviations of $Dom(X_1, X_2)$ from 1 lead to errors in the estimation of Hav, 22 smaller datasets (each of 10 objects) were randomly generated. For each object x of these artificial data sets the Hav-value based on the scheme given in Eq. 12 was calculated, HavDom(x) and the exact value, Havexact, based on the lattice theoretical method presented by De Loof et al. (2006, 2011, 2012). The deviation was calculated:

$$Eps(x) := | Haxexact(x) - HavDom(x) | \tag{13}$$

For each dataset a final value *epsav* was determined:

$$epsav := \sum Eps(x)/n \quad \text{with } n = | X | \tag{14}$$

the quantity epsav being the average error related to any single object. In Fig. 1 the scatterplot, together with the regression equation is shown.

Figure 1 confirms that the deviations epsav will be rather large, when $Dom(X_1, X_2)$ becomes small values. It is clear that the way, how the partitioning of X into two subsets X_1 and X_2 is selected, plays an important role. However, aiming at an efficient method for the calculation of Hav, the principles were:

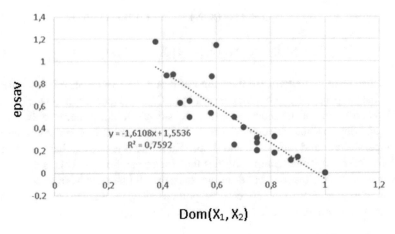

Fig. 1 Scatterplot of *epsav* vs *Dom(X₁, X₂)* and the regression equation with $R^2 \approx 0.76$

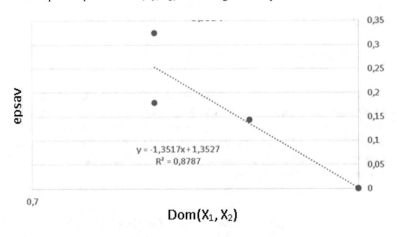

Fig. 2 *epsav* vs *Dom(X₁, X₂)* , with *Dom(X₁, X₂)* ≥ 0.8

1. To select the X_1, X_2 in that manner that they have approximately the same number of elements
2. To find a selection that maximizes $Dom(X_1, X_2)$.

The principle (1) was a priori considered as more important than the principle (2). From Fig. 1 it becomes clear that obviously in the specific considered randomly generated case (for details, see below) the deviations *epsav* require $Dom(X_1, X_2) \geq 0.8$.

When the regression is restricted to those pairs of values $(Dom(X_1, X_2)$, *epsav*), where $Dom(X_1, X_2) \geq 0.8$, then the result is (more or less trivially) better, see Fig. 2.

The regression equation based on 6 pairs (the pair (1,0) is realized three times) has the striking structure:

$$epsav = a * (1 - Dom\,(X_1, X_2)\,) \text{ for } 0.8 < Dom\,(X_1, X_2) \le 1 \qquad (15)$$

with the coefficient of determination, $R^2 = 0.88$ and the coefficient a around 1.35. This statistical result indicates that the relevant quantity is the deviation Δ:

$$\Delta := 1 - Dom\,(X_1, X_2) \qquad (16)$$

Hence, Eq. 15 expresses proportionality between the error $epsav$ and Δ. The crucial value 0.8 for separating relevant $Dom(X_1,X_2)$-values from irrelevant ones, may vary from case to case and is open for future research. Furthermore, Fig. 2 shows that the average error related to single objects is less than 0.25 and the deviations from the regression line will be larger the smaller the value $Dom(X_1, X_2)$ is.

3.2 Application to Real Data Set

The estimation method needs the following steps:

1. Defining X_1, X_2 and the Hasse diagram for the full set X
2. Calculation of $Dom(X_1, X_2)$
3. Providing the data for X_1 and X_2
4. Application of the lattice theoretical method: (a) for X, (b) for X_1, (c) for X_2
5. Performing the calculations due to Eqs. 8 and 9
6. Inspecting $epsav$ to check the quality of the results

Up to now there is no program performing all 6 steps. However, for steps (1)–(3) the program package PyHasse (see for details Bruggemann et al. 2014) was extended by the new module DomRkav. Its graphical user interface is shown in Fig. 3. The data is found in Table 1.

The corresponding Hasse diagram is shown in Fig. 4.

Some remarks concerning Fig. 4 may be useful here:

- Hg and As are minimal elements, they cause the least deviation from a natural state
- Fe and Zn are maximal elements, they are most problematic because the deviation of the natural state is very high.
- Incomparabilities, such as for Zn and Fe show that the loading of the lichens in general is high, however with some geographical differentiation.

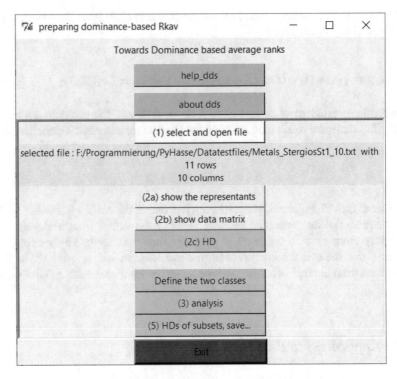

Fig. 3 Graphical user interface for the new PyHasse module DomRkav

Fig. 4 Hasse diagram of the
metals, according to their
deviations from their natural
state in the studied stations

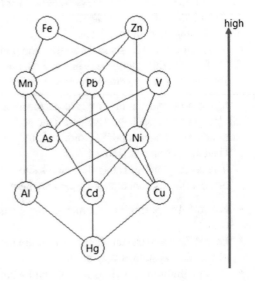

In order to perform the calculation scheme based on Eqs. 8 and 9 the first step is to select X_1 and X_2, i.e. the partitioning of set X.

Step 1:

The sets X_1 and X_2 are:

set X_1: Fe, Zn, Mn, Pb, V, As
set X_2: Ni, Al, Cd, Cu, Hg

Step2:

The Dom-matrix is:

	X_1	X_2
X_1:	0.361	0.733
X_2:	0.0	0.48

Remark 1:

Although $Dom(X_1, X_2) = 0.733$ is less 0.8 the next calculation steps are documented, just for a demonstration.

Remark 2:

The partitioning selected above is not the only possible one. For example, the metalloid As is a minimal element. Why not assign As to X_2? Let $X_2' = X_2 \cup \{As\}$ and $X_1' = X_1 - \{As\}$. Indeed the value of $Dom(X_1', X_2') = 0.833$ is better than that of $Dom(X_1, X_2)$ and correspondingly $epsav = 0.395$. The disadvantage is that $(X_2', \leq)$ leads due to its symmetry to a very high degree of degeneracy: Fe$\cong$ Zn, Pb $\cong$ V and As $\cong$ Al $\cong$ Cd $\cong$ Cu. Therefore we continue with the partitioning of X into X_1 and X_2 as given above.

Step 3: Calculation of the averaged ranks by the lattice-theoretical method (De Loof et al., 2006) due to X_1 and X_2 of step 1.

Figure 5 shows the Hasse diagrams of the two subsets.

Fig. 5 The two Hasse diagrams due to X_1 and X_2

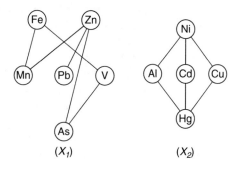

Table 2 Summarizing the results of step 5 and step 5. Wether $x \in X1$ or $\in X2$ is indicated by a membership function. If $x \in X1$ then the value in the corresponding column $= 1$, otherwise 0. Similarly for $x \in X_2$

Object x	Havexact	X_1	X_2	Hav($...,X_1$)	Hav($...,X_2$)	HavDom	Eps(x)
Al	3,581	0	1		3	3	0,581
As	4,246	1	0	1,597		6,597	2,351
Cd	3,36	0	1		3	3	0,36
Cu	3,36	0	1		3	3	0,36
Fe	10,487	1	0	5,455		10,455	0,032
Hg	1,135	0	1		1	1	0,135
Mn	7,175	1	0	2,364		7,364	0,189
Ni	6,559	0	1		5	5	1,559
Pb	7,169	1	0	2,636		7,636	0,467
V	8,693	1	0	3,576		8,576	0,117
Zn	10,236	1	0	5,273		10,273	0,037

$\sum Eps(x) = 6.188$
$epsav = 0.563$
$Dom(X_1, X_2) = 0.733$

Step 4: Application of Eqs. 8 and 9.
Step 5: Check for the accuracy of the results.

The remaining steps 4 and 5 are summarized in the following Table 2. The column below X_1 and X_2 is the membership function, indicating whether or not the metal belongs to X_1 or to X_2.

The value of $epsav = 0.563$ deviates from the value obtained from Eq. 15; ($epsav_{eq.15} = 0.36$). However the value of $Dom(X_1, X_2)$ is not within the range of applicability of Eq. 15. As to be expected, the measure of deviation, epsav, indicates a bad approximation. Due to pretty large deviations (in terms of epsav the final weak order shows two inversions:

Exact: Hg < Cd $\cong$ Cu < Al < As < Ni < Pb < Mn < V < Zn < Fe
Approx.: Hg < Cd $\cong$ Cu $\cong$ Al < Ni < As < Mn < Pb < V < Zn < Fe

As it is often the case, different methods coincide, when extremal ranking positions are to be detected. This empirical finding is found here as well, i.e., Hg Cd, Cu, as well as V, Zn and Fe coincide in their positions at the beginning or the end of the ranking sequence. The other positions in a ranking sequence are usually determined by many factors. Therefore, here different methods will lead to different ranking positions. Here, indeed, some other metals change their position (As, Ni) and (Mn, Pb), when the exact, lattice theoretical method is compared with the approximation, suggested here. The reasons for the inversion Mn, Pb is that Mn "sees" four vertices order theoretically less than Mn, whereas Pb only "sees" three vertices. In the approximation however, both are minimal elements, so that for both metals the Eq. 8 gives the same summand $|X_2|$, being 4. A similar argument holds for the pair (As, Ni).

4 Discussion

4.1 Lichen Biomonitoring/Bioaccumulation Matrices as Multi-indicator Systems

Undoubtedly, any data pre-processing, as done here, are of high importance not only in chemical risk assessment and management but also broader, in the decision-making process of environmental policy. However, here is not the place to discuss in depth the pre-processing, defined by Nimis and Bargagli (1999), nevertheless, we think that here some words may be helpful:

In biomonitoring techniques of air quality with native lichens, an approach to the interpretation of data of native lichens is the so-called "naturality/alteration scales" based on thresholds identifying classes of increasing element concentrations, and obtained by the meta-analysis of a large set of bioaccumulation data. The method by Nimis and Bargagli (1999) defines seven classes of element concentrations. These classes are built up on hundreds of data points collected in Italy between the 1980s and the 1990s. The seven class scale refer to (1) very high naturality, (2) high naturality, (3) middle naturality, (4) low naturality/alteration, (5) middle alteration, (6) high alteration and (7) very high alteration based on the percentile distributions of element concentrations in lichens (Nimis et al. 2000).

Recently a paper was published, where the data pre-processing of data (is examined under the methodological background of partial order theory, see Fattore et al. (2019).

4.2 Applicability of the Proposed Method

The quantity *epsav*, Eq. 14 is an average value and is – as mentioned already above – related to a single object. The domain of validity for Eq. 15 is given by $0.8 \leq Dom(X_1, X_2) \leq 1.0$.

If $Dom(X_1, X_2) \to 0.8$ the deviations *Eps* (Eq. 13) become quickly large as Figs. 1 and 2 (randomly generated data) show. Consequently in the following paragraph we investigate reasons for large deviations of Eps.

4.3 Reasons for Large Deviations

First of all, a dissection of a poset $(X, \leq)$ into two subposets $(X_1, \leq)$ and $(X_2, \leq)$ leads to more symmetry in the resulting graphs of the subposets (as already mentioned above). Hence, the degeneracy of *Hav*-values is increased. Even if the enhanced degree of ties is accepted, there can be large deviations, which result from structures like the one shown in Fig. 6.

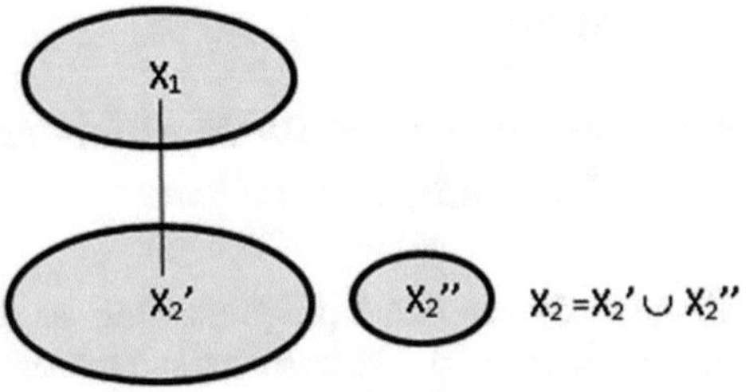

Fig. 6 X_1 dominates fully X_2' but not X_2. A typical situation causing deviations

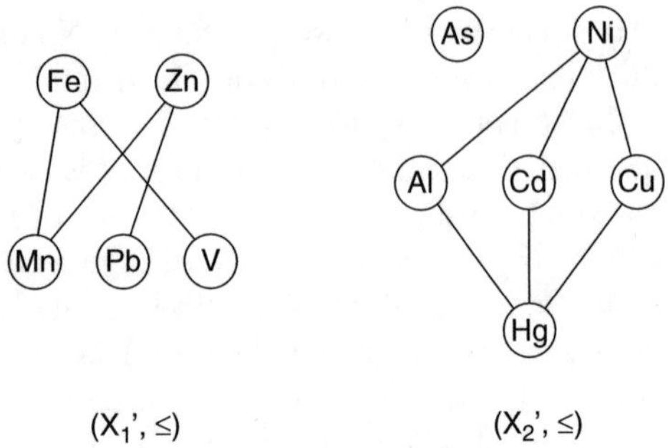

$X_2 = X_2' \cup X_2''$

Fig. 7 A variant for partitioning of set X

$(X_1', \leq)$ $(X_2', \leq)$

Considering Fig. 6 a situation, similar to that, causing eq. 12, arises. For $x \in X_1$, $Hav(x, X) = Hav(x, X_1) + |X_2|$ is an overestimation, because by constructing the linear extensions, the elements of X_2'' can also be located above the elements of X_1, whereas by $Hav(x, X) = Hav(x, X_1) + |X_2'|$ an underestimation follows. Based on remark 2 (see above) the two Hasse diagrams are shown, when $X_1' = X_1 - \{As\}$ and $X_2' = X_2 \cup \{As\}$ (Fig. 7)

The element Fe "sees" the same number of lower neighbours as Zn. Similarly, Pb and V have one upper neighbour. Therefore the exact method delivers $Havexact(Fe, X_1') = Havexact(Zn, X_1')$ as well as $Havexact(Pb, X_1') = Havexact(V, X_1')$. In Fig. 7, the subposet $(X_2', \leq)$ has also symmetries, leading to: Al $\cong$ Cd $\cong$ Cu with respect to $Havexact$.

4.4 Conclusive Consideration

By applying the dominance matrix and based on this, the calculation scheme seems to be attractive for an estimation method of Hav. However, the requirement of very high values of $Dom(X_1, X_2)$ seems to be too restrictive to justify to propose this method as a general approximation method. Thus, up to our actual knowledge this new procedure will not be practically feasible in comparison with exact results. When, however, a first check is wanted, for example to start from this a refinement procedure, then the scheme based on Eqs. 8 and 9 may be useful. When this line of research is to be followed, then

- A catalogue could be aimed, where structures are gathered, which typically lead to strong deviations
- As a candidate for a better approximation the method by Bubley and Dyer (1999), may be selected and modified in that manner that the weak order as a result of Eqs. 8 and 9 is a starting linear extension.

Summarizing, we hope that the present study has revealed some mathematical ideas which may be of interest and attract new research by scholars of the mathematical chemistry scene.

References

Annoni, P., Bruggemann, R., & Carlsen, L. (2014). A multidimensional view on poverty in the European Union by partial order theory. *Journal of Applied Statistics, 42*, 535–554.

Atkinson, M. D., & Chang, H. W. (1986). Extensions of partial orders of bounded width. *Congressus Numerantium, 52*, 21–35.

Bruggemann, R., & Carlsen, L. (2011). An improved estimation of averaged ranks of partially orders. *MATCH Communications in Mathematical and in Computer Chemistry, 65*, 383–414.

Bruggemann, R., & Halfon, E. (1997). Comparative analysis of nearshore contaminated sites in Lake Ontario: ranking for environmental hazard. *Journal of Environmental Science and Health, A32*(1), 277–292.

Bruggemann, R., & Patil, G. P. (2011). *Ranking and prioritization for multi-indicator systems – Introduction to partial order applications*. New York: Springer.

Bruggemann, R., & Voigt, K. (2008). Basic principles of hasse diagram technique in chemistry. *Combinatorial Chemistry & High Throughput Screening, 11*, 756–769.

Bruggemann, R., & Voigt, K. (2011). A new tool to analyze partially ordered sets – Application: Ranking of polychlorinated biphenyls and alkanes/alkenes in River main, Germany. *MATCH Communications in Mathematical and in Computer Chemistry, 66*, 231–251.

Bruggemann, R., & Voigt, K. (2012). Antichains in partial order, example: Pollution in a German region by lead, cadmium, zinc and sulfur in the herb layer. *MATCH Communications in Mathematical and in Computer Chemistry, 67*, 731–744.

Bruggemann, R., Halfon, E., Welzl, G., Voigt, K., & Steinberg, C. (2001). Applying the concept of partially ordered sets on the ranking of near-shore sediments by a battery of tests. *The Journal for Chemical Information and Computer scientists, 41*, 918–925.

Bruggemann, R., Sørensen, P. B., Lerche, D., & Carlsen, L. (2004). Estimation of averaged ranks by a local partial order model. *The Journal for Chemical Information and Computer scientists, 44*, 618–625.

Bruggemann, R., Voigt, K., Restrepo, G., & Simon, U. (2008). The concept of stability fields and hot spots in ranking of environmental chemicals. *Environmental Modelling & Software, 23*, 1000–1012.

Bruggemann, R., Carlsen, L., Voigt, K., & Wieland, R. (2014). PyHasse software for partial order analysis: Scientific background and description of selected modules. In R. Bruggemann, L. Carlsen, & J. Wittmann (Eds.), *Multi-indicator systems and modelling in partial order* (pp. 389–423). New York: Springer.

Bubley, R., & Dyer, M. (1999). Faster random generation of linear extensions. *Discrete Mathematics, 201*, 81–88.

Buonocore, E., Mellino, S., De Angelis, G., Liu, G., & Ulgiati, S. (2018). Life cycle assessment indicators of urban wastewater and sewage sludge treatment. *Ecological Indicators, 94*, 13–23.

Carlsen, L. (2008a). Hierarchical partial order ranking. *Environmental Pollution, 155*, 247–253.

Carlsen, L. (2008b). Partial ordering and prioritising polluted sites. In M. Pavan & R. Todeschini (Eds.), *Data handling in science and technology* (Vol. 27, pp. 97–109). Elsevier B.V.

Carlsen, L. (2013). Assessing chemicals using partial order ranking methodology. *Advances in Combinational Chemistry & High Throughput Screening, 1*, 3–35.

Carlsen, L. (2018). Happiness as a sustainability factor. The world happiness index: A posetic – based data analysis. *Sustainability Science, 13*, 549–571.

Carlsen, L., & Bruggemann, R. (2011). Risk assessment of chemicals in the river Main (Germany): Application of selected partial order ranking tools. *Statistica & Applicazioni*, Special Issue, 2011, 125–140.

Carlsen, L., & Bruggemann, R. (2014a). Partial order methodology: A valuable tool in chemometrics. *Journal of Chemometrics, 28*, 226–234.

Carlsen, L., & Bruggemann, R. (2014b). An analysis of the "failed states index" by partial order methodology. *Journal of Social Structure, 14*, 1–31.

Davey, B. A., & Priestley, H. A. (1990). *Introduction to lattices and order*. Cambridge: Cambridge University Press.

De Loof, K., De Meyer, H., & De Baets, B. (2006). Exploiting the lattice of ideals representation of a poset. *Fundamenta Informaticae, 71*, 309–321.

De Loof, K., De Baets, B., & De Meyer, H. (2011). Approximation of average ranks in posets. *MATCH Communications in Mathematical and in Computer Chemistry, 66*, 219–229.

De Loof, K., Rademaker, M., Bruggemann, R., De Meyer, H., Restrepo, G., & De Baets, B. (2012). Order theoretical tools to support risk assessment of chemicals. *MATCH Communications in Mathematical and in Computer Chemistry, 67*, 213–230.

Demiray, A. D., Yolcubal, I., Akyol, N. H., & Cobanoglu, G. (2012). Biomonitoring of airborne metals using the lichen *Xanthoria parietina* in Kocaeli Province, Turkey. *Ecological Indicators, 18*, 632–643.

Fattore, M., & Maggino, F. (2014). Partial orders in socio-economics. A practical challenge for poset theorists or a cultural challenge for social scientists? In R. Bruggemann, L. Carlsen, & J. Wittmann (Eds.), *Multi-indicator systems and modelling in partial order* (pp. 197–214). New York: Springer.

Fattore, M., Arcagni, A., & Maggino, F. (2019). *Optimal scoring of partially ordered data, with an application to the ranking of smart cities*. Smart Statistics for Smart Applications Book of Short Papers SIS 2019 – ISBN 9788891915108.

Ferretti, M., Innes, J. L., Jalkanen, R., Saurer, M., Schäffer, J., Spiecker, H., & von Wilpert, K. (2002). Air pollution and environmental chemistry–what role for tree-ring studies? *Dendrochronologia, 20*, 159–174.

Frati, L., Caprasecca, E., Santoni, S., Gaggi, C., Guttova, A., Gaudino, S., Pati, A., Rosamilia, S., Pirintsos, S. A., & Loppi, S. (2006). Effects of NO_2 and NH_3 from road traffic on epiphytic lichens. *Environmental Pollution, 142*(1), 58–64.

Galassi, S., Provini, A., & Halfon, E. (1996). Risk assessment for pesticides and their metabolites in water. *International Journal of Environmental Analytical Chemistry, 65*, 331–344.

Grisoni, F., Consonni, V., Nembri, S., & Todeschini, R. (2015). How to weight Hasse matrices and reduce incomparabilities. *Chemometrics and Intelligent Laboratory Systems, 147*, 95–104.

Grönlund, S. E. (2019). Indicators and methods to assess sustainability of wastewater sludge management in the perspective of two systems ecology models. *Ecological Indicators, 100*, 45–54.

Halfon, E., & Reggiani, M. G. (1986). On ranking chemicals for environmental hazard. *Environmental Science & Technology, 20*, 1173–1179.

Hasse, H. (1967). *Vorlesungen über Klassenkörpertheorie*. Marburg: Physica-Verlag.

Helm, D. (2003). Bewertung von Monitoringdaten der Umweltprobenbank des Bundes mit der Hasse-Diagramm-Technik. *UWSF – Z Umweltchem Ökotox, 15*, 85–94.

Jin, C., Nan, Z., Wang, H., & Jin, P. (2017). Plant growth and heavy metal bioavailability changes in a loess subsoil amended with municipal sludge compost. *Journal of Soils and Sediments, 17*, 2797–2809.

Klein, D. J., & Ivanciuc, T. (2006). Directed reaction graphs as posets. In R. Bruggemann & L. Carlsen (Eds.), *Partial order in environmental sciences and chemistry* (pp. 35–57). Berlin: Springer.

Loppi, S., & Pirintsos, S. A. (2003). Epiphytic lichens as sentinels for heavy metal pollution at forest ecosystems (central Italy). *Environmental Pollution, 121*, 327–332.

Loppi, S., Pirintsos, S. A., & De Dominicis, V. (1999). Soil contribution to the elemental composition of epiphytic lichens (Tuscany, central Italy). *Environmental Monitoring and Assessment, 58*, 121–131.

Newlin, J., & Patil, G. P. (2010). Application of partial order to stream channel assessment at bridge infrastructure for mitigation management. *Environmental and Ecological Statistics, 17*, 437–454.

Nimis, P. L., & Bargagli, R. (1999). *Linee-guidaper l'utilizzo come bioaccumulatori di metalli in traccia*. In: Proceedings of Workshop Biomonitorraggio della qualita dell' ariasul terratorio nazionale, Roma 26–27, Giugno 1998, ANPA-Serie Atti, pp. 279–287

Nimis, P. L., Lazzarin, G., Lazzarin, A., & Skert, N. (2000). Biomonitoring of trace elements with lichens in Veneto (NE Italy). *Science of the Total Environment, 255*, 97–111.

Patil, G. P., & Taillie, C. (2004). Multiple indicators, partially ordered sets, and linear extensions: Multi-criterion ranking and prioritization. *Environmental and Ecological Statistics, 11*, 199–228.

Pavan, M., & Todeschini, R. (2004). New indices for analysing partial ranking diagrams. *Analytica Chimica Acta, 515*, 167–181.

Pirintsos, S. A., & Loppi, S. (2008). Biomonitoring atmospheric pollution: The challenge of times in environmental policy on air quality. *Environmental Pollution, 151*, 269–271.

Pirintsos, S. A., Matsi, T., Vokou, D., Gaggi, C., & Loppi, S. (2006). Vertical distribution patterns of trace elements in an urban environment as reflected by their accumulation in lichen transplants. *Journal of Atmospheric Chemistry, 54*, 121–131.

Pirintsos, S., Bariotakis, M., Kalogrias, V., Katsogianni, S., & Bruggemann, R. (2014). Hasse diagram technique can further improve the interpretation of results in multielemental large-scale biomonitoring studies of atmospheric metal pollution. In R. Bruggemann, L. Carlsen, & J. Wittmann (Eds.), *Multi-indicator systems and modelling in partial order* (pp. 237–251). New York: Springer.

Pudenz, S., & Heininger, P. (2006). Comparative evaluation and analysis of water sediment data. In R. Bruggemann & L. Carlsen (Eds.), *Partial order in environmental sciences and chemistry* (pp. 111–151). Berlin: Springer.

Quintero, N. Y., Bruggemann, R., & Restrepo, G. (2018). Mapping posets into low dimensional spaces: The case of uranium trappers. *MATCH Communications in Mathematical and in Computer Chemistry, 80*, 793–820.

Restrepo, G., & Bruggemann, R. (2008). Dominance and separability in posets, their application to isoelectronic species with equal total charge. *Journal of Mathematical Chemistry, 44*, 577–602.

Restrepo, G., Bruggemann, R., Weckert, M., Gerstmann, S., & Frank, H. (2008a). Ranking patterns, an application to refrigerants. *MATCH Communications in Mathematical and in Computer Chemistry, 59*, 555–584.

Restrepo, G., Weckert, M., Bruggemann, R., Gerstmann, S., & Frank, H. (2008b). Ranking of refrigerants. *Environmental Science & Technology, 42*, 2925–2930.

Rocco, C. M., & Tarantola, S. (2014). Evaluating ranking robustness in multi-indicator uncertain matrices: An application based on simulation and global sensitivity analysis. In R. Bruggemann, L. Carlsen, & J. Wittmann (Eds.), *Multi-indicator systems and modelling in partial order* (pp. 275–292). New York: Springer.

Simon, U., Bruggemann, R., & Pudenz, S. (2004). Aspects of decision support in water management – Example Berlin and Potsdam (Germany) I – Spatially differentiated evaluation. *Water Research, 38*, 1809–1816.

Simon, U., Bruggemann, R., Behrendt, H., Shulenberger, E., & Pudenz, S. (2006). METEOR: A step-by-step procedure to explore effects of indicator aggregation in multi criteria decision aiding – Application to water management in Berlin, Germany. *Acta Hydrochimica et Hydrobiologica, 34,* 126–136.

Sørensen, P. B., Mogensen, B. B., Gyldenkaerne, S., & Rasmussen, A. G. (1998). Pesticide leaching assessment method for ranking both single substances and scenarios of multiple substance use. *Chemosphere, 36,* 2251–2276.

Sørensen, P. B., Mogensen, B. B., Carlsen, L., & Thomsen, M. (2000). The influence on partial order ranking from input parameter uncertainty – Definition of a robustness parameter. *Chemosphere, 41,* 595–600.

Trotter, W. T. (1992). *Combinatorics and partially ordered sets, dimension theory.* Baltimore: The Johns Hopkins University Press.

Voigt, K., Welzl, G., & Bruggemann, R. (2004a). Data analysis of environmental air pollutant monitoring systems in Europe. *Environmetrics, 15,* 577–596.

Voigt, K., Bruggemann, R., & Pudenz, S. (2004b). Chemical databases evaluated by order theoretical tools. *Analytical and Bioanalytical Chemistry, 380,* 467–474.

Winkler, P. (1982). Average height in a partially ordered set. *Discrete Mathematics, 39,* 337–341.

Problem Orientable Evaluations as *L*-Subsets

Adalbert Kerber and Rainer Bruggemann

1 Evaluation

Evaluations using sequences of parameter values can be considered as sets over a *lattice* (cf. (Bruggemann and Kerber 2018; Bruggemann et al. 2011; Kerber 2006, 2017; Pollandt 1997)). An example is G. Restrepo's evaluation (Restrepo 2008) of 40 refrigerants using the parameters *ODP* (ozone depletion potential), *GWP* (general warming potential) and *ALT* (atmospheric life time). The parameters were normalized, i.e. their values were elements of the interval $[0, 1]$ and therefore the lattice containing these triples of parameter values is $L = [0, 1]^3$. With a refrigerant *ref* he associated the following triple of real numbers between 0 and 1:

$$(ODP(ref); GWP(ref); ALT(ref)) \in [0, 1]^3.$$

Examples of such triples that he obtained are

refrigerants	values of (ODP; GWP; ALT)
ref_1	(0.19607843; 0.31621622; 0.01406219)
ref_2	(0.16078431; 0.72432432; 0.0312497)
ref_3	(0.00980392; 0.12027027; 0.00374969)
ref_4...	(0.00431373; 0.00513514; 0.00040594)

A. Kerber
Department of Mathematics, University of Bayreuth, Bayreuth, Germany
e-mail: kerber@uni-bayreuth.de

R. Bruggemann (✉)
Leibniz-Institute of Freshwater Ecology and Inland Fisheries, Berlin, Germany
e-mail: brg_home@web.de

This evaluation associates, e.g. with the refrigerant ref_4, the *truth value* $tv(ref_4$ *has* $(ODP;GWP;ALT)) = (0.00431373;0.00513514;0.00040594)$.

As the entries of this table are in $L = [0,1]^3$, this evaluation is a mapping

$$\mathrm{E}: \{ref_1, ref_2, \dots\} \times \{(ODP; GWP; ALT)\} \to [0, 1]^3,$$

with, e.g., the value

$$\mathrm{E}\left(ref_4, (ODP; GWP; ALT)\right) = (0.00431373; 0.00513514; 0.00040594).$$

These values are elements of the lattice $L = [0,1]^3$, and *hence we may consider such an evaluation as an L-subset of the set of refrigerants*. This way of analysis implies that we have no more crisp sets (an element is a member of a set: "yes" or "no"), but fuzzy sets, where the membership can be any number between 0 and 1. The general case reads as follows:

1.1   Definition

An *evaluation* E of objects $o_i \in O$ w.r.t. attributes $a_k \in A$ and over L is a mapping

$$\mathrm{E}: O \times A \to L : (o_i, a_k) \to \mathrm{E}((o_i, a_k)) = tv\,(o_i \text{ has } a_k),$$

i.e. we consider it as an *L-subset* E of $O \times A$, containing (o_i, a_k) with the *truth value* $tv(o_i$ has $a_k) \in L$.

1.2   Basic theory

Evaluations of objects o_i w.r.t. attributes a_k.

– Consider

$$L^{O \times A} := \{\mathrm{E} \mid \mathrm{E} : O \times A \to L\},$$

the set of all *L–subsets* of $O \times A$, for a given lattice L. In case $L = [0,1]^3$, an L-subset of $O \times A$ is an association of triples of parameter values to the pairs $(o,a) \in O \times A$.

By this generalization of the evaluation *we can choose a set theory and its logic* over L and this *allows problem–orientation*. Hereby we adopt the fuzzy notion of membership functions (Pollandt 1997) as the set theoretical operations such as subset-set relation, inclusion, set differences or union are not necessarily related to crisp sets.

− On *L–subsets* S, S' of a set *X* we introduce *L–inclusion* as follows:

$$S \subseteq_L S' \Longleftrightarrow \forall x \in X : S(x) \leq S'(x).$$

− Intersections of two such *L*-subsets can be defined, using *t-norms*
   $\tau : L \times L \to L$, mappings with *symmetry, monotony, associativity* and *side condition* $\tau(x, 1_L) = x$. They yield τ–*intersections* I on L^X with *M* and *N* as arbitrary membership functions (Pollandt 1997).

$$I(x) = (M \cap_\tau N)(x) = \tau(M(x), N(x)).$$

One of the most important *t*–norms is:

− The *standard* norm *s*, defined as

$$s(x, y) = x \wedge y.$$

Other *t*-norms are the drastic norm, the algebraic product and the bounded difference, see (Kerber 2006, 2017)
We use a notion of truth, based on τ and its residuum:

− $\tau^* : L \times L \to L$ is a *residuum* of τ, iff

$$\tau(x, y) \leq v \Longleftrightarrow x \leq \tau^*(y, v).$$

In this case τ is called a *residual t*–norm.
As an example of a residuum for $L = [0, 1]$ we select the standard norm, s(α,β)

$$s^*(\alpha, \beta) = \begin{cases} 1, & if\ \alpha \leq \beta \\ \beta, & otherwise \end{cases}$$

This means that we have choices, and that we can use a *problem orientation*:

− Choose a suitable lattice *L* as set of values; pick a suitable residual *t*-norm τ obtaining a set theory. Its residuum τ^* gives the corresponding logic, i.e. a quantification of the subset-set-relation. Apply that to E $\in L^{O \times A}$, the evaluation considered, and get a basis of the implications (see below)!

1.3 Exploration

For the exploration of the evaluation E we can use that object *o has* attribute *a* if and only if E(*o,a*) > 0. We put

$$A'(o) = \tau^*(A \Rightarrow E) = \bigwedge_{a \in A} \tau^*(A(a), E(o, a)),$$

and we evaluate $A \in L^A$ *implies* $B \in L^A$ in E by:

$$\tau^* (A \Rightarrow B) = \bigwedge_{o \in O} \tau^* \left(A'(o), B'(o) \right).$$

$A \Rightarrow B$ *holds* in E if and only if $\tau^*(A \Rightarrow B) = 1$, i.e., iff $A' \subseteq_L B'$. Defining *pseudo-contents* (Ganter and Wille 1996), by

$$P \neq= P'' \text{ and for each pseudo} - \text{content } Q \subset_L P : Q'' \subseteq_L P,$$

we get the *Duquenne/Guigues-basis* (Duquenne 1987) which implies every attribute implication following from E,

$$P = \left\{ P \Rightarrow \left(P'' \backslash P \right) \mid P \text{ pseudo} - \text{content} \right\}.$$

1.4   Example

Adding substructures, *Cl-*, *F-*, *Br-*, *I*-atoms, and using simplified binary parameters *nODP* *, *nGWP* *, *nALT* *, ... , we obtain for an arbitrary subset of refrigerants (see for the complete set (Restrepo 2008)) in order not to get too huge outputs:

E	nODP*	nGWP*	nALT*	nC	Cl	F	Br	I	ether	CO_2	NH_3
1	1	0	0	0	1	1	0	0	0	0	0
2	0	1	0	0	1	1	0	0	0	0	0
6	0	0	0	1	1	1	0	0	0	0	0
7	0	0	0	1	1	1	0	0	0	0	0
8	0	1	1	0	0	1	0	0	0	0	0
16	0	0	0	1	0	0	0	0	0	0	0
21	0	0	0	0	0	0	0	0	0	1	0
22	1	0	0	0	1	1	1	0	0	0	0
23	0	1	1	1	0	1	0	0	0	0	0
29	0	1	1	1	0	1	0	0	1	0	0
32	0	0	0	0	1	0	0	0	0	0	0
33	1	0	0	1	1	1	0	0	0	0	0
35	1	0	1	1	1	1	0	0	0	0	0
36	0	0	0	0	0	1	0	1	0	0	0
37	0	0	0	1	0	0	0	0	1	0	0
38	0	0	0	0	0	0	0	0	0	0	1
39	0	0	0	1	0	1	0	0	1	0	0
40	0	0	0	1	0	1	0	0	1	0	0

The Duquenne/Guigues basis of it yields all what follows, it can be obtained online, using CONEXP–1.3 (Yevtushenko 2000). We find the implications

$\{nODP'\}$	$\Longrightarrow \{Cl,F\}$
$\{nGWP'\}$	$\Longrightarrow \{F\}$
$\{nALT'\}$	$\Longrightarrow \{F\}$
$\{nC,Cl\}$	$\Longrightarrow \{F\}$
$\{nALT^*,Cl,F\}$	$\Longrightarrow \{nODP^*,nC\}$
$\{nGWP',nC,F\}$	$\Longrightarrow \{nALT^*\}$
$\{Br\}$	$\Longrightarrow \{nODP^*,Cl,F\}$
$\{I\}$	$\Longrightarrow \{F\}'$
$\{ether\}$	$\Longrightarrow \{nC\}$
$\{nALT',nC,F,ether\}$	$\Longrightarrow \{nGWP^*\}$

Summarizing we suggest: In order to explore an evaluation of objects $o \in O$ according to given attributes $a \in A$ do the following:

- Choose a suitable set theory, i.e. a residual τ and its τ^*,
- use Bruggemann's̈ PyHasse (Bruggemann et al. 2014), to avoid the tedious manually operations,
- evaluate, using CONEXP (Yevtushenko 2000) if it is binary, the Duquenne/ Guigues basis
- evaluate, using CONEXP (Yevtushenko 2000) if it is binary, the Duquenne/ Guigues basis (Duquenne 1987), a set of *hypotheses* on possibly interesting bigger sets $\Omega \supset O$ of objects. Try to prove (or at least to check) these!

2   A Further Example

Eight regions along river Rhine in the southwest of Germany were monitored with respect to pollution by the chemical elements lead, *Pb*, cadmium, *Cd*, zinc, *Zn* and sulfur, *S*. As the middle range transport was of main interest, the herb layer was more closely investigated by the environmental protection agency of Baden Württemberg. The regions, together with their labels and the total concentrations˙(mg/kg dry mass) of *Pb*, *Cd*, *Zn* and *S* are shown in the following table (Table 1)

Table 1 Measured regional pollution

Region	Label	Pb	Cd	Zn	S
Lorrach	1	1	0.04	21	1540
Westwards of Freiburg	24	1.7	0.18	39	1740
Heidelberg	52	2	0.23	36	4030
South of Mulheim	10	1	0.03	29	1780
Offenburg	31	1.1	0.15	28	1740
Mannheim	56	1	0.11	34	1970
Westwards of Mulheim	19	0.8	0.01	18	4030
Rastatt	43	0.5	0.11	39	4030

Table 2 Discretized regional pollution

Regions	nPb	nCd	nZn	nS
1	0	0	0	0
10	0	0	0	0
24	1	1	1	0
31	1	1	0	0
19	0	0	0	1
43	0	1	1	1
52	1	1	1	1
56	1	1	1	0

We coarsen the data to values 0 or 1: Let $mw(j)$ be the arithmetic mean value of j-th chemical element (Pb or Cd or Zn or S) taken over all eight regions, indicate by $q(i,j)$ the total concentration of the j-th chemical element of region i and put

$$q_b(i, j) = \begin{cases} 0 \; if \; q(i, j) > mw(j) \\ 1 \; otherwise \end{cases}$$

It is completely clear that the results are depending on outliers and the distribution of the data, because the mean value is statistically not a robust measure. We suppress the corresponding analysis, as these data serve only as a demonstration. The mean values are:

$$Pb : 1.1375; \; Cd : 0.1075; \; Zn : 30.5; \; S : 2607.5.$$

Here is an input for the program CONEXP by Yevtushenko (Yevtushenko 2000). We write (instead of $q_b(-,j)$) nPb, nCd, etc. for easier understanding, write 1 for × and 0 for the empty cell, and find (Table 2)

The program CONEXP (Yevtushenko 2000) delivers some implications, which should be considered as geochemical hypotheses. We list implications as outcome of *CONEXP* (Yevtushenko 2000). How often the premises are realized in the (transformed) data matrix is given by the numbers in <>-brackets:

1. $< 4 > nPb \Longrightarrow nCd$;
2. $< 4 > nZn \Longrightarrow nCd$;
3. $< 2 > nCd,nS \Longrightarrow nZn$.

Within the transformation, and the selection of regions the hypotheses would be: High pollution by lead implies a high pollution by cadmium. This is plausible as often by mining activities lead and cadmium are simultaneously found. Similarly Zn also implies Cd, and finally, when the pollution by Cd and by S is high, then the hypothesis is that in this case Zn is also highly polluting. It should be realized that an implication of the form 'S implies...' is not found. These hypothetic implications can a posterior easily be verified by checking Table 2, whereas it is often difficult to detect implications from reading the table. However, the reader should still be aware that the implications are derived from binary data which originally were continuous in concept. In another contribution within this book (Bruggemann, Kerber) the mathematical framework given above is reformulated for programming purposes and may be helpful for a further understanding of the "problem orientable evaluations as L-subsets".

References

Bruggemann, R., & Kerber, A. (2018). Fuzzy logic and partial order; first attempts with the new PyHasse-program L eval. *MATCH, Communications in Mathematical and in Computer Chemistry, 80*, 745–768.

Bruggemann, R., Kerber, A., & Restrepo, G. (2011). Ranking objects using fuzzy orders with an application to refrigerants. *MATCH Communications in Mathematical and in Computer Chemistry, 66*, 581–603.

Bruggemann, R., Carlsen, L., Voigt, K., & Wieland, R. (2014). PyHasse software for partial order analysis: Scientific background and description of selected modules. In R. Bruggemann, L. Carlsen, & J. Wittmann (Eds.), *Multi-indicator systems and modelling in partial order* (pp. 389–423). New York: Springer.

Duquenne, V. (1987). Contextual implications between attributes and some representation properties for finite lattices. In B. Ganter, R. Wille, & K. E. Wolff (Eds.), *Beiträge zur Begriffsanalyse* (pp. 213–239). Mannheim: BI Wissenschaftsverlag.

Ganter, B., & Wille, R. (1996). *Formale Begriffsanalyse Mathematische Grundlagen*. Berlin: Springer.

Kerber, A. (2006). Contexts, concepts, implications and hypotheses. In R. Bruggemann & L. Carlsen (Eds.), *Partial order in environmental sciences and chemistry* (pp. 355–365). Berlin: Springer.

Kerber, A. (2017). Evaluation and exploration, a problem-oriented approach. *Toxicological & Environmental Chemistry, 99*, 1270–1282.

Pollandt, S. (1997). *Fuzzy-Begriffe*. Berlin: Springer.

Restrepo, G. (2008). *Assessment of the environmental acceptability of refrigerants by discrete mathematics: Cluster analysis and Hasse diagram technique*. Doctoral thesis, University of Bayreuth, Germany. Available via http://opus.ub.uni-bayreuth.de/frontdoor.php?sourceopus=393&la=de

Yevtushenko, S. A. (2000). *System of data analysis concept explorer*. Proceedings of the 7th National Conference on Artificial Intelligence, Russia, K II-2000, pp. 127–134. Software: http://sourceforge.net/projects/conexp

Evaluations as Sets over Lattices – Application Point of View

Rainer Bruggemann and Adalbert Kerber

1 Introduction

This contribution is based on Bruggemann and Kerber 2018 and on the theory developed there (in the following referred as BK). The essential point is that an evaluation, as well as a subsequent exploration starts not always with an ordinal data matrix or even with a data matrix built of binary data, but on data continuous in concept, for example: data taken from $[0,1]^m$, as described for m=3 in BK. Therefore, it is of interest how far the powerful and elegant method of Formal Concept Analysis (Ganter and Wille 1996) can be applied. Especially, without the crucial and often arbitrary scaling method which was developed in the school of Wille. The mathematical concept of an alternative method, avoiding the scaling procedure (Ganter and Wille 1989) is explained in BK. Here the methods are examined in more detail with respect to typical tasks in decision support systems and taking the point of view of a statistician. I.e. a special focus was set on the point, how the lattice theoretical method can be compared with conventional statistical methods, as for example correlation analysis.

A technical problem arises: The lattice theoretical method in BK is selected as one of the modules of the PyHasse software, see BK, because the method is basically simple however manually performed, very tedious. (For the PyHasse software itself, see e.g. Bruggemann et al. 2014). Therefore, the mathematical notation is to be replaced by a notation, which also can be used in the programming language Python.

R. Bruggemann (✉)
Leibniz-Institute of Freshwater Ecology and Inland Fisheries, Berlin, Germany
e-mail: brg_home@web.de

A. Kerber
Department of Mathematics, University Bayreuth, Bayreuth, Germany
e-mail: kerber@uni-bayreuth.de

© The Author(s), under exclusive license to Springer Nature Switzerland AG 2021
R. Bruggemann et al. (eds.), *Measuring and Understanding Complex Phenomena*,
https://doi.org/10.1007/978-3-030-59683-5_7

2   Materials and Methods

2.1   Notation

Table 1 shows the most important issues.

2.2   The Nature of Mapping A (and B), Standard t-norm

In BK the mappings A and B are introduced as subsets of $[0,1]^m$,m being $|Q|$, in order to be most general and (as will be shown later) to calculate properly derivations of A and B, resp.. Both mappings are the basis for formulating an implication A $\Rightarrow$ B, i.e. to calculate as to how far A implies B. In data exploration and in Formal Concept Analysis one wants to find out, as to how far a subset of Q implies another subset of Q. An example based on refrigerants is shown in BK, where {nODP*} $\Rightarrow$ {Cl, F} (*if there is an ozone depletion potential observed (nODP* = 1), then (within the given set of refrigerants), the chemicals have Cl- and F substituents*). Both sets {ODP*} and {Cl, F} are crisp subsets of Q = {nODP*, nGWP*, nALT*, nC,Cl, F, Br, J, ether, CO_2, NH_3}. For more details, see Sect. 4.1 or Kerber, Bruggemann, this volume.

In other words: Starting a data exploration in terms of finding out as to how far implications can be established among subsets of Q needs the formulation of mapping A and B as subsets of $\{0,1\}^{|Q|}$. For example the subsets with $|Q| = 2$ would be {(0,0)} or {(0,1),(1,0)}, etc. Note that a deepened analysis (here not considered) would require of $[0,1]^{|Q|}$. Then a subset with $|Q| = 2$ could be {(0.2,0.7), (0.8,0.01)}, etc.

Table 1 Notation

Issue	Notation	Remark
t-norm	t	in BK: τ
Standard t-norm	s	in BK: s ; in the following text only the standard norm will be applied.
Residuum of s	s* in this text	resid in Python programs
Object set	X	
Set of indicators	Q	in former papers also called IB (information basis)
ith object	x(i)	in BK: o
jth indicator	q(j)	in BK: a
Entry of a data matrix	x(i,j)	in BK: ε(o,a). It is assumed that $0 \leq x(i,j) \leq 1$
Mapping A	A	A
Mapping B	B	B

2.3   Role of Mapping A, Standard Norm and the Data Matrix

A central role in the data exploration applying data continuous in concept plays the "has"-relation:

$$\text{An object } x \in X \text{ has an indicator } q \in Q \qquad (1)$$

Because the data are supposed to be continuous, the term "has" must be interpreted as a degree, i.e. a truth value (tv) of the "has"-relation. If tv $= 0$ then object x has not the indicator q, if tv $= 1$, then the object has certainly the indicator q, hence in general tv $\in [0,1]$. Statement (1) in terms of A, Q and x(i,j) is given by Eq. 2:

$$A'(x) = \wedge_{q(j)\in Q1 \subset Q} s * (A(q(j)), x(i,j)) \qquad (2)$$

Therein is Q1 a crisp subset of Q. The meet-operation in Eq. 2 can in the case of data continuous in concept replaced by the Min-operation. When Q1 is a singleton, say Q1 $= \{q(1)\}$ the evaluation of Eq. 2 is very easy:

$$A'(x) = \text{Min} \{s * (A(q(j)), x(i,j))\}.$$

As shown in BK the residuum of the standard norm is given by Eq. (3):

$$s * (\alpha, \beta) := \begin{cases} 1 & if \alpha \leq \beta \\ 0 & otherwise \end{cases} \qquad (3)$$

α, β being real numbers ≥ 0.

A is a tuple of length $|Q|$. A(q(j)) means, the value of the tuple A at position j. Let us select a 1 at the jth position and 0 otherwise, in order to describe the crisp subset $\{q(j)\}$. Then, the residuum of the standard norm delivers for x(i) everywhere, where A(q(j)) $= 0$ the value 1, and only in the jth position the value x(i,j). Hence a singleton Q1 $= \{q(j)\}$ selects just the entry x(i,j) and if all x $\in X$ are considered just the jth column of the data matrix.

Consequently, the subset Q1, with several indicators, say $\{q(j1), q(j2), q(j3)\}$ delivers for the object x(i) first the entries x(i,j1), x(i,j2) and x(i,j3) and A'(x) is 0 besides at the positions j $= j1$, j $= j2$, j $= j3$) where the actual values of x(i,j) are to be inserted (which nevertheless can also be 0) and after this selection the minimal value among the set of entries $\{x(i,j1), x(i,j2), x(i,j3)\}$ is to be found.

In Fig. 1 the situation, due to Eq. (2) is schematically shown, assuming that $|Q| = 7$ and A $= (0,1,0,0,1,0,0)$ describing the crisp subset $\{q(2), q(5)\}$.

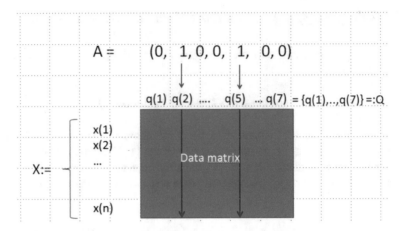

Fig. 1 Scheme for the evaluation of Eq. (2), assuming 7 indicators and a data matrix with values in [0,1]. For one object x(i) Eq. 2 selects just x(i,2) and x(i,5). For each row, the minimal value is to be selected

2.4 Implication Between Two Disjoint Singletons of Q

Let Q1 be {q(j*)} and Q2 = {q(j**)}. The truth-value tv of an implication can be calculated, by evaluating Eq. (4):

$$tv\,(Q1 \Rightarrow Q2) = Min_{x \in X}\left\{s * \left(A^{'}(x), B^{'}(x)\right)\right\} \tag{4}$$

A'(x) and B'(x) quantify, as to how far x has q(j*) and q(j**), resp. From Sect. 2.3 it is known that the has – relation for Q1 selects just the column x(i, j*), whereas the has-relation of Q2 selects the column x(i, j**). Figure 2 shows schematically the procedure.

3 Towards a Statistical Approach

3.1 Role of Subsets of X

As should be clear from the above, the truth values of implications will rarely be 1. Hence it is meaningful to check whether or not subsets of X will modify the truth values.

Proposition

$$X1 \subseteq X \Rightarrow tv\,(q\,(j*) \Rightarrow q\,(j**))\,|_X \leq tv\,(q\,(j*) \Rightarrow q\,(j**))|_{X1} \tag{5}$$

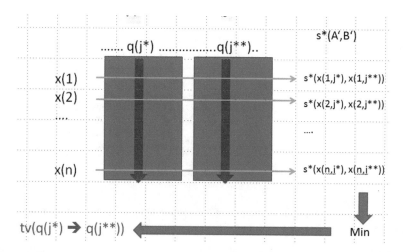

Fig. 2 Procedure to determine the truth value of the implication q(j*) ⇒ q(j**). The blue blocks symbolize the data matrix x(i,j) , j = 1,...,|Q| and I = 1,...,n

in Eq. 5 the notation ... $|_X$ and ... $|_{X1}$ means that the values of tv are taken once from the set of X and once from the set X1.

Proof The tv-value is taken from Min{s*(x(i,j*)),x(i,j**) } according to Eqs. (2) and (4). The Min-value is obtained from X; let x(i*, j') be this minimum. If X1 ⊂ X, and x(i*,j') is not in X1, then necessarily tv(X1) ≥ tv(X). Otherwise tv(X1) = tv(X).

Corollary By a proper selection of X1 the truth value can be enlarged.

3.2 Role of Transposed Data Matrix x(j,i) j = 1, ...,|Q|, i = 1, ...,n

The question is, when can be guaranteed that the implication between two disjoint singletons yield tv = 1, applying the standard norm.

Proposition

$$x(i, j_1) \leq x(i, j_2) \quad \text{for all } i = 1, \ldots, n \implies \text{tv}(\{q(j_1)\} \implies [q(j_2)]) = 1 \qquad (6)$$

Proof A(q(j₁)) induces x(i,j₁), B(q(j₂)) induces x(i,j₂). From each pair (x(i,j₁), x(i,j₂)) the residual standard norm has to be taken. Due to: x(i,j₁) ≤ x(i,j₂) for all i = 1,...,n the residual standard nor equals 1 for x(i), hence the Min-value, taken over all x(i) equals 1, and tv(q(j₁) ⇒ q(j₂)) is therefore 1.

Corollary 1 The test for: x(i,j₁) ≤ x(i,j₂) for all i = 1,...,n means that a partial order can be defined among the indicators. Because: x(i,j₁) ≤ x(i,j₂) for all

Fig. 3 Application of the
product order for the
transposed data matrix

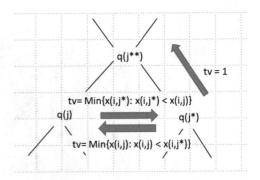

Fig. 3 Application of the product order for the transposed data matrix

$i = 1, \ldots, n$ $q(j_1) \leq q(j_2)$ is fulfilled. Checking $x(i,j_1) \leq x(i,j_2)$ for all $i = 1, \ldots, n$ means to investigate the partial order by examining the transposed data matrix.

Corollary 2 If $q(j) \parallel q(j^*)$ (i.e. if $(q(j,1), q(j,2), \ldots, q(j,n))$ incomparable with $(q(j^*,1), q(j^*,2), \ldots, q(j^*,n))$, n being the number of objects, then the minimal values of all $x(i, j^*)$ is to be checked, for which $x(i,j^*) < x(i,j)$ to establish the truth value for the implication $q(j) \Rightarrow q(j^*)$, and analogously the minimal values of all $x(i,j^*)$ for which is found: $x(i,j) \leq x(i,j^*)$-

Figure 3 shows this result schematically (instead of $q(j1)$ and $q(j2)$, resp. it is used $q(j^*)$ and $q(j^{**})$, resp.

Taking the scheme in Fig. 3 literally, then also $tv(q(j) \Rightarrow q(j^{**})) = 1$ is valid.

3.3 Implications and Correlation

As already stated, data continuous in concept can also be analyzed with simple statistical tools, such as the (Spearman or Pearson) correlation analysis. However, it is difficult, to find a theoretical relation between tv-values of an implication and the correlation coefficient. In order to get an idea how the tv-values and the correlation coefficients could be related, a fictitious data set was analyzed. Table 2 shows the data.

The term z in Table 2 stands for values from 0.1 to 1 in 0.1 steps, so that in practice 10 data matrices are analyzed, which only differ in the value $x(11,q2)$.

It is clear that correlation analysis is from its very nature a symmetric analysis, whereas the truth values of implications depend on the direction of the implication, i.e. whether $q1 \Rightarrow q2$, or $q2 \Rightarrow q1$.

Furthermore, the truth values of implications depend on the lowest possible value of z, i.e. the tv cannot be considered as a statistical robust measure. The correlation coefficient (Pearson) and the two truth values are calculated for each of the ten possible data matrices and the result is shown in Fig. 4.

Table 2 A fictitious data set with 11 objects $x(1), \ldots, x(11)$ and two indicators $q(1)$ and $q(2)$

	q(1)	q(2)
x(1)	0	0
x(2)	0.1	0.1
x(3)	0.2	0.2
x(4)	0.3	0.3
x(5)	0.4	0.4
x(6)	0.5	0.5
x(7)	0.6	0.6
x(8)	0.7	0.7
x(9)	0.8	0.8
x(10)	0.9	0.9
x(11)	1.0	z

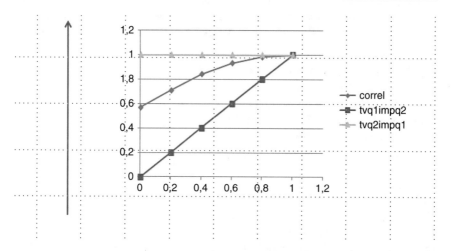

Fig. 4 The values of the abscissa are the values of z, whereas the ordinate is either the Pearson correlation coefficient or the two truth values

In that specific case, with a very special family of data matrices, the Pearson correlation coefficient has an upper limit by the truth value of q2 $\Rightarrow$ q1, and a lower limit by the truth value of q1 $\Rightarrow$ q2.

This example shows that considering the implication based on indicators continuous in concept as an approach for a correlation is misleading: The correlation aims at a more or less good co-monotony in the two indicators, whereas the truth value, say of q1 $\Rightarrow$ q2 depends on the value of z and in the case of the implication q2 $\Rightarrow$ q1, where the truth value equals 1 for all z only confirms that all data of q2 are larger than those of q1, when all objects (here x1 to x11) are considered.

4   Two Examples

4.1   Example: Refrigerants

Refrigerants are chemicals, which reduce the temperature during a controlled evaporation process in various machines (Restrepo et al. 2008a, b; Bruggemann et al. 2011). In BK the data were transformed to binary ones, and based on this, implications were deduced. In Table 3 a subset of 4 refrigerants is shown, all having chlorine and fluorine as substituent. The data set selected is small, just for demonstration and to avoid too large outputs.

The chemicals are listed in Table 4.

Only the implications between a premise as singleton are analyzed, and only the standard norm and its residuum resp. is applied. As can be seen from Table 3, the data are in the interval [0,1], and not in binary form. We have the implications:

(1) F, implies Cl, with truth-value 1.0
(2) Cl, implies F, with truth-value 1.0
(3) nC, implies F, with truth-value 1.0
(4) nC, implies Cl, with truth-value 1.0
(5) nC, implies Cl, F, with truth-value 1.0
GWP, implies F, with truth-value 1.0
GWP, implies Cl, with truth-value 1.0
GWP, implies Cl, F, with truth-value 1.0
ODP, implies F, with truth-value 1.0
ODP, implies Cl, with truth-value 1.0
ODP, implies Cl, F, with truth-value 1.0
ODP, implies GWP, with truth-value 1.0
ODP, implies GWP, F, with truth-value 1.0
ODP, implies GWP, Cl, with truth-value 1.0
ALT, implies F, with truth-value 1.0
ALT, implies Cl, with truth-value 1.0
ALT, implies GWP, with truth-value 1.0

Table 3 Four refrigerants (with labels "1","2",...) and a reduced set of indicators

	ALT	ODP	GWP	nC	Cl	F
"1"	0.01	0.2	0.32	0.0	1.0	1.0
"2"	0.03	0.16	0.72	0.0	1.0	1.0
"6"	0.0	0.02	0.05	1.0	1.0	1.0
"7"	0.01	0.01	0.15	1.0	1.0	1.0

ALT atmospheric lifetime, *ODP* ozone depletion potential, *GWP* general warming potential, *Cl* presence of chlorine in the molecule, F analogously, *F* analogously, *nC* at least one C-C-bond

Table 4 Chemicals, labeled by "1", "2", ...

Label	Chemical
"1"	CCl_3F
"2"	CCl_2F_2
"6"	$C_2H_3Cl_2F$
"7"	$C_2H_2ClF_2$

The first five implications are trivial, as one can see from the molecular formulas, because Cl and F are always present. Nevertheless, a series of implications is found, where the truth value equals 1. However, it should be clear that the restriction of only 4 chemicals is very severe, especially when originally 40 refrigerants were part of a partial order driven analysis (Restrepo et al. 2008a). That usually the truth values are rarely 1 will be shown in the next section.

4.2 Eight German Regions Along the River Rhine

The Environmental Protection Agency of the German state Baden-Württemberg initialized a large monitoring study, concerning the pollution by lead, Pb, cadmium, Cd, zinc, Zn and sulfur, S. Different targets were investigated, so for example the herb layer (see for a general overview: Bruggemann et al. 1998, 1999). The herb layer is of interest, because it supports to identify middle range transport phenomena. The total concentrations were measured in mg/kg dry mass. Here, in this section, eight regions along the river Rhine, from the German part of Basel, until Karlsruhe are used as example. Originally the interest in these regions was to clarify whether or not a trend can be observed, following downstreams (Bruggemann et al. 1997). The regions are labelled by {1,10,24,31,19,43,52,56}.

Here implications are investigated between two singletons, once again, only the standard norm and its residuum are selected.

```
standard-norm
premises and conclusions: only one indicator
Analysis
concerning the set of objects as follows
X= {1, 10, 24, 31, 19, 43, 52, 56},
S, implies Zn, with truth-value 0.0
S, implies Cd, with truth-value 0.0
S, implies Pb, with truth-value 0.0
Zn, implies S, with truth-value 0.0
Zn, implies Cd, with truth-value 0.091
Cd, implies Zn, with truth-value 0.476
```

As can be seen, the truth values of these implications are remarkably low, showing that the transformation into binary values must be considered with care. More details and the role of the selection of another t-norm can be inspected in Bruggemann and Kerber 2018. Furthermore it is of interest to state that based on binary transformed data Zn implies Cd, whereas here the implication Zn implies Cd (however with a small truth value) is found.

5 Discussion

The example, discussed in Sect. 4.2 shows remarkably that accepting a continuous range of data and implications based no more on a pure binary point of view will also lead in general to truth values less 1. The main problematic point is that in the binary case a "1" stands not for the maximum of a possible range [0, 1] but for the presence of an indicator, so to say, in a close interpretation of the has-relation. In the case of a continuous scale also the has-relation has truth values and the truth values in consequence can also take values between 0 and 1. At which truth value can we speak of a has-relation? Therefore, one of the main tasks in the future is, to provide tools, how a coarsening of truth values can be performed. Is for example 0.476 for the implication Cd $\Rightarrow$ Zn big enough to establish contextually that Cd implies Zn?

The very simple and still pretty arbitrary example for a correlation analysis shows that the generation of hypotheses can better be based on correlation coefficients than on the truth values of implications, if the nature of the data allows a correlation analysis (either Spearman or Pearson, just to denote two famous methods).

Nevertheless, the theoretical concept behind implications based on continuous data opens another tool, which is worth to be examined further. The tasks for the future are:

How can the theoretical framework, presented in BK and partially in this chapter be embedded into the general Formal Concept Analysis, where the concepts, i.e. pairs, for which (A')' = A is valid, play an important role in deriving implications. Here up to now, there was no mentioning of concepts, although the theoretical framework is general enough, to establish concepts even for data continuous in concept. Nevertheless, first approaches indicate that even with data continuous in concept, the requirement that the "second derivative" of A (=A" = (A')'), has to be equal to A itself works well as a method to find concepts. The fact that the second derivative is to be formed, makes a posteriori understandable, that in BK the formulation of A as a set $[0,1]^{|Q|}$ was selected: The first derivative may deliver a tuple of length |Q| whose components are indeed values taken from [0,1].

The number of implications in the binary case can be large. If -as shown in the former section- truth values are to be accepted which are not equal 1, then still the number of implications increases dramatically. Therefore the construction of a basis, the Duquenne, Guigues basis (Duquenne 1987), is urgently needed. So, even if actually the results are not that convincing, the theoretical framework needs still future work before a final conclusion about its usefulness in case of non-binary data can be done.

References

Bruggemann, R., & Kerber, A. (2018). Fuzzy logic and partial order; first attempts with the new PyHasse-Program L_eval. *Communications in Mathematical and in Computer Chemistry (MATCH), 80*, 745–768.

Bruggemann, R., Kaune, A., Komossa, D., Kreimes, K., Pudenz, S., & Voigt, K. (1997). Anwendungen der Theorie partiell geordneter Mengen in Bewertungsfragen. *DGM, 41*(5), 205–209.

Bruggemann, R., Voigt, K., Kaune, A., Pudenz, S., Komossa, D., & Friedrich, J. (1998). *Vergleichende ökologische Bewertung von Regionen in Baden- Württemberg GSF-Bericht 20/98*. Neuherberg: GSF.

Bruggemann, R., Pudenz, S., Voigt, K., Kaune, A., & Kreimes, K. (1999). An algebraic/graphical tool to compare ecosystems with respect to their pollution. IV: Comparative regional analysis by Boolean arithmetics. *Chemosphere, 38*, 2263–2279.

Bruggemann, R., Kerber, A., & Restrepo, G. (2011). Ranking objects using fuzzy orders, with an application to refrigerants. *MATCH Communications in Mathematical and in Computer Chemistry, 66*, 581–603.

Bruggemann, R., Carlsen, L., Voigt, K., & Wieland, R. (2014). PyHasse software for partial order analysis. In R. Bruggemann, L. Carlsen, & J. Wittmann (Eds.), *Multi-indicator systems and modelling in partial order* (pp. 389–423). New York: Springer.

Duquenne, V. (1987). Contextual implications between attributes and some properties of finite lattices. In: B. Ganter, R. Wille, & K. E. Wolff (Eds.), *Beiträge zur Begriffsanalyse* (pp. 213–239). BI Wissenschaftsverlag.

Ganter, B., & Wille, R. (1989). Conceptual scaling. In F. Roberts (Ed.), *Applications of combinatorics and graph theory to the biological and social sciences* (pp. 139–167). New York: Springer.

Ganter, B., & Wille, R. (1996). *Formale Begriffsanalyse Mathematische Grundlagen*. Berlin: Springer.

Restrepo, G., Weckert, M., Bruggemann, R., Gerstmann, S., & Frank, H. (2008a). Ranking of refrigerants. *Environmental Science and Technology, 42*, 2925–2930.

Restrepo, G., Bruggemann, R., Weckert, M., Gerstmann, S., & Frank, H. (2008b). Ranking patterns, an application to refrigerants. *MATCH Communications in Mathematical and in Computer Chemistry, 59*, 555–584.

Part II
Indicators for Special Purposes

Indicators for Sustainability Assessment in the Procurement of Civil Engineering Services

Nora Pankow, Rainer Bruggemann, Jan Waschnewski, Regina Gnirss, and Robert Ackermann

1 Introduction

In order to face global challenges, the UN passed a joint solution also known as Agenda 2030. UN member states declare themselves ready to implement the Agenda 2030 and its set of 17 sustainable development goals (SDG) on a national level (UN 2015). Germany applied the ideas of the Agenda 2030 in the *German Sustainability Strategy* and adjusted the SDGs from the global perspective to a German perspective with specific indicators for the different goals (Bundesregierung 2016). The strategy also states that for an inclusion of a sustainable development a further implementation into economy, academics and civil society is necessary (Bundesregierung 2016).

The rising awareness for sustainability, increases the demand for companies to include and monitor different objectives not just from governmental but also from customer side (Jasch 2009; Koplin 2006). Consequently, mandatory indicators must be expanded. The Berliner Wasserbetriebe (BWB) provide Berlin and parts of the state Brandenburg with drinking water and are responsible for the wastewater

N. Pankow (✉) · R. Ackermann
Institute of Environmental Science & Technology, Technische Universität Berlin, Berlin, Germany
e-mail: robert.ackermann@tu-berlin.de

R. Bruggemann
Leibniz-Institute of Freshwater Ecology and Inland Fisheries, Berlin, Germany

J. Waschnewski · R. Gnirss
Berliner Wasserbetriebe, Berlin, Germany
e-mail: jan.waschnewski@bwb.de; regina.gnirss@bwb.de

© The Author(s), under exclusive license to Springer Nature Switzerland AG 2021
R. Bruggemann et al. (eds.), *Measuring and Understanding Complex Phenomena*,
https://doi.org/10.1007/978-3-030-59683-5_8

treatment (BWB 2016). The main goal is to ensure the water quality for present and the future generation (BWB 2018). This involves the restoration, renovating, reconstructing and enhancing of plants. The engineering and purchasing of different engineering projects already require various legal, technical and company internal guidelines (Senatsverwaltung für Stadtentwicklung und Umwelt Berlin 9/5/2016; BWB 2014). The implementation of sustainability indicators needs to be fitted for this complex process. For an enterprise the implementation of indicators given in the Agenda 2030 can therefore be problematic.

This paper exhibits an option to incorporate sustainability indicators into the procurement process to support decision-makers. The inclusion of the engineering phase in the procurement process is seen as essential to secure a more sustainable company. Thus a sustainable procurement can only be realised when the engineering phase and purchasing phase also consider sustainability indicators (Koplin 2006; Wutke 2016).

For this, the top-down approach of the Agenda 2030 was linked to the problem-oriented bottom-up approach (Coen 2000). Using the bottom-up approach the specific demands in the procurement of the urban water management can be considered.

2 Method

The procurement process at the Berliner Wasserbetriebe (BWB) consists of several phases involving different departments. The main departments are planning and construction (PB) as well as purchasing (EK) (Pankow 2018). The PB is the biggest division and oversees systems planning which includes the environmental laws, technical standards and economic indicators. The decision making for different variants during the engineering process is usually based on economic indicators. In consultation with the PB the EK oversees the tendering process. As a state authority the BWB is bound to the procurement law of the EU and Germany and the environmental procurement directive of the Senate Department for the Environment, Transport and Climate Protection of Berlin (Europäische Union 2014; Senat Berlin; VgV 2016; Senatsverwaltung für Stadtentwicklung und Umwelt Berlin 9/5/2016).

Indicators that have already been included in procurement process of the BWB are defined as "status quo indicators". The status quo indicators should be complemented with sustainability indicators, those indicators are called "target indicators". However, the indicators must be assessed for their applicability and their relevance in the specific field for procurement of civil engineering. In a technical workshop with employees of relevant departments, indicators were selected. Afterwards the selected indicators had to be tested on their causal relationships in the procurement process.

2.1 Development of Indicators

The goals defined in the German Sustainability Strategy address different problem areas in Germany and were not developed for a specific sector or a company. For example: The German SDG 6 (ensure availability and water sustainable management of water and sanitation for) contains three indicators which are focusing on water quality (phosphor and nitrate) and on access to drinking water. But for the procurement of civil engineering projects in the public sector other aspects need to be considered (e.g. cost of the plant, safety of worker). Thus, indicators of the German Sustainability Strategy need a specification before they can be applied in the assessment of the procurement process Instead of relying solely on the Sustainable Development (SDG)-indicators this approach will, as a first step, take other indicators into consideration. Those indicators should add technical, water- and BWB-related aspects to ensure the sectoral focus. By adding other indicators, a base-set of indicators is created (see Table 1). This base-set with over 150 indicators contains various kinds of aspects, which are of different relevance in the civil engineering service in the BWB.

For a practical application of the Multi-indicator system (MIS) the number of *target indicators* needs to be decreased. Consequently, the next step should be to narrow down the base set to indicators with a specific relevance in the civil engineering service.

To identify the indicators with the highest relevance a workshop with the main contributing departments in the BWB was held. This collective selection round together with *PB*, *EK* and the department of strategy could highlight the most relevant topics. Each participant was given the same number of votes. With this workshop a new set of indicators was created, which are here defined as "theoretical target indicators".

Since the indicators are from different sources, most of them have different degree of detail and may contextually overlap. Every indicator from the set of the *theoretical target indicators* was therefore reassessed. To ensure the practicability every indicator should have a direct causal relation or a short chain of causation (Pankow 2018; Mischke 2017). The theoretical target indicators could be reduced to

Table 1 overview of frameworks for a base-set of indicators

National framework	- Sustainable Development Goals of Germany
Technical framework	- VDI-Indicators of the technical assessment guideline - NaCoSi-Indicators for the sustainability controlling of residential water management systems
Regional framework	- Administrative provision procurement and environment (VwVBU) - Order and procurement regulations of Berlin
Legal framework	- Procurement regulations - Environmental regulations
Company intern framework	- Sustainability indicators of the company

Table 2 Practical target
indicators

I1	Tariff stability
I2	Air pollutants (health)
I3	Worker safety
I4	Competence management
I6	Free space loss
I8	Acceptance
I9	Affected sources of water
I10	Energy intensity
I11	Power consumption
I13	Greenhouse Gas emissions
I14	Other significant air emissions
I15	Biodiversity of water
I16	Waste
I17	Secondary raw materials
I18.1	Investment costs
I18.2	Operating cost
I19	Spending research and development
I20	Creativity
I21	Environmental management external
I22	Innovation and adaptability
I23	Robustness

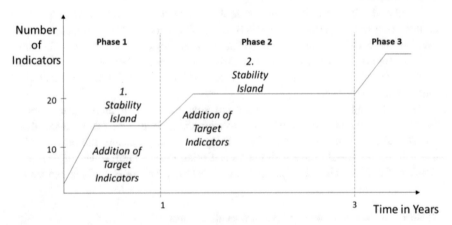

Fig. 1 Implementation of Indicators in an iterative process

"practical target indicators". The *practical target indicators* are a set of 21 indicators
in total (see Table 2). For the assessment of the case examples the indicators were
divided into the three sustainability dimensions.

Since the concept of sustainability is in transition over influenced by culture and
knowledge a regular reassessment is vital. That is why an incremental implementation of indicators is proposed (see Fig. 1). By adding indicators over time, the MIS
can be readjusted, and new indicators can be added if necessary.

2.2   Application to Case Examples

By applying the MIS to case examples, the applicability and practicability was reviewed. For the utilization in the procurement process the MIS must be used by *PB* and *EK*. As stated by both departments the time exposure and data availability are defining limitations for the practicability in the procurement process (Pankow 2018).

The selected indicators were tested on four case examples at the BWB. The goal was to identify the most sustainable variant of three given options. For further discussion the data was evaluated with a multi-criteria decision analysis using three different tools: Dashboard of Sustainability, Value Analysis and Partial Order.

As previously explained the MIS requires an incremental implementation and must be reassessed over time. In this work the partial order was used for an analysis of the MIS.

2.2.1   Multi-criteria Decision Analysis

There is a broad variety of multi-criteria decision analysis (MCDA). MCDA aims to systematically analyse a complex problem to support the decision-making process (Wilkens 2012).

The most prominently used method is a value benefit analysis. The indicators form the assessment system to which each variant is allocated a *partial use value* which can be summed to a *total value* for each variant (Zangemeister 1971). The use value allocates a certain weight for each indicator. The weight can be derived by different methods (Sartorius et al. 2017; Müller-Herbers 2007). In this work the weights were assigned by the direct ranking method. The direct ranking method gives each indicator a certain ranking between the indicators. Depending on the rank a weight is appointed (Žižović et al. 2017).

The second MCDA used is the Dashboard of Sustainability (DS). With a DS the disaggregated results of a sustainability assessment should be easier to communicate with non-expert users in decision-making (Traverso et al. 2012). The Dashboard of Sustainability is a free software, developed by the UN for the assessment of the Millennium Development Goals of cities and countries (Sachs 2012; Saltelli et al. 2005). The Excel-tool, provided by the UN, can be readjusted with individual indicators and presents the results in a coloured dashboard (O'Conner 2003, 2012). The indicators are normalized, and the variants are compared directly to each other (Scipioni et al. 2009). The tool provides the option to create an own dashboard and edit parameters (Seidel-Schulze and Grabow 2007). The variants are conjugated to a colour range from red (critical) over yellow (medium) to green (very good) (O'Conner 2003).

The third method is partial order. Partial order is a non-parametric method where objects are related by the $\leq$-relation. In addition to this relation, a partially ordered set is subject to the three axioms of reflexivity, antisymmetry and transitivity

(Bruggemann and Patil 2011). With the partial order, objects are characterized by attributes (or in this case indicators). The sorting of variants or objects using the $\leq$-relation results in a rating network with characteristic structures, a so-called Hasse Diagram (HD) (Steinberg 2002). A HD displays the objects and their $\leq$-relation by a directed graph (chain). In a poset not all objects are comparable, so they are not connected by a graph (antichain). The program PyHasse was used to create those HDs (Bruggemann 2018).

During the course of this work PyHasse showed a low validity for the assessment of the variants because of the high number of indicators and low number of objects (Pankow 2018; Bruggemann et al. 2014). The focus of this work was to derive a MIS to assess variants in decision-making. Therefore, the indicators should also be assessed in this process. For the assessment of the indicators, the indicators were treated as objects and the variants as attributes. When the indicators are treated as objects, the variants can be used to characterize the indicators. For the assessment of the indicators the case examples were analysed in two different approaches. In the first approach each case examples forms a matrix and is analysed individually. Therefore, each case example was considered individually. The other approach is to analyse all case example in its entirety in one matrix.

The examination tools for both approaches were identical. The indicators must be normalized to allow forthcoming analysis. The goal of the partial order is the mapping of indicators on a metric scale. Indicators should therefore not dominate other indicators due to a large measure-unit. To avoid this kind of dominance a normalization can be appropriate. After the normalization the HDs for all case examples and the entirety of all case examples were created.

The characteristic structures formed in the HDs, especially the subsets in the HD were of interest. With the PyHasse module "sepanal15_4" the separability of subsets from the HD can be investigated (Restrepo and Bruggemann 2008). The module provides a degree of separability and the attributes that cause the separability between two subsets. The result can be displayed graphically in a so-called tripartite graph.

All indicators were associated with a sustainability dimension (social, ecological or economic). If one dimension dominates the other dimensions, it would point to an unbalanced set of indicators. The module "dds_12" in PyHasse analyses the dominance of an assigned group of attributes. With *dds_12* a dominance histogram is generated. The histogram shows the distribution of the normalized dominance of objects. In addition to the histogram, the module calculates the dominance matrix, the separability matrix and the degree of separability.

2.2.2 Case Example

The case examples were previous projects at the BWB. Three of those examples were projects in different wastewater treatment plants.

The first case example is the exhaust air treatment in the wastewater treatment plant (WWTP) 1. In the inlet area of the mechanical wastewater treatment at the

WWTP 1, the maximum workplace concentration and the applicable occupational exposure limits of hydrogen sulphide concentration have been exceeded. To stay within the threshold value, the sand trap should be covered, and the exhaust air sucked off. For the treatment of the exhaust air, there are three different variants: co-treatment in the aeration, fume scrubber and UV treatment.

The second example is the renewal of the digester chambers for the WWTP 2. The WWTP is operated with a mechanical and biological stage. The sludge from the wastewater treatment are fed to sludge digestion. The variants for this case example were the renovation of the existing septic tanks, an addition of mixed sludge dewatering and the construction of the septic tanks.

The third example is the treatment of process water in WWTP 3. During the dewatering of the sludge, the process water with high ammonium content is produced. This process water is returned to the main sewage stream. Due to lower effluent criteria of ammonium in the WWTP the process water must be treated separately. Consequently, three variants were assessed: Stripping and acid wash, activated sludge process or deammonification in a sequencing batch reactor.

The fourth example is taken from the sewer system rehabilitation. The so-called Cured-in-place pipe (CIPP) is one of several trenchless rehabilitation methods where a resin-saturated felt tube is inserted into a damaged pipe. The liner is inserted using water or air pressure and dries in place. In this case study a product with different properties was examined.

3   Results

3.1   Variant Analysis

With the value benefit analysis all case studies could be evaluated. The specific value for each indicator was determined by the direct ranking method, which allowed a flexible assessment process. Due to the high number of indicators a consideration of the three sustainability dimensions simplifies the results. The dimension view shows the high and low rated dimensions of the variants as it can be seen in Fig. 2 for the case example exhaust air treatment in the WWTP1. A high rating indicates a positive assessment. In this case example each variant has a high value in one dimension. Variant 3 has the highest total value but is just in one dimension the highest rated variant. Variant 2 has the highest rating in the ecological dimension. In the economic and social dimension, the value is the lowest. Variant 1 is like variant 2, in the social dimension the value is high but the other two dimensions are lower than the other variants.

In Fig. 3 the DS for the same case example can be seen. The created graphic shows the three sustainability dimensions. The colours mark the different variants and therefore the "good" and "bad" aspects of each variant are easy to distinguish. The same case example shows slightly different results as in the value benefit

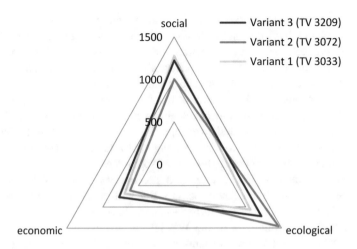

Fig. 2 Net diagram of the variant analysis of WWTP 1

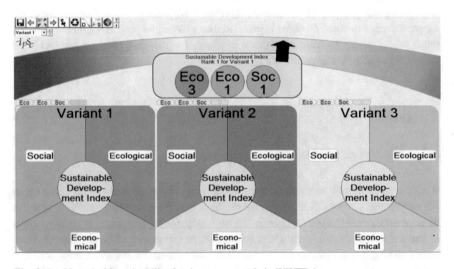

Fig. 3 Dashboard of Sustainability for the case example in WWTP 1

analysis. Variant 1 displays better results in the social and economic dimension. The variant 2 has better result in the ecological dimension. The variant 3 shows a light green colour in the ecological dimension, but medium to critical rankings in the social and economic dimension.

The created HD showed complete antichains for all three variants in all case examples (see Fig. 4), which implies that each variant has specific advantages and disadvantages (which could be expected, after inspecting Fig. 3).

Fig. 4 HD for the case
example WWTP 1

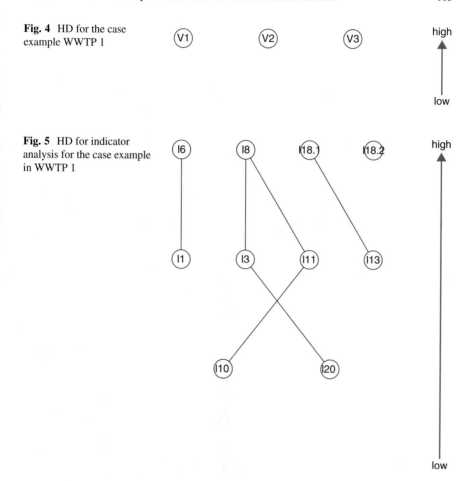

Fig. 5 HD for indicator
analysis for the case example
in WWTP 1

3.2 Indicator Analysis

First, HDs for all case examples and the entirety of all case examples were
developed, whereby now the indicators are considered as objects (vertices in the
HD) evaluated by the variants. Each of the HD formed certain subsets, for example
as shown below for the exhaust air treatment in WWTP1 (see Fig. 5). The chains
I6 > I1 und *I8 > I3 > I10, I8 > I3 > I20, I8>I11>I10, I8>I11>I20* and *I18.1
> I13* were identified. The indicators I16, I22 and I23 are equivalent to I8 and
are therefore not visible on the HD. The object I18.2 is an isolated object. The
degree of separation among those subsets was analysed with the PyHasse module
"sepanal15_4". The comparison of the subsets C1 (I6, I1) with C2 (I8,I3,I10), C3
(I8,I3,I20), C4 (I8,I11,I10), C5 (I8,I11,I20) and the component C6 (I18.1,I13) with
C4 shows a high degree of separation in the variants 2 and 3. The indicator group can
be separated by these two variants in a present separation . The scatter plot of Fig. 6

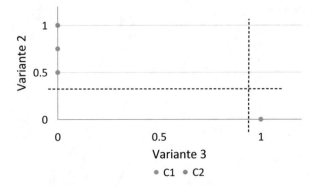

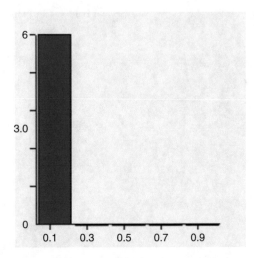

Fig. 6 Scatter plot of two components for the case example in WWTP 1

Fig. 7 Dominance histogram
of the sustainable dimensions

illustrates this separation by means of the separation of components C1 and C2 by variant 3 and variant 2. All components were showing a high degree of separation (Fig. 6).

The dominance was analysed with the PyHasse modul "dds12". In each HD the indicators were assigned to one sustainability dimension (see Fig. 8). The indicators of the same dimension are coloured in orange, blue and green. Indicators from the social dimension are marked orange, the ecological dimension green, and the economic dimension blue. Based on this allocation, the investigation of the dominance between the three dimensions with the module *dds12* was carried out. The dominance histogram for the case example WWTP 1 shows a low degree of dominance (see Fig. 7). The dominance analysis for the other case examples and the entirety of all case examples showed similar results.

Fig. 8 HD with the assigned dimensions (orange: social; green: ecological; blue: economical)

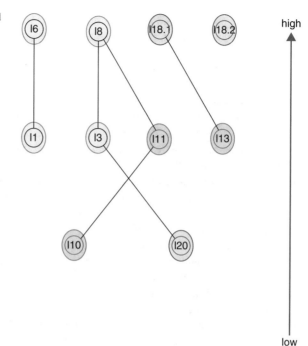

4   Conclusion and Outlook

To secure water quality and access to water for next generations sustainable development is a vital component of the BWB company strategy. Therefore, a stronger focus on implementation of sustainability in the various company departments and especially their practical implementation is of interest. Due to the different protection goals, it is necessary to adapt sustainability to a problem area. This work dealt with the development of sustainability indicators in procurement of engineering services, while considering the specific requirements in urban water management.

The German Sustainability Strategy has not been able to fully adhere to the goals of the Agenda 2030 but has partially incorporated them into the sustainability strategy. Consequently, these objectives cannot be transferred to the water sector. Beyond the German Sustainability Strategy, additional framework conditions were included for the recording of the indicators and the development of a multi-indicator system. On the one hand, this has allowed for more specific indicators for the water sector. On the other hand, this also led to a high number of indicators with different degrees of detail. The support provided by the employees was an important basis in the determination of the requirements and the relevance of the indicators. The participation of employees significantly reduced number of indicators from over 150 to 21.

It has also proved useful to review the causal relationship. In the next stages of the indicator development, this step should be included to ensure the relevance in the specific department.

The trial based on case examples has shown a solid data availability in the sewage treatment sector. The variants evaluated with the utility analysis and the dashboard of sustainability were practical in their application and displayed easy to interpret graphics. Due to the direct ranking method the value benefit analysis has a subjective value attitude. A stronger focus on the weighting method as suggested by Sartorius et al. (2017) with the involvement of experts and decision-makers is recommended. The Dashboard of Sustainability simplified the results with its colour representation even more than the value benefit analysis. The colour assignment made the identification of "good" and "bad" rated variants easy. The high degree of simplification is accompanied by a loss of transparency. The direct comparison of variants could lead to wrong conclusion especially for non-expert users.

The partial order and the program PyHasse could not be used for the variant evaluation. The small number of variants and the high number of indicators lead to many incompatibilities. Examples with more than 3 variants could show comparability and hence better interpretable results.

The program PyHasse, however, showed usability for the assessment of the MIS. The study of separability points to differing indicators. The dominance analysis has shown that no sustainability dimension dominates. This suggests a diverse set of indicators that covers different aspects. However, the indicator rating used with PyHasse could be further developed. Since the number of variants was too small to statistically secure the results.

As clarified at the beginning of this work, the MIS can be developed in an iterative process. As a result, the changes of social, legal, technical or scientific nature occurring over time can be included. Furthermore, this procedure offers the possibility to test the set of indicators with current obtained data. Once a stable number of indicators has been reached, the analyses already performed by PyHasse can be reapplied and expanded.

During the work, further questions related to the indicators and the evaluation methods have emerged: The evaluation of qualitative indicators has so far been subjective. For further application it is appropriate to develop a rating scheme to facilitate the evaluation of these indicators.

The causal relationship and the quality of the indicators were only briefly addressed in this paper. Further work requires a closer examination of the indicators in terms of their causal relationship to a specific field in civic engineering and their assessability. As a result, the evaluation of sustainability by indicators and thus the decision could be hedged better.

In this work three multicriteria evaluation methods were considered. In the next phases, further multi-criteria evaluation methods could be tested by case studies within the implementation phases. The testing of new methods could also lead to the creation of a proprietary algorithm that can be used for planning and purchasing.

In the future, this set of indicators for the sewer network and the sewage treatment sector can be applied and expanded on other departments as well, so that the company can apply the same pool of indicators.

References

Bruggemann, R. (2018). *PyHasse*. Available online at https://www.pyhasse.org/start, updated on 2/14/2018.

Bruggemann, R., & Patil, G. P. (2011). *Ranking and prioritization for multi-indicator systems. introduction to partial order applications* (Environmental and Ecological Statistics, 5). New York: Springer Science+Business Media, LLC.

Bruggemann, R., Carlsen, L., & Wittmann, J. (Eds.) (2014). *Multi-indicator systems and modelling in partial order*. New York, s.l.: Springer New York. Available online at https://doi.org/10.1007/978-1-4614-8223-9

Bundesregierung. (2016). Deutsche Nachhaltigkeitsstrategie. Neuauflage 2016. Edited by Die Bundesregierung. Available online at https://www.bundesregierung.de/Content/DE/_Anlagen/2017/01/2017-01-11-nachhaltigkeitsstrategie.pdf;jsessionid=AC20D6F985F308DE 387921C12C264936.s1t1?__ blob=publicationFile&v=5, checked on 1/20/2017.

BWB. (2014). *Einkaufs- und Vergabehandbuch*. Berlin: Berliner Wasserbetriebe (BWB).

BWB. (2016). *Nachhaltigkeitsbericht 2016*. Berlin: Berliner Wasserbetriebe (BWB).

BWB. (2018). *Verhaltenskodex. Grundsätze für ein verantwortungsvolles und rechtmäßiges Handeln*. Berlin: Berliner Wasserbetriebe (BWB). Available online at http://www.bwb.de/content/language1/downloads/Verhaltenskodex_BWB_web.pdf, checked on 2/28/2018.

Coen, R. (2000). Konzeptionelle Aspekte von Nachhaltigkeitsindikatorensystemen. In *TA-Datenbank-Nachrichten 9*, pp. 47–53. Available online at https://www.tatup-journal.de/downloads/2000/tadn002_coen00a.pdf, checked on 2/26/2018.

Europäische Union. (2014). RICHTLINIE 2014/24/EU DES EUROPÄISCHEN PARLAMENTS UND DES RATES vom 26. Februar 2014 über die öffentliche Auftragsvergabe und zur Aufhebung der Richtlinie 2004/18/EG. 2014/24/EU. Source: das Amtsblatt der Europäischen Union (ABl.). Available online at http://eur-lex.europa.eu/legal-content/EN/ALL/?uri=CELEX:32014L0024

Jasch, C. (2009). EVANAB – systematische Evaluierung von Nachhaltigkeitsberichten mit dem EVANAB Tool und Strategieanalyse österreichischer Nachhaltigkeitsberichte : Erstellung eines Tools für die Evaluierung von Nachhaltigkeitsberichten und für die Begleitung des Erstellungsprozesses. hrsg. im Rahmen der Programmlinie Fabrik der Zukunft, Impulsprogramm Nachhaltig Wirtschaften. With assistance of Richard Tuschl, Andreas Windsberger. Edited by Wien : Bundesministerium für Verkehr, Innovation und Technologie. Wien (Berichte aus Energie- und Umweltforschung). Available online at www.nachhaltigwirtschaften.at

Koplin, J. (2006). *Nachhaltigkeit im Beschaffungsmanagement. Ein Konzept zur Integration von Umwelt- und Sozialstandards*. Dissertation. Universität Oldenburg, Oldenburg. Wirtschaftswissenschaften.

Mischke, M. (2017). *Bewertung der Nachhaltigkeit chemischer Substanzen. Die Methode' SusDec' als schutzgutbezogenes Nachhaltigkeitsindikatorensystem*. Wiesbaden: Springer Fachmedien Wiesbaden. Available online at https://ebookcentral.proquest.com/lib/gbv/detail.action?docID=4789926

Müller-Herbers, S. (2007). *Methoden zur Beurteilung von Varianten*. Arbeitspapier. Stuttgart: Universität Stuttgart, Institut für Grundlagen der Planung.

O'Conner, J. (2003). *The dashboard manuel*. United Nations. Available online at www.webalice.it/jj2006/MdgDashboard.htm, checked on 2/14/2018.

O'Conner, J. (2012). *Dashboard of sustainability*. Version V50.4: CGSDI. Available online at https://web.archive.org/web/20150508004640/http://esl.jrc.ec.europa.eu/envind/dashbrds.htm, checked on 10/17/2017.

Pankow, N. (2018). *Nachhaltige Indikatoren für die Planung nd Vergabe von Ingenieurbauleistungen*. Masterarbeit. Berlin: Technische Universität Berlin. Sustainable Engineering.

Restrepo, G., & Bruggemann, R. (2008). Dominance and separability in posets, their application to isoelectronic species with equal total nuclear charge. *J Math Chem, 44*(2), 577–602. https://doi.org/10.1007/s10910-007-9331-x.

Sachs, J. D. (2012). From millennium development goals to sustainable development goals. *The Lancet, 379*(9832), 2206–2211. https://doi.org/10.1016/S0140-6736(12)60685-0.

Saltelli, A., Giovannini, E., Tarantola, S., Saisana, M., Hoffman, A., & Nardo, M. (2005). OECD Statistics Working Papers (2005/03).

Sartorius, C., Lévai, P., Nyga, I., Sorge, C., Menger-Krug, E., Niederste-Hollenberg, J., & Hillebrand, T. (2017). Multikriterielle Bewertung von Wasserinfrastruktursystemen am Beispiel des TWIST-Modellgebietes Lünen. *DWA KA Korrespondenz Abwasser, 64*(11), 999–1007.

Scipioni, A., Mazzi, A., Mason, M., & Manzardo, A. (2009). The dashboard of Sustainability to measure the local urban sustainable development. The case study of Padua Municipality. *Ecological Indicators, 9*(2), 364–380. https://doi.org/10.1016/j.ecolind.2008.05.002.

Seidel-Schulze, A., & Grabow, B. (2007). *Nutzung von Urban Audit-Daten – eine Arbeitshilfe des KOSIS-Verbundes Urban Audit*. Berlin: Deutsches Institut für Urbanistik. Available online at http://edoc.difu.de/edoc.php?id=BIRJ3S6N, checked on 12/8/2017.

Senat Berlin: Berliner Ausschreibungs-und Vergabegestz. BerkAVG, revised 7/8/2010. Source: Gesetz-und Verordnungsblatt für Berlin *66*(17), 399–401. Available online at http://www.hfk.de/w/files/fachbeitraege/berlinerausschreibungsundvergabegesetz.pdf

Senatsverwaltung für Stadtentwicklung und Umwelt Berlin (9/5/2016). *Verwaltungsvorschrift Beschaffung und Umwelt*. VwVBU, revised 2/23/2016. Source: KNBBE. Available online at http://www.nachhaltige-beschaffung.info/DE/DokumentAnzeigen/dokumentanzeigen.html?idDocument=1231&view=knbdownload, checked on 2/1/2017.

Steinberg, C. (2002). *Nachhaltige Wasserwirtschaft. Entwicklung eines Bewertungs- und Prüfsystems*. Berlin: Erich Schmidt (Initiativen zum Umweltschutz, 36).

Traverso, M., Finkbeiner, M., Jørgensen, A., & Schneider, L. (2012). Life cycle sustainability dashboard. *Journal of Industrial Ecology, 16*(5), 680–688. https://doi.org/10.1111/j.1530-9290.2012.00497.x.

UN. (2015). *Die Transformation unserer Welt: die Agenda 2030 für nachhaltige Entwicklung*. New York: United Nations.

VgV. (2016). *Der Deutsche Bundestag zur Verordnung über die Vergabe öffentlicher Aufträge* (Vergabeverordnung). VgV, revised 4/12/2016. Source: 703-5-5. Available online at https://www.gesetze-im-internet.de/bundesrecht/vgv_2016/gesamt.pdf

Wilkens, I. (2012). *Multikriterielle Analyse zur Nachhaltigkeitsbewertung von Energiesystemen. Von der Theorie zur Praktischen Anwendung*. Dissertation. Technische Universität Berlin, Berlin. Fakultät III- Prozesswissenschaften, checked on 2/14/2017.

Wutke, S. (2016). *Entwicklung eines Gestaltungsmodells zur Berücksichtigung von Nachhaltigkeit bei der Ausschreibung und Vergabe logistischer Leistungen im Straßengüter-verkehr*. Berlin: Universitätsverlag der TU Berlin (Schriftenreihe Logistik der Technischen Universität Berlin, 36).

Zangemeister, C. (1971). *Nutzwertanalyse in der Systemtechnik. Eine Methodik zur multidimensionalen Bewertung und Auswahl von Projektalternativen*. Teilw. zugl.: Berlin, Univ., Diss., 1970. 2. Aufl. München: Wittemann.

Žižović, M., Damljanović, N., & Žižović, M. R. (2017). Multi-criteria decision making method for models with the dominant criterion. *Filomat, 31*(10), 2981–2989. Available online at http://www.jstor.org/stable/26195029

Dependent Indicators for Environmental Evaluations of Desalination Plants

Ghanima Al-Sharrah and Haitham M. S. Lababidi

1 Introduction

Availability of data is essential for effective environmental assessment study. In desalination studies, environmental data are not commonly available (Roberts et al. 2010). Most of reported data in environmental assessment of desalination operations are qualitative (low, moderate, above limits, etc.), incomplete and in most cases inconsistent. For instance, salinity and ion concentrations of the discharged brine are not enough to assess the environmental impact of the desalination facilities. Temperature of the brine discharge, for example, is an important variable that has a direct effect on other variables that affect the marine life, such as the amount of dissolved oxygen. Another example is salinity, which is frequently reported by environmental engineers in different methods that are not directly comparable. It may be reported as mass fraction of dissolved salt (in ppm or g/kg), conductivity (in Siemens per meter, S/m), or as TDS, which is expressed as total dissolved solids or total dissolve salts (Boerlage 2011).

The selection of appropriate measures of environmental performance for desalination depends on the nature of the environmental concerns, the type and quantity of available information, and the degree of accuracy required in the representation. Different environmental indicators are suitable for different stages of process development, design or operation. Some indicators are general and can be applied to a wide range of processes and industries, while others are more specific to the unit under consideration. Selection of environmental indicators is not an easy task. Prior to environmental assessment, indicators should be screened to eliminate irrelevant

G. Al-Sharrah (✉) · H. M. S. Lababidi
Chemical Engineering Department, College of Engineering & Petroleum, Kuwait University, Safat, Kuwait
e-mail: g.sharrah@ku.edu.kw; haitham.lababidi@ku.edu.kw

© The Author(s), under exclusive license to Springer Nature Switzerland AG 2021
R. Bruggemann et al. (eds.), *Measuring and Understanding Complex Phenomena*,
https://doi.org/10.1007/978-3-030-59683-5_9

and redundant ones. For this reason, it is important to study the relationships between indicators and their expected impacts on the final decisions.

In general, most of the indicators are in some way or another related to other properties. Values of indicators are normally measured using online sensors or determined by chemical Lab analysis of collected samples. In many occasions, Lab analyses are also necessary to calibrate the sensors. Moreover, most of the indicators are related to one or more physical or chemical properties. The relationship between the indicator and its related properties may be expressed in the form of simple equation derived from basic principles. In many cases, this relationship is correlated using experimental data. For both cases, the relationship between the indicator and other variables and parameters is referred to as a "model". The output of the model is basically the value of the indicator, while the inputs to the model are the dependent variables, which are usually known as "primary variables".

A "dependent indicator" is an indicator that is correlated to other indicators or to primary variables. The correlations are expressed as models, which are mathematical formulations that describe how the value of the dependent indicators change with changes in their corresponding variables. Methods for developing the correlations (models) are generally classified as statistical and conceptual. Statistical methods are basically applied to numerical data and result in correlations described as simple to use formulations (Imam et al. 1993). Conceptual methods are oriented towards qualitative data and rules. Environmental assessments can benefit from the correlations that may exist among indicators (Sutherland et al. 2016). These correlations are also known as dependencies.

Examples of reported dependencies between environmental indicators include the correlation between dissolved oxygen and pH (Makkaveev 2009), CO_2 emission and energy consumption (Omri 2013) and the watershed indicator that relates percentages of old forest to interior forest (Sutherland et al. 2016). Dependent environmental indicators can be modeled as linear, multiple-linear or simple non-linear models (Piegorsch and Bailer 2005). In an attempt to rank chemicals according to their hazardous effects (such as threshold limit value or lethal dose), Al-Sharrah (2011) defined the dependency between the indicator Y_1 and another two indicators, X_1 and X_2, using the multiple-linear model represented by Eq. (1).

$$Y_1 = \rho X_1 + \sqrt{\left(1 - \rho^2\right)} X_2 \qquad (1)$$

Where ρ is a correlation parameter that can be determined using multiple-linear least-squares regression. In Eq. (1), the indicator Y_1 is proportionally correlated to the indicators X_1 and X_2 with corresponding coefficients ρ and $\sqrt{\left(1 - \rho^2\right)}$, respectively. The model suggests that Y_1 is equally correlated to X_1 and X_2 for $\rho = 0.5$, correlated to X_2 more than X_1 for $0 \leq \rho < 0.5$, and to X_1 more than X_2 otherwise.

2   Dependent Indicators for Seawater and Brine

Correlation between indicators can be simply identified through expert knowledge or mathematically using statistical analysis of collected data. Dependent indicators are not necessarily redundant, and dependencies through correlations do not necessarily mean causations (Guttman 1977), but may represent different concepts. Hence, exclusion of dependent indicators may affect the overall decision, and it is possible only when set against the analysis of influence on decision.

In the desalination field, environmental indicators are defined to characterize water or air releases. For airborne releases, general dependencies include emissions of greenhouse gases resulting from the required power generation (Younos 2005). Indicators that are usually considered for assessing the impacts to air pollution are basically related to the amount and type of the burned fuel. For the waterside, the indicators are defined to describe the effect of the rejected brine on the marine environment. Table 1 lists a number of examples on the main dependencies between environmental indicators. It provides also examples of primary variables that may be used in correlating the values of dependent indicators.

3   Ranking with Correlated Data

As mentioned earlier, models that are used in correlating environmental data may be linear, multiple-linear, non-linear, or other more complex forms (Piegorsch and Bailer 2005). The proposed approach is to study the dependencies in environmental indicators and utilize these dependencies in refining the available data. The aim is to improve the environmental decision making when ranking is applied. The dependencies can be effectively used in completing missing data as well as excluding redundant indicators. The key challenge is identifying the dependencies and determining the correlation models.

In this work, the Copeland score (Copeland 1951) ranking method is used. It is more than a half-century-old voting procedure, which is simply based on pairwise comparisons of candidates. It is one of many vote-aggregation systems that social-choice theorists have invented in their attempts to determine the most appropriate systems for a variety of voting situations. The Copeland rule selects the object with the largest Copeland score, which is the number of times an object beats other objects minus the number of times that object loses to other alternatives when the objects are considered in pairwise comparisons. Using the concept of partially ordered sets and social choice theory, the Copeland score ranking methodology was applied outside of its usual political environment (voting) by Al-Sharrah (2011) to rank objects in science and engineering applications. This method assumes neither linearity nor any mathematical relationship among indicators and is therefore defined as a non-parametric method. The method is presented next.

Table 1 Examples of dependent indicators and primary variables used in characterizing brine discharge of desalination plants

No	Dependent indicator	Primary variables	Model	References
1	Electric conductivity	Chlorine (Cl^-) and sodium (Na^+) ions	Linear	Fondriest (2015) and Sharp and Culberson (1982)
2	Total hardness, expressed as equivalent $CaCO_3$	Ca^{2+} and Mg^{2+} ions	$[CaCO_3] = 2.5[Ca^{2+}] + 4.1[Mg^{2+}]$	Venkateswarlu (1996) and LENNTECH (2016)
3	Langelier Saturation Index (LSI)	Total dissolved solids (TDS), concentrations of calcium (Ca^{2+}) and bicarbonates (HCO_3), and water temperature	The LSI is expressed as the difference between the actual system pH and the saturation pH_s ($LSI = pH - pH_s$). The saturation pH_s a log function of the primary variables	Alvarez-Bastida et al. (2013)
4	Density, viscosity	Temperature Salinity	Empirical model valid for salinities between 0 and 160 ppt and temperature between 10 and 180 °C at a pressure of 1 atm	El-Dessouky and Ettouny (2002)
5	Dissolved oxygen	pH	Non-linear	Makkaveev (2009)
6	Dissolved oxygen	Temperature Salinity	Non-linear	Lewis (2013)
7	Total alkalinity (At)	Total amount of calcium carbonate	$A_T = [HCO_3^-] + 2[CO_3^{-2}]$ (mmol/l)	Danoun (2007)
8	Carbonate (CO_3^{-2} and HCO_3^-),	Total alkalinity	$[CO_3^{-2}] = 0.6$ At (mg/l) $[HCO_3^-] = 1.22$ At(mg/l)	California Environmental Protection Agency (2016)

A data matrix containing a set of objects (e.g., chemicals, projects, universities, etc.) and their corresponding indicators (e.g., lethal dose, profit, number of courses, etc.), is used to rank the objects according to a desired aim (e.g., most hazardous, more sustainable, best performance, etc.). The Copeland method is applied by comparing one indicator at a time for each pair of objects. This is followed by counting the number of "greater than" results between indicators (add +1) and the number of "less than" results (add -1). The total sum of comparisons is set as an element in a *comparison matrix*. Detailed description of this method is given by Al-Sharrah (2011). A simple example is illustrated in Table 2, where the Copeland rank is applied for four objects ($O_1 \ldots O_4$) using three indicators ($I_1 \ldots I_3$) for the

Table 2 Copeland rank using (a) three indicators and (b) two indicators

		Indicator			(a)	Indicator		(b)
		I_1	I_2	I_3	Copeland rank	I_1	I_2	Copeland rank
Objects	O_1	1	3	6	1	1	3	0
	O_2	0.5	2	4	−5	0.5	2	−4
	O_3	5	1	2	−4	5	1	−1
	O_4	5	5	10	8	5	5	5

first case and two indicators (I_1 and I_2) for the second case. The third indicator may be assumed dependent indicator and can be correlated to indicators I_1 or I_2. In this case, I_3 may be excluded from the ranking procedure, in the second case, to test the effect of removal of dependent indicators. Overall ranking results do not consider the numerical value of the obtained rank, however, it concentrates on the relative position of objects (i.e., which one is more important). The results of the example in Table 2 show that the objects are ranked similarly in both cases, from most important to least important: O_4, is the first, followed by O_1, O_3, then O_2 is the last. The similarity in the ranks of the two cases (removing a dependent indicator) cannot be taken as a general result. This is studied in more detail in the subsequent sections.

Four types of models were used in representing dependencies between studied data. The models are defined as follows:

3.1 Linear

$$Y_1 = \alpha_0 + \alpha_1 X_1 + \varepsilon_1 \tag{2}$$

where Y_1 is the dependent indicator and X_1 is another environmental indicator or a primary variable. Here, Y_1 is linearly correlated to X_1. α_0 and α_1 are the correlation coefficients, and ε_1 is the residual error.

3.2 Multiple Linear

$$Y_1 = \beta_1 X_1 + \beta_2 X_2 + \varepsilon_1 \tag{3}$$

Here, the dependent indicator Y_1 is a function of two indicators, X_1 and X_2. β_1 and β_2 are the correlation coefficients. Equation (1) is one form of the multiple linear models.

3.3   Non-linear

$$Y_1 = \frac{\gamma_1 X_1}{1 + \gamma_2 X_2} + \varepsilon_1 \tag{4}$$

3.4   Complex

$$Y_1 = \delta_1 e^{-\delta_2 X_1} (1 - \delta_3 X_2) + \varepsilon_1 \tag{5}$$

Equations (4) and (5) are examples of the nonlinear and complex models, respectively. Assessment of the proposed approach was carried out using the following steps:

1. Original Data: A sequences of uncorrelated normally distributed random indicators $X_1, X_2 \ldots X_n$ for hypothetical objects $Obj_1, Obj_2 \ldots Obj_m$ are generated.
2. Extended Data: A correlation model is selected (Eqs. 2, 3, 4, and 5), and the value of the dependent indicator Y_1 is evaluated for all objects.
3. Decision ranking: Ranking is performed using the original and extended data (like the example in Table 2). This results in ordering the objects (assigned a numerical rank) from top to bottom to represent the most and the least important object.
4. Comparison: The rankings from original and extended data are compared using the Spearman's rank correlation coefficient (SRCC).

All the above steps and the simulation runs were implemented using MATLAB. This included generation of the datasets, evaluation of dependency models, application of Copeland ranking, and finally the SRCC value. Random datasets included sizes of up to 10 objects and 10 indicators, which is a suitable size for a typical environmental assessment problem. The comparison is made using SRCC, the most widely used measure of correlation or association between ranks. In statistics, the SRCC is a non-parametric measure of rank correlation (statistical dependence between the pairs rankings of two methods). It assesses how well the relationship between two rankings can be described using a monotonic function. Naturally, the SRCC between two variables will be high when observations have similar ranking between the two variables (correlation of 1) and low when observations have dissimilar or fully opposed ranking between the two variables (correlation of -1).

The analysis started with 64 datasets. Each data item is ranked and then re-ranked after the addition of four different dependent indicators one after another (a total 320 ranking runs). The results show that the rankings are not profoundly affected by the addition of a correlated indicator, no matter whether they were linear and nonlinear. The average SRCC scores that resulted for linear, multi-linear, non-linear, and complex correlations are 0.892, 0.894, 0.892, and 0.857, respectively. The multi-linear having the highest SRCC means that if indicators exist in a decision-making

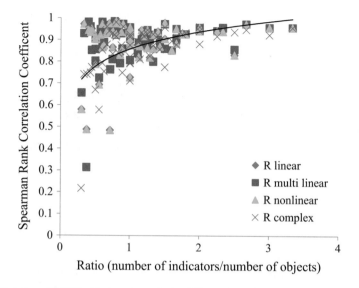

Fig. 1 Variation of SRCC with data size ratio for different dependency models

exercise with dependencies, their exclusion will slightly affect the ranking of objects if several of them are correlated multi-linearly. This is also very helpful in cases of incomplete datasets since the models can help in filling the gaps.

Results with low SRCC values were obtained when the ratio of the number of indicators to the number of objects is low and vice versa. In other words, if the number of indicators is high, excluding an indicator or using a dependency model to predict missing data will not affect the ranking results. On the other hand, all indicators must be considered for cases where number of indicators is small, whereas it is acceptable to find dependencies to fill the gaps of missing data. The results of the randomly simulated data are reported in Fig. 1, which plots the SRCC values against the ratio of the number of indicators to the number of objects (x-axis), for all generated datasets for the four different dependency models. The figure shows a monotonic increase in SRCC with values as low as 0.22.

4 Case Studies

One of the biggest challenges is the validation of any proposed approach. Validation covers both, applicability of the proposed methodology on real-world problem as well as accuracy of the outcomes. Professionals must make sure that the methodology behaves as expected in practice. Even strategies with high statistical significance may occasionally fail, and they often do (Harris 2015).

Validation procedures usually follow three steps (Varshney et al. 2013): (1) prospective validation, which occurs before the methodology is used, (2) concurrent validation, which occurs simultaneously with the implementation, and (3) retrospective validation, which occurs after establishment of the methodology and it is usually applied to real-world data. In this study, the prospective validation of the proposed approach has been already applied using the random data case presented above (see Fig. 1).

Three case studies were carried out to demonstrate and validate the proposed approach. All three case studies are based on actual environmental measurements, reported in literature, for operational desalination plants in the GCC. The first two cases studies are considered for the concurrent validation, while the third case study is for retrospective validation.

4.1 Case Study 1

The first case study is based on data reported by Bu-Olayan and Bivin (2006). Their study included trace metal levels in seawater from five sites in Kuwait's Bay, where many desalination plants exist. The case study covers concentrations of five metals during harmful and non-harmful algal blooms for both seawater samples and ctenophore samples in two seasons, summer and winter. The resulted data matrix consists of five objects (sites I to V) and 40 indicators (concentrations of five metal in two samples in two seasons in two blooms). All data sets are provided, i.e. there are no missing data for this case study. Another good point is that the five metal concentrations were found to be linearly correlated to a high degree. Hence, data reduction was possible, and it was quite representative to use the concentration of only one of the metals for the environmental ranking of the sites. This led to a reduced data matrix of five objects (sites) with eight indicators (one metal concentration in two samples in two seasons in two blooms). Applying Copeland ranking for both the original and reduced data matrices gave the same decisions. The sites were ranked, from the most contaminated to the lowest, as: Site III (Khadma), Site IV (Towers), Site V(Salmia), Site II (Doha) and Site I (Subiyah). Hence, excluding dependent indicators was safe given the dependences and large number of indicators.

4.2 Case Study 2

The next case study is based on data reported by Mohamed et al. (2005) for six desalination plants in the GCC. Assessment of the plants (objects) is carried out using 27 indicators, which are listed in Table 3. As shown in the table, there are several missing data (indicated as 'NA'), which should be completed if a decision is to be made about which of the six plants is affecting the environment the most. Given the relatively high number of indicators, it is recommended to analyze the data and safely reduce the number of indicators as much as possible. One option is to exclude those indicators with incomplete data. This option is not acceptable for the data presented in Table 3 because it will result in eliminating almost two thirds of the indicators.

Another option is to exclude indicators that are not highly relevant to the environmental objective. For instance, SiO_2 is considered by the World Health Organization (WHO) as a safe chemical for marine environment, mainly because it is a dietary requirement for various organisms. Hence, SiO_2 (indicator 14 in Table 3) can be safely excluded from the environmental analysis.

The data presented in Table 3 still has many missing values some adjustments are needed before testing the assessment methodology presented above. The procedure starts by studying the dependencies between indicators in order to complete missing data or otherwise reduce the number of indicators. Examples of using dependencies in estimating missing values for indicators are:

a) Measurements for the concentration of carbonate (CO_3^-), which is indicator 15 in Table 3, is missing. But as indicated in Table 1, carbonate can be estimated using the total alkalinity (At), i.e. $[CO_3^{-2}] = 0.6$ At. Furthermore, alkalinity measurements are available for three plants (indicator 25 in Table 3).

b) Permanent water hardness (indicator 26 in Table 3) is expressed as equivalent of $CaCO_3$ and this property is usually related to compounds with calcium and magnesium ions (Ca^{++} and Mg^{++} ions). It is therefore possible to estimate the missing total hardness measurement for plant (d) by applying the correlation in Table 1, i.e. $[CaCO_3] = 2.5[Ca^{2+}] + 4.1[Mg^{2+}]$.

c) Langelier Saturation Index (LSI) (indicator 22 in Table 3) is a calculated property that reflects the stability of calcium carbonate in water. It estimates the saturation level of calcium carbonate and indicates the extent of scale deposition on heat transfer surfaces. Values of the LSI indicator is missing for the last three plants in Table 3. The exact procedure for evaluating the LSI is relatively complex. The alternative was to estimate the missing values of the LSI indicator using an online LSI calculator reported by LENNTECH (2016). Required data include acidity (pH), calcium ion (Ca^{++}), bicarbonate (HCO_3^-), and total dissolved solids (TDS), which are indicators 7, 1, 16 and 9 in Table 3, respectively, together with brine temperature. Since the brine temperature is not reported, the LSI was evaluated at 40 °C, which is considered a typical brine temperature taken from different studies (see for example Dawoud and Al Mulla 2012 and Kotb 2015).

To check the validity of the assessment steps presented above, the data provided in Table 3 is divided into three datasets. The aim is to test the effect of indicator dependencies, data sizes, and model format when ranking with the Copeland method. The selected datasets are presented in Table 4. The first dataset includes three objects (plants a, b and c) and 23 indicators, with no missing data. One of the excluded indicators (SiO_2) is irrelevant, while the rest (Carbonate, LSI and Hardness) are initially excluded from the analysis due to incomplete data. Applying the Copeland ranking procedure on the (3×23) dataset that includes the three plants a, b and c (Alssadanat, Umm Alquain and Hamriyah) results in normalized ranks of 0.227, 1.0 and zero, respectively. The next step is to extend the original

Table 3 Chemical composition of rejected brine from six inland desalination plants in the GCC (Mohamed et al. 2005)

No	Parameter	Alssadanat, Oman (a)	Umm Alquain, UAE (b)	Hamriyah, Sharjah, UAE (c)	Saja'a Sharjah UAE (d)	Buwaib Saudi Arabia (e)	Salboukh Saudi Arabia (f)
1	Ca^{++}, mg/l	923	202	173	188	573	404
2	Mg^{++}, mg/l	413	510	311	207	373	257
3	Na^{++}, mg/l	2780	3190	1930	4800	2327	1433
4	K^{++}, mg/l	81.5	84.5	50.7	60	NA	NA
5	Sr^{++}, mg/l	28.2	21.1	14.2	40	NA	NA
6	Sum cation, meq/l	203.06	192.98	119.48	NA	NA	NA
7	pH	7.21	7.54	7.66	7.95	4.1	4.5
8	Electrical conductivity, mS/cm	16.8	14.96	127.41	NA	NA	NA
9	TDS, mg/l	10,553	10,923	7350	12,239	10,800	6920
10	NO_3, mg/l	7.2	27.4	15.9	NA	143	142
11	F^-, mg/l	0	1.6	1.3	8.0	NA	NA
12	Cl^-, mg/l	4532	4108	2933	4860	2798	1457
13	SO_4, mg/l	1552	2444	1537	2400	4101	2840
14	SiO_2, mg/l	NA	164.09	133.71	120	NA	NA
15	Carbonate (CO_3^-), mg/l	NA	NA	NA	NA	NA	NA
16	Bicarbonate (HCO_3^-), mg/l	466	656	753	NA	NA	NA
17	N^-	1.6	6.2	3.6	NA	NA	NA
18	Sum anions, meq/l	167.88	198.05	127.41	NA	NA	NA
19	Ion balance	9.48	4.02	−3.21	NA	NA	NA
20	SAR	19.12	27.2	20.3	NA	NA	NA
21	SER	59.55	71.91	70.27	NA	NA	NA
22	LSI	1.24	1.04	1.26	NA	NA	NA
23	R.I.	4.73	5.46	5.14	NA	NA	NA
24	Total ion, mg/l	10781	11245	7719	NA	NA	NA
25	Total alkalinity	380	538	617	NA	NA	NA
26	Total hardness	4041	2630	1730	NA	2968	2066
27	Fe, meq/l	0.06	0.08	0.05	NA	65.5	NA

Table 4 SRCC between ranks of original and extended data based on data in Table 3 for different data subsets

No	Objects	Original indicators (Table 3)	Ratio[+]	Extended indicators	Type of correlation	SRCC[*]
1	a,b,c	All indicators except: • SiO_2 • Carbonate • LSI • Hardness	23/3 = 7.67	Carbonate	Linear	0.9996
				LSI	Complex	0.9996
				Hardness	Multi-linear	0.9966
2	a,b,c	• Ca^{++} • Mg^{++} • pH • Electrical Conductivity • Bicarbonate • Alkalinity	6/3 = 2	Carbonate	Linear	1
				LSI	Complex	0.9449
				Hardness	Multi-linear	1
3	a,b,c,d,e,f	• Ca^{++} • Mg++ • pH • Na^+	4/6 = 0.67	Carbonate	Linear	0.7314
				LSI	Complex	0.9112
				Hardness	Multi-linear	0.9580

[+]Ratio = number of indicators/number of objects
[*]SRCC between objects' ranks using original and extended indicators

dataset by adding one additional dependent indicator at a time. Starting with the carbonate indicator, which is 'linearly' correlated to the total alkalinity indicator, as described above, the extended dataset consists now of 3 objects and 24 indicators. The new Copeland normalized ranks of the extended dataset are 0.235, 1 and 0. The normalized ranks of the original (23×3) and extended (24×3) datasets are then compared by evaluating their SRCC value (see Table 4). Similarly the original dataset is extended to include the LSI and total hardness indicators.

The same procedure was applied for the second and third subsets, as shown in Table 4. The second subset included 6 indicators and 3 objects, while the third subset included 4 indicators and all 6 objects.

Results shown in Table 4 indicate that decisions on excluding or including dependent indicators are highly related to the ratio of the number of indicators to the number of objects. When the number of indicators is twice or higher the number of objects, then dependent indicators in the dataset that have problems of quality (incomplete, inaccurate, etc.) can be safely discarded from the analysis. In other words, the effect on the quality of the decisions obtained from the Copeland ranking method will be minimal. The first two subsets in Table 4 would result in excellent

Table 5 Characteristics of brine water from five different desalination plants in the GCC (Dawoud and Al Mulla 2012)

No.	Environmental Indicators	Abu-Fintas Qatar	Ajman	Um-Quwain	Qidfa I Fujairah	Qidfa II Fujairah
		(a)	(b)	(c)	(d)	(e)
1	Temperature, °C	40	30.6	32.5	32.2	29.1
2	pH	1.2	0.5	0.3	0.03	0.99
3	EC, μS/cm	NR[*]	16.5	11.3	77	79.6
4	Ca, ppm	1350	312	173	631	631
5	Mg, ppm	7600	413	282	2025	2096
6	Na, ppm	NR	2759	2315	17,295	18,293
7	HCO_3, ppm	3900	561	570	159	149.5
8	SO_4, ppm	3900	1500	2175	4200	4800
9	Cl, ppm	29,000	4572	2762	30,487	31,905
10	TDS, ppm	52,000	10,114	8275	54,795	57,935
11	Total hardness, ppm	NR	NR	32	198	207
12	Free Cl_2, ppm	Trace	NR	0.01	NR	NR
13	SiO_2, ppm	NR	23.7	145	1.02	17.6
14	Langelier SI (LSI)	NR	0.61	0.33	NR	NR

[*]*NR* not reported

ranking quality because they have high value of SRCC due to their high indicators to objects ratio, which is almost 8 times for the first case and twice for the second. In contrast, subset 3 showed lower SRCC values because the number of indicators is lower than the number of objects. The SRCC values of third case (see Table 4) indicate that extending the dataset with the hardness indicator would result in a rank quality better than the LSI and carbonate indicators.

4.3   Case Study 3

The retrospective validation of the proposed methodology will be carried out here using the brine discharge data reported by Dawoud and Al Mulla (2012) for five desalination plants in the GCC. The datasets listed in Table 5 consists of 14 environmental indicators and 5 objects. The aim is to assess the level of the environmental deterioration caused by these desalination plants. This will be attained by ranking the plants with respect to their anticipated environmental impacts.

The first validation step is to prepare the datasets for the ranking process, mainly by attempting to complete the missing data. Like the previous case study, it is safe to exclude the SiO_2 indicator from the analysis. Next, models listed in Table 1 will be used to find estimates for missing data in Table 5, which are electrical conductivity (EC), Na concentration, total hardness, free chlorine concentration and

LSI (indicators 3, 6, 11, 12 and 14). The procedure used in estimating the values of these indicators, except free chlorine (indicator 12) is described in Al-Sharrah et al. (2017). For plant (a), the missing values for EC, Na concentration, total hardness and LSI were estimated as 72 μS/cm, 16,385 ppm, 734 ppm and 2.5, respectively. For plant (b), the total hardness was estimated as 45 ppm. Moreover, the LSI values for plants (d) and (e) were estimated as -0.63 and 0.29, respectively (Al-Sharrah et al. 2017).

The remaining missing values are the free chlorine concentration (indicator 12) for all plants except Um-Quwain (plant c). The importance of this indicator and whether it can be accurately estimated through dependencies will be studied first. The final decision will be dependent on the ability to proceed with the analysis in case the indicator has been excluded. As demonstrated in the previous case studies, the indicators to objects ratio is a reliable measure for exclusion of indicators.

The source of free chlorine in the brine is disinfection of feed seawater, which is essential to prevent biofouling in desalination processes. Chlorine is toxic to the aquatic organisms; hence, it is crucial that chlorination is tightly controlled to achieve the desired task with reduced harm to the marine life. The concentration of free chlorine is dependent on several factors related to the chemical and physical conditions of seawater and the flushing ability of the coastal zone (Hamed et al. 2017). Estimation of the free chlorine concentration is not possible with available dependencies and correlations. In effect, free chlorine can be only determined by chemical analysis. Hence, the free chlorine indicator cannot be used for this problem, and the question is whether it can be safely excluded from further analysis.

Going back to the data in Table 5, the number of indicators after excluding the SiO_2 indicator is 13 and the number of objects is 5, which results in the ratio of indicators to objects equals to 2.6 prior to excluding the free chlorine indicator and 2.4 after exclusion. Therefore, exclusion of the free chlorine indicator may be justified because the ratio is greater than two.

Two scenarios will be presented for ranking the environmental performance of the five plants by the Copeland method. The first scenario utilizes the 12 remaining indicators (excluding SiO_2 and free Cl_2), while the second one considers the indicators with complete data only (excluding indicators 3, 6, 11, 12, 13 and 14). Hence, the ranking will be performed using 12 × 5 and 8 × 5 datasets. The ranking results for the two datasets are shown in Table 6. The reported ranks are normalized in the range from one to zero, where a higher rank indicates more serious environmental impact.

The ranking results in Table 6 indicate that the ranks for both cases are the same. The SRCC value calculated for the 8 indicators with respect to 12 indicators is 0.985, which is reasonably high. The results show also that the worst plant in terms of environmental performance due to brine discharge is Abu-Fintas while the best is Um-Quwain. These are clear cuts in a sense that identical results have been obtained from complete and reduced sets of indicators. However, such clarity in decisions cannot be always achieved with low number of indicators. For instance, the ranks of Ajman and Um-Quain (plants b and c) can be compared in more confidence when using 12 indicators. In fact, for the 8 indicators case, it is not realistic to conclude

Table 6 Copeland normalized ranks for data in Table 5 for datasets of 12 and 8 indicators

	Desalination plants	Case 1 12 Indicators	Case 2 8 Indicators
(a)	Abu-Fintas	1	1
(b)	Ajman	0.17	0.05
(c)	Um-Quwain	0	0
(d)	Qidfa I	0.64	0.47
(e)	Qidfa II	0.77	0.69
	SRCC w.r.t. 12 indicators	–	0.985

that Um-Quain is the "worst" plant and Ajman is not when the difference between them is 0.05. Better resolution is provided for the case of 12 indicators allowing more confidence in making decision.

Further validation was carried out by comparing the ranking results of the Qidfa I and II desalination plants with similar results reported by Rustum et al. (2020). The two plants are in the same area in United Arab Emirates (UAE) but use different technologies. Qidfa I uses Reverse Osmosis (RO), while Qidfa II uses Multi-Stage Flash (MSF). Rustum et al. (2020) demonstrated the use of fuzzy modeling in "sustainability" ranking of typical desalination plants in UAE. One of the studied sustainability components "Reject stream characteristics", which is equivalent to the characteristics of the rejected brine considered in this case study. The reported results indicated that the normalized ranks for Qidfa I and Qidfa II are 0.7 and 0.6, respectively. In their formulation, the higher the rank is the better the object is in terms of sustainability. Converting from "sustainable" to "environmental deterioration" objectives will result in the normalized ranks of 0.3 and 0.4 for Qidfa I and Qidfa II, respectively. The relative rank is 1.33, which means that Qidfa II is 1.3 times (33% more) adverse to the environment compared to Qidfa I. This is conceptually true because it is a known fact that RO brine has less pollutants than that of the MSF. These results are comparable to the results obtained in Table 6. Moreover, the relative ranks for the 12 and 8 indicators cases are 1.2 and 1.47, respectively. In conclusion, our proposed approach gave consistent ranking results when compared with other more complex ranking methods.

5   Conclusions

Seawater desalination is vital in arid regions and for many countries it is the main source of potable water. However, they pose real environmental challenges, which should be addressed to mitigate the impacts on the environment. An important step in this direction is to quantify and assess the impacts of different desalination technologies and plants to enable decision makers to take environmental issues when considering new plants. Assessment is usually based on indicators that characterize the technologies or plants on common basis. There are many environmental indicators that are specific for desalination. The most important one that are widely used have been discussed in this study. The main difficulty in handling environmental indicators is dealing with incomplete datasets, primarily because industries are usually interested in measuring operational parameters that are directly related to the product quality and profitability, not those related to environmental performance. An effective approach has been proposed for filling missing values of the indicators. The approach is based on identifying dependencies between indicators and utilizing such dependencies in correlating the missing values using a number of proposed models. In case there are no dependencies, exclusion of indicators is possible with little risk of incorrect decisions only when the numbers of indicators are relatively high compared to the numbers of studied objects. Results derived from the case studies showed that exclusion of indicators would not have drastic effect on the assessment or decision-making as far as the number of indicators is more than twice the number of objects.

Acknowledgments This work is supported by the Kuwait-MIT Centre for Natural Resources and the Environment CNRE, which is funded by the Kuwait Foundation for Advancement of Sciences (KFAS) Project no. P31475EC01.

References

Al-Sharrah, G. (2011). The Copeland method as a relative and categorized ranking tool. *Statistica & Applicazioni Special Issue*, 81–95.

Al-Sharrh, G., Lababidi, H. M. S., & Al-Anzi, B. (2017). Environmental ranking of desalination plants: The case of the Arabian Gulf. *Toxicological & Environmental Chemistry, 99*(7–8), 1054–1070.

Alvarez-Bastida, C. V., Martínez-Miranda, M., Solache-Ríos, G., de Oca, F.-M., & Trujillo-Flores, E. (2013). The corrosive nature of manganese in drinking water. *Science of the Total Environment, 447*, 10–16.

Boerlage, S. (2011, November–December). Measuring seawater and brine salinity in seawater reverse osmosis. *Desalination & Water Reuse*, 26–30.

Bu-olayan, A., & Bivin, V. (2006). Validating ctenophore pleurobrachia pileus as an indicator to harmful algal blooms (HABs) and trace metal pollution in Kuwait Bay. *Turkish Journal of Fisheries and Aquatic Science, 6*, 1–5.

California Environmental Protection Agency, State Water Resources Control Board. (2016). https://www.waterboards.ca.gov/drinking_water/certlic/drinkingwater/documents/drinkingwaterlabs/AlkalinityConversions.pdf. Accessed 2 Aug 2018.

Copeland, A. (1951). *A "reasonable" social welfare function*. Mimeographed notes from a Seminar on Applications of Mathematics to the Social Sciences. University of Michigan.

Danoun, R. (2007). *Desalination plants: Potential impacts of brine discharge on marine life*. Project Report, The Ocean Technology Group, University of Sydney.

Dawoud, M., & Al Mulla, M. (2012). Environmental impacts of seawater desalination: Arabian gulf case study. *International Journal of Environment and Sustainability, 1*(3), 22–37.

El-Dessouky, H., & Ettouny, H. (2002). *Fundamentals of salt water desalination*. Amsterdam: Elsevier Science.

Fondriest: Environmental Learning Center. (2015). *Conductivity, salinity & total dissolved solids*. http://www.fondriest.com/environmental-measurements/parameters/water-quality/. Accessed 5 June 2016.

Guttman, L. (1977). What is not what is statistics. *The Statistician, 26*(2), 81–107.

Hamed, M. A., Moustafa, M. E., Soliman, Y. A., El-Sawy, M. A., & Khedr, A. I. (2017). Trihalomethanes formation in marine environment in front of Nuweibaa desalination plant as a result of effluents loaded by chlorine residual. *Egyptian Journal of Aquatic Research, 43,* 45–54.

Harris, M. (2015). *Fooled by Technical Analysis: The perils of charting, backtesting and data-mining.* Price Action Lab. http://www.priceactionlab.com/Blog/the-book/. Accessed 1 Apr 2020.

Imam, I. F., Michalski, R. S., & Kerschberg, L. (1993). *Discovering attributes dependence in databases by integrating symbolic learning and statistical analysis techniques.* In Proceedings of Knowledge Discovery in Database Workshop. AAAI-93, pp. 264–275.

Kotb, O. A. (2015). Optimum numerical approach of a MSF desalination plant to be supplied by a new specific 650 MW power plant located on the Red Sea in Egypt. *Ain Shams Engineering Journal., 6*(1), 257–265.

LENNTECH. (2016). *Langelier saturation index calculator.* Water Treatment Solutions. http://www.lenntech.com/calculators/langelier/index/langelier.htm. Accessed 3 May 2016.

Lewis, M. E. (2013). *National field manual for the collection of water-quality data, field Measurements: Dissolved oxygen.* http://water.usgs.gov/owq/FieldManual/Chapter6/Archive/6.2_v2.0.pdf. Accessed 25 May 2018.

Makkaveev, P. N. (2009). The features of the correlation between the pH values and the dissolved oxygen at the Chistaya Balka test area in the northern Caspian sea. *Oceanology, 49*(4), 466–472.

Mohamed, A. M. O., Maraqa, M., & Al Handhaly, J. (2005). Impact of land disposal of reject brine from desalination plants on soil and groundwater. *Desalination, 182,* 411–433.

Omri, A. (2013). CO_2 Emissions, energy consumption and economic growth nexus in MENA Countries: Evidence from simultaneous equations models. *Energy Economics, 40,* 657–664.

Piegorsch, W. W., & Bailer, A. J. (2005). *Analyzing environmental data.* Chichester: Wiley.

Roberts, D. A., Johnston, E. L., & Knott, N. A. (2010). Impacts of desalination plant discharges on the marine environment: A critical review of published studies. *Water Research, 44*(18), 5117–5128.

Rustum, R., Kurichiyanil, A. M. J., Forrest, S., Sommariva, C., Adeloye, A. J., Zounemat-Kermani, M., & Scholz, M. (2020). Sustainability ranking of desalination plants using Mamdani fuzzy logic interference system. *Sustainability, 12*(2), 631–653.

Sharp, J. H., & Culberson, C. H. (1982). The physical definition of salinity: A chemical evaluation. *Limnology and Oceanography, 27*(2), 385–387.

Sutherland, G., Waterhouse, L., Smith, J., & Saunders, S. (2016). Developing a systematic simulation-based approach for selecting indicators in strategic cumulative effects assessments with multiple environmental valued components. *Ecological Indicators, 61*(2), 512–525.

Varshney, P., Shah, M., Patel, P., & Rohit, M. (2013). Different aspects involved in process validation. *Innovare Journal of Sciences, 1*(2), 16–19.

Venkateswarlu, K. S. (1996). *Water chemistry: Industrial and power station water treatment.* New Delhi: New Age International Publisher.

Younos, T. (2005). Environmental issues of desalination. *Journal of Contemporary Water Research & Education, 132,* 11–18.

Introduction into Sampling Theory, Applying Partial Order Concepts

Bardia Panahbehagh and Rainer Bruggemann

1 Introduction

In sampling theory, for estimating some population parameters, each survey based on a strategy involves two stages; sampling stage and estimation stage (Hajek 1959). In the sampling stage, we indicate how it is supposed to select the sample units and in the estimation stage, estimators will be proposed for estimating the respective parameters.

The development of sampling theory is based on efficiency; i.e. on the search for high precision, low cost, etc. Searching is mostly based on the two principles; Randomization and Representation.

- **Randomization:** Based on randomization, the sampling design should select the sample units at random such that all the population units have chances to be selected. We know such designs as probability sampling designs. In contrast to probability sampling designs, we have non-probability or selective sampling designs in which the sampling is based on many factors including personal or expert-oriented ideas, the population situations, budgets of projects, etc. Probability sampling designs have at least two advantages relative to their non-probability versions; first, according to the randomize bases of the designs, we can make inferences about the estimators, second, we decrease the chance of facing a bias sample result of personal factors of the sampler mind (Sarndal et al. 2003, page 8).

B. Panahbehagh (✉)
Faculty of Mathematical Sciences and Computer, Kharazmi University, Tehran, Iran
e-mail: panahbehagh@khu.ac.ir

R. Bruggemann
Leibniz-Institute of Freshwater Ecology and Inland Fisheries, Berlin, Germany
e-mail: brg_home@web.de

- **Representation:** The exact definition of the representative sample has always been the subject of various discussions in the statistical literature (for some interesting discussions see Kruskal and Mosteller 1979a,b,c, 1980; Rao 2005; Dumicic 2011; Tille and Wilhelm 2017). In fact, the definition of a representative sample depends on our purpose. For example, if we are going to estimate the population density of the respective variable, then a simple random sample (SRS) would be a representative sample; if we are going to estimate the population total, a sample proportional to size, which is not a miniature of the population would be a representative sample (Tille and Wilhelm 2017) and if we are going to know if any member in our population has a specified disease or not, then a non-probability sample with many non-responses, containing just a few diseased people, could be a representative sample.

In the way of searching for efficient strategies, based on the two principals, randomization and representation, auxiliary variables have an important role in the past, present, and probably future of sampling theory (Rao and Fuller 2017). Some examples of the roles of auxiliary variables in the design stage are ranked set sampling (McIntyre 1952; Chen et al. 2004; Bouza-Herrera and Al-Omari 2018), judgment post-stratified sampling (MacEachern et al. 2004), balanced sampling (Deville and Tille 2004) and for estimation stage are ratio and regression estimators (Cochran 1953; Deng and Chhikara 1990), calibration estimator (Deville and Sarndal 1992).

Among the strategies based on auxiliary variables, some of them like balanced sampling, regression sampling, etc, need almost complete information about the population of the auxiliary variables. For example, in balanced sampling, it is assumed that we know the auxiliary variables for all the population units, before starting the procedure of sampling, and in the regression estimator, it is assumed that the population means of the auxiliary variables are known.

However some strategies just need partial information about the auxiliary variables like ranked set sampling (RSS) and judgment post-stratified sampling. In these designs just it is assumed that we know or even measure easily auxiliary variables for the sampled units and based on such information a more representative sample will be achievable.

Here we are going to discuss the later kind of strategies (that just need partial information about the auxiliary variables) intending to reduce costs and enhance the precision for multivariate variables. Then this chapter proceeds as follows; in Sect. 2 for using the information of ranks of data in the univariate case, we discuss RSS design and an economic version of RSS introduced by Panahbehagh et al. (2018), in Sect. 3 based on partial order set theory we discussed RSS for multivariate cases based on a research of Panahbehagh (2020) and the chapter will be finished with a conclusion in Sect. 4.

2 Univariate Sampling Based on Ranks of Data

Using ranks of the main variables, based on auxiliary variables, it is possible to take a representative sample from the target population. Panahbehagh (2020) showed that if we have complete information about the ranks of all the main variables we can estimate the population total with a great efficiency relative to SRS. Knowing the ranks of all the main or even auxiliary variables is very ambitious but it would be reasonable to assume that if we take a sample, before measuring the main variable, we can rank these sample units based on some easy to measure auxiliary variables. If it is possible, then it is better first to take an initial large sample, rank them based on some easy to measure auxiliary variable, and then select a good, more representative sample and measure the final sample units exactly based on the main variable. This idea goes back to McIntyre (1952). He introduced RSS for estimating a kind of crops, without extending theory. Takahasi and Wakimoto (1968) extended the theory of RSS and showed that this design is more efficient (more precise) than SRS.

2.1 Ranked Set Sampling

The idea of RSS is simple and beautiful. As a simple example assume we are going to estimate the mean height of the students in a college based on a sample of size 3. In SRS we select 3 students randomly and estimate the mean population based on the sample mean. In RSS, because it is easy and inexpensive to rank the students based on their heights, we first select 3 students, sort it, and select the shortest student as the first sample unit. For the second sample unit, we select another SRS sample of size 3, sort them and select the middle student and for the last sample unit we select another SRS and select the tallest student as the final sample and then measure the heights of these three students exactly and estimate the population mean based on them (see Fig. 1a).

Many different kinds of research, theoretically and practically showed that with considering precision, RSS is more efficient than SRS in many different problems (for some complete reviews of RSS see Chen et al. 2004; Bouza-Herrera and Al-Omari 2018). But yet many researchers are reluctant to use RSS for gathering their sample because different versions of RSS produce non-iid (independent and identical) samples and then conventional inferences that are extended based on iid samples assumption are not applicable for RSS samples. That's why MacEachern et al. (2004) introduced judgment post stratified sampling (JPS).

The idea of JPS is using the information of ranks of data just in the estimation stage and not the design stage. Then we can have an SRS (that is an iid sample) but take advantage of the ranks of the observations. Following the example of RSS, for estimating the mean height of the student, assume we are going to select a JPS sample of size 3. We select 3 students by SRS as the main sample and for indicating rank for each of the 3 observations, we select 2 students by SRS (as an auxiliary

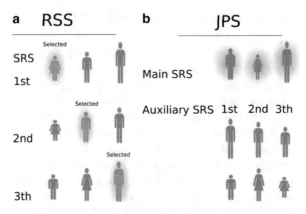

Fig. 1 Procedures of selection an RSS sample (**a**) and a JPS sample (**b**) of sizes three. In RSS three sets of independent SRS of size three should be selected and each set should be sorted based on their heights by eyes. The highlighted persons are selected as the final sample and should be measured exactly. For JPS an SRS as the main sample is selected and for each of its observations, we should select an SRS of size two to indicate the rank of the respective observation in the respective sample. For example in (**b**), for the first observation (the man in left-up of (**b**)) we select two persons (indicated by 1st auxiliary SRS) and we indicate the rank of the respective person in 1st auxiliary SRS and then we should allocate rank 2 to him. We proceed the same until the ranks of all the main SRS sample are indicated

sample) and indicate the rank of the respective observation in the respective sample (see Fig. 1b). Then based on ranks, we post-stratified the sample and estimate the mean height of the population based on the conventional estimator in stratified sampling (Sarndal et al. 2003). MacEachern et al. (2004) showed that the efficiency of JPS is between SRS and RSS and goes to RSS as the size of the sample is increasing.

2.2 An Unbalanced Ranked Set Sampling to Reduce the Costs

RSS is an efficient sampling strategy concerning precision. But with considering cost, it would be inefficient because of needing too many initial samples. To clarify this situation, suppose that X is the variable of interest (main variable) with probability density function f_μ, expectation $E(X) = \mu$ and variance $V(X) = \sigma^2 < \infty$ and we are to estimate μ and the variance of the estimator with an RSS of size m. We can suppose further that there is an auxiliary variable (used for ranking) with finite expectation and variance, and suppose that this auxiliary variable has a reasonable correlation with the main variable X and then we can use this auxiliary variable for ranking X. Here we assume perfect ranking (i.e. ranking based on X itself and not using an auxiliary variable) with this guarantee that all the results are also valid for the case of using an auxiliary variable for ranking. If we use an

Table 1 Selecting a RSS of size m. $X_{(h)i}$ for $h, i = 1, 2, \ldots, m$ is h-th order statistic based on X. The highlighted units will be selected as the final sample for full measurement and the other units will be used just for ranking

		Ranks			
	Set	1 (Smallest)	2 (Second smallest)	$\cdots$	m (Largest)
Sets	1	$X_{(1)1}$	$X_{(2)1}$	$\cdots$	$X_{(m)1}$
	2	$X_{(1)2}$	$X_{(2)2}$	$\cdots$	$X_{(m)2}$
	$\vdots$	$\vdots$	$\vdots$	$\ddots$	$\vdots$
	m	$X_{(1)m}$	$X_{(2)m}$	$\cdots$	$X_{(m)m}$

auxiliary variable for ranking, the density probability function of the order statistics should be defined based on the auxiliary variable instead of the main variable.

Now in the case of perfect ranking, we should select m independent sets each of size m based on SRS, sort each of them based on X and in the first set we select the smallest unit, in the second set we select the second smallest unit and so on until in the last set we select the largest unit as the final sample (see Table 1). As we can see in Table 1, to have an RSS of size m we need to select an initial sample of size m^2. Also to have an RSS of size $n_\bullet = d \times m$, instead of selecting $d \times m$ sets of size $d \times m$, it is recommended to implement d times (cycles) an RSS of size m, because it would be much more complicate to sort $d \times m$ units relative to sorting m units (Table 2).

As a cost-efficient RSS, Wang et al. (2004) proposed L-Tuple RSS (LTR) as follows; to have an LTR of size m, first we need to select $C_t^{m'}$ (the number of t-combinations from a given set of m' units) and then select t units from each set (resulting in $t \times C_t^{m'} = m$ final sample units) identified by mutually different ranks. For example for $m' = 5$ and $t = 2$, first select $C_2^5 = 10$ sets of size 5, sort each of them based on X and then select the units with ranks 1 and 2 from the first set and units with ranks 1 and 3 from the second set and so on until selecting units with ranks 4 and 5 from the last set which results to $m = 10 \times 2 = 20$ final sample size. In LTR setting m' and t, and also restriction of dependency between m, m' and t is challenging (for more details see Panahbehagh et al. (2018)).

To overcome these disadvantages of RSS and LTR, Panahbehagh et al. (2018), presented an easy to implement and calculate, unbalanced and cost efficient version of RSS as Virtual Stratified Sampling Using Ranked Set Sampling (VSR). The idea of VSR is very simple, to have a VSR of size $n_\bullet = d \times m$ we need to select K sets ($K > d$) of size m, sort each set based on X, resulting a post-stratified initial sample and then select a SRS of size d from each stratum. For its unbalanced version, it is enough to set $K > \max_{h=1}^m d_h$ and then select a SRS of size d_h (say s_h) from h-th stratum (see Table 3). Now it is possible to estimate μ unbiasedly by $\widehat{\mu}_{VSR}$ as

$$\widehat{\mu}_{VSR} = \frac{1}{m} \sum_{h=1}^m \bar{X}_{(h)} = \frac{1}{m} \sum_{h=1}^m \frac{1}{d_h} \sum_{i \in s_h} X_{(h)i}$$

Table 2 Selecting a RSS of size $d \times m$ in the case of d cycles of a RSS of size m. $X_{(h)ij}$ for $h, i = 1, 2, \ldots, m$ and $j = 1, 2, \ldots, d$ is h-th order statistics in i-th set for j-th cycle. The highlighted units will be selected as the final sample for full measurement and the other units will be used just for ranking

Cycle	Set	Ranks			
		1	2	$\cdots$	m
1	1	$X_{(1)11}$	$X_{(2)11}$	$\cdots$	$X_{(m)11}$
	2	$X_{(1)21}$	$X_{(2)21}$	$\cdots$	$X_{(m)21}$
	$\vdots$	$\vdots$	$\vdots$	$\ddots$	$\vdots$
	m	$X_{(1)m1}$	$X_{(2)m1}$	$\cdots$	$X_{(m)m1}$
2	1	$X_{(1)12}$	$X_{(2)12}$	$\cdots$	$X_{(m)12}$
	2	$X_{(1)22}$	$X_{(2)22}$	$\cdots$	$X_{(m)22}$
	$\vdots$	$\vdots$	$\vdots$	$\ddots$	$\vdots$
	m	$X_{(1)m2}$	$X_{(2)m2}$	$\cdots$	$X_{(m)m2}$
	$\vdots$	$\vdots$	$\vdots$	$\ddots$	$\vdots$
d	1	$X_{(1)1d}$	$X_{(2)1d}$	$\cdots$	$X_{(m)1d}$
	2	$X_{(1)2d}$	$X_{(2)2d}$	$\cdots$	$X_{(m)2d}$
	$\vdots$	$\vdots$	$\vdots$	$\ddots$	$\vdots$
	m	$X_{(1)md}$	$X_{(2)md}$	$\cdots$	$X_{(m)md}$

Table 3 Selecting a VSR of size $n_{\bullet} = \sum_{m=1}^{m} d_h$. $X_{(h)i}$ for $h, i = 1, 2, \ldots, m$ is h-th order statistics in i-th set. The highlighted units will be selected as the final sample (assuming) for full measurement and the other units will be used just for ranking

	Set	Ranks			
		1 (1-th Stratum)	2 (2-th Stratum)	$\cdots$	m (m-th Stratum)
Sets	1	$X_{(1)1}$	$X_{(2)1}$	$\cdots$	$X_{(m)1}$
	2	$X_{(1)2}$	$X_{(2)2}$	$\cdots$	$X_{(m)2}$
	3	$X_{(1)3}$	$X_{(2)3}$	$\cdots$	$X_{(m)3}$
	$\vdots$	$\vdots$	$\vdots$	$\ddots$	$\vdots$
	K	$X_{(1)K}$	$X_{(2)K}$	$\cdots$	$X_{(m)K}$

with

$$V(\widehat{\mu}_{\text{VSR}}) = \frac{\sigma^2}{Km} + \frac{1}{m^2} \sum_{h=1}^{m} \frac{1 - \frac{d_h}{K}}{d_h} \sigma_h^2 \tag{1}$$

where $\mu_h = E(X_{(h)})$ and $\sigma_{(h)} = V(X_{(h)})$ are the mean and variance of the h-th order statistics based on f_μ. As we can see in equation (1), if $\frac{d_h}{K} \longrightarrow 0$, then

$V(\widehat{\mu}_{\mathrm{VSR}}) \longrightarrow V(\widehat{\mu}_{\mathrm{RSS}})$ where $\widehat{\mu}_{\mathrm{RSS}}$ is the conventional estimator in the conventional RSS (for more details about definition of $\widehat{\mu}_{\mathrm{RSS}}$ and its variance see Panahbehagh et al. 2018). As we can see, VSR is a kind of RSS that is executable with any K larger than d. For example if we set $m = 5$, to have a ranked sample of size $n_{\bullet} = d \times m = 3 \times 5 = 15$, we can implement a VSR with selecting $Km = 4 \times 5 = 20$ units while for a conventional RSS we need $d \times m^2 = 3 \times 25 = 75$ units. Then VSR is a cost efficient, easy to implement version of RSS. Panahbehagh et al. (2018) also showed that if we define the efficiency based on high precision and low cost simultaneously, VSR can be more efficient than RSS and LTR. In next section we extend such idea to multivariate variables.

3 Multivariate Sampling Based on Ranks of Data

When we have just one variable, it is easy to rank and sort the sample units. For example if we have two persons, it is straightforward to sort them based on their heights or their weights. But it would be complicated or even impossible if we want to sort them based on the two variables, height and weight simultaneously (see Fig. 2).

In this section we will make a connection between sampling theory and partial order set theory. For this purpose first we discuss few version of multivariate RSS, involving multiple variable VSR.

3.1 A Multivariate Ranked Set Sampling

In the case of multivariate variables (Patil et al. 1994) considered one of the variables as the main variable and sorted the units based this main variable. With this approach, the strategy is efficient for estimating the mean of the main variable and efficiency for the other variables depends on their correlations with the main one. To consider all the variables (say R) simultaneously, often, however, too many initial sample are taken, and then running a R-layer procedure to consider each variable in a layer such that at the last we have all combinations of all the ranks for all the variables together (for more details about such strategies see Al-Saleh and Zheng 2002; Chen and Shen 2003; Arnold et al. 2009). For example with $R = 2$ and $m = 5$, we need $5^4 = 625$ initial units to perform such designs that indicates inefficiency of them with considering cost.

After presenting a multivariate version of VSR based on the method of Patil et al. (1994), in the next subsection, based on partial order set theory, a simple and cost efficient version of multivariate RSS introduced by Panahbehagh (2020) will be presented.

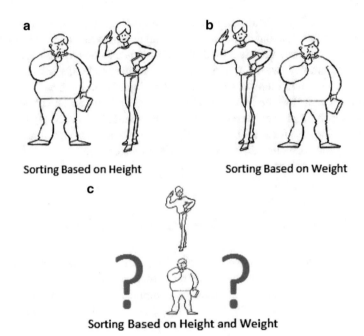

Fig. 2 Sorting two persons based on their heights (**a**) or based on their weights (**b**) and sorting them based on their heights and weights simultaneously (**c**)

3.2   Multivariate VSR

For the demonstration of the multivariate VSR (MVSR), we consider a bivariate case. The results can be easily extended to more than 2 variables. Assume there is a 2 dimensional variable $\mathbf{Z} \sim f_{\boldsymbol{\mu}}$ with $E(\mathbf{Z}) = \boldsymbol{\mu}$, where $\mathbf{Z} = (X, Y)$ and $\boldsymbol{\mu} = (\mu_x, \mu_y)$ also $Var(X) = \sigma_x^2$, $Var(Y) = \sigma_y^2$ and $Cov(X, Y) = \rho_{x,y}\sigma_x\sigma_y$ that are all finite. Here we are going to estimate $\boldsymbol{\mu}$.

Now to have a MVSR of size $n_\bullet = \sum_{m=1}^m d_h$ first we select an iid sample of $\mathbf{Z}_i$s of size m from $f_{\boldsymbol{\mu}}$ and sort it according to X in m columns and repeat this, K times and then select a SRS of size d_h from h-th column (see Table 4). Please note that in Table 4, $\mathbf{Z}_{(h)i} = (X_{(h)i}, Y_{[h]i})$, and $X_{(h)i}$ is the h-th order statistics in the i-th set with $\mu_{x(h)}$ and $\sigma_{x(h)}^2$, and $Y_{[h]i}$ is concomitant variable with respect to $X_{(h)i}$ in i-th set with $\mu_{y[h]}$ and $\sigma_{y[h]}^2$ as the mean and variance respectively. Also here again it is possible to use an auxiliary variable instead of the main variable for ranking. At the last, it is possible to estimate the elements of $\boldsymbol{\mu}$, unbiasedly by

$$\widehat{\mu}_{x.\text{MVSR}} = \frac{1}{m}\sum_{h=1}^m \bar{X}_{(h)} = \frac{1}{m}\sum_{h=1}^m \frac{1}{d_h}\sum_{i\in s_h} X_{(h)i},$$

Table 4 Selecting a MVSR of size $n_\bullet = \sum_{m=1}^{m} d_h$. In this table $\mathbf{Z}_{(h)i} = (X_{(h)i}, Y_{[h]i})$, and $X_{(h)i}$ is the h-th order statistics in the i-th set with $\mu_{x(h)}$ and $\sigma^2_{x(h)}$, and $Y_{[h]i}$ is concomitant variable with respect to $X_{(h)i}$ in i-th set with $\mu_{y[h]}$ and $\sigma^2_{y[h]}$ as the mean and variance respectively. The highlighted units will be selected (assuming) as the final sample for full measurement and the other units will be used just for ranking

		Ranks			
	Set	1 (1-th Stratum)	2 (2-th Stratum)	$\cdots$	m (m-th Stratum)
Sets	1	$\mathbf{Z}_{(1)1}$	$\mathbf{Z}_{(2)1}$	$\cdots$	$\mathbf{Z}_{(m)1}$
	2	$\mathbf{Z}_{(1)2}$	$\mathbf{Z}_{(2)2}$	$\cdots$	$\mathbf{Z}_{(m)2}$
	3	$\mathbf{Z}_{(1)3}$	$\mathbf{Z}_{(2)3}$	$\cdots$	$\mathbf{Z}_{(m)3}$
	$\vdots$	$\vdots$	$\vdots$	$\ddots$	$\vdots$
	K	$\mathbf{Z}_{(1)K}$	$\mathbf{Z}_{(2)K}$	$\cdots$	$\mathbf{Z}_{(m)K}$

$$\widehat{\mu}_{y.\text{MVSR}} = \frac{1}{m} \sum_{h=1}^{m} \bar{Y}_{[h]} = \frac{1}{m} \sum_{h=1}^{m} \frac{1}{d_h} \sum_{i \in s_h} Y_{[h]i}$$

with

$$V(\widehat{\mu}_{x.\text{MVSR}}) = \frac{\sigma^2_x}{Km} + \frac{1}{m^2} \sum_{h=1}^{m} \frac{1 - \frac{d_h}{K}}{d_h} \sigma^2_{x(h)},$$

$$V(\widehat{\mu}_{y.\text{MVSR}}) = \frac{\sigma^2_y}{Km} + \frac{1}{m^2} \sum_{h=1}^{m} \frac{1 - \frac{d_h}{K}}{d_h} \sigma^2_{y[h]}$$

and unbiased estimators of the variances as

$$\widehat{V}(\widehat{\mu}_{x.\text{MVSR}}) = \frac{K-1}{m(mK-1)} \sum_{h=1}^{m} \frac{1}{d_h(d_h-1)} \sum_{i \in s_h} (X_{(h)i} - \bar{X}_{(h)})^2 + \frac{1}{m(mK-1)} \sum_{h=1}^{m} (\bar{X}_{(h)} - \widehat{\mu}_{x.\text{MVSR}})^2,$$

$$\widehat{V}(\widehat{\mu}_{y.\text{MVSR}}) = \frac{K-1}{m(mK-1)} \sum_{h=1}^{m} \frac{1}{d_h(d_h-1)} \sum_{i \in s_h} (Y_{[h]i} - \bar{Y}_{[h]})^2 + \frac{1}{m(mK-1)} \sum_{h=1}^{m} (\bar{Y}_{[h]} - \widehat{\mu}_{y.\text{MVSR}})^2.$$

With equal size MVSR ($d_h = d, h = 1, 2, \ldots, m$) it is easy to show that

$$V(\widehat{\mu}_{x.\text{MVSR}}) = \frac{1}{dm} (\sigma^2_x - \frac{(1 - \frac{d}{K})}{m} \sum_{h=1}^{m} (\mu^1_{x(h)} - \mu_x)^2),$$

$$V(\widehat{\mu}_{y.\text{MVSR}}) = \frac{1}{dm} (\sigma^2_y - \frac{(1 - \frac{d}{K})}{m} \sum_{h=1}^{m} (\mu_{y[h]} - \mu_y)^2)$$

and if we assume that X and Y are linked by a linear regression model:

$$Y_i = \mu_y + \rho_{x,y}\frac{\sigma_y}{\sigma_x}(X_i - \mu_x) + \varepsilon_i$$

where ε is a random variable independent from X, then

$$V(\widehat{\mu}_{y.\text{MVSR}}) = \frac{1}{dm}(\sigma_y^2 - \frac{(1 - \frac{d}{K})}{m}\rho_{x,y}^2 \sum_{h=1}^{m}(\mu_{y(h)} - \mu_y)^2),$$

which shows that MVSR is an efficient designs for estimating the population mean for the main variable and efficiency of the other variable is dependent upon its correlations with the main one. Then MVSR just consider one of the variables. In the next subsection, based on partial order set theory (Poset), we will show that, it is easy to present a strategy to consider all the variables simultaneously.

3.3   Ranked Set Sampling Based on Poset

First briefly we introduce Poset and Linear Extensions (LE).

3.3.1   Poset, Linear Extensions, and Hasse Diagram

The application of partial order set theory for ranking has been described by Bruggemann and Carlsen (2011). In this theory, we have a set containing m units each of them with R variables, with a binary relation between the units. To compare two units of the set, if all variables of the first unit are equal or bigger (smaller) than the second one, then the first unit is better ($\geq$) (worse ($<$)) than the second one, otherwise the two units are not comparable. Linear extensions (LEs) are different projections of the partial order into a complete order that respect all the relations in the partial order set. I.e. linear extensions are the result of order preserving mappings. Therefore a relation $a < b$ in a Poset is preserved in all linear extensions. Also, a Hasse diagram is a graphical representation of the relation of units of a Poset with an implied upward orientation. A point is drawn for each unit of the Poset and joined with the line segment according to the following rules:

- If $a < b$ in the Poset, then the point corresponding to a appears lower in the drawing than the point corresponding to b.
- The two points a and b will be joined by line segment iff $a < b$ or $b < a$ and there is no other element, z for which is $a < z < b$ or $b < z < a$.

For an example of constructing all the LEs and plotting Hasse diagram, consider a set of $m = 5$ units with $R = 2$ variables as presented in Table 5 and Fig. 3.

Table 5 Units and all the their possible LEs

Set	X	Y	LE1	LE2	LE3	LE4	LE5	LE6	LE7	LE8
a	0	1	d	d	d	e	d	d	d	e
b	2	1	c	c	e	d	b	b	e	d
c	1	2	b	e	c	c	c	e	b	b
d	3	3	e	b	b	b	e	c	c	c
e	0	4	a	a	a	a	a	a	a	a

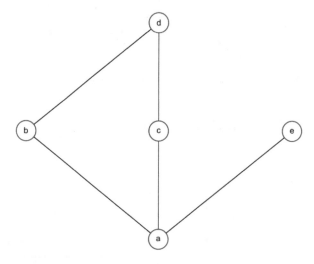

Fig. 3 Hasse diagram of the units presented in Table 5

Now it is possible to present two strategies:

- ranking based on all the LEs (POC),
- ranking based on a random sample of LEs (POR),

which both of them (with a reasonable sample size) consider all the variables in ranking procedure simultaneously (Panahbehagh 2020). Indeed in POC based on rounded mean height (ranks of the units in the respective LEs) of the units we put them in the strata. To illustrate procedures of POC and POR, we consider a simple example, a set with $m = 5$ and $R = 2$ (see Table 6). As we can see in Table 6, there is a complete order between the units $\{a, c, d, e\}$ and $\{b\}$ is incomparable with $\{c, d, e\}$. Then we have 4 LEs and based on them we can calculate mean height (MH) and after rounding them we can decide about putting them in the strata. With repeating this procedure for K sets, we will have an unequal size post-stratified initial sample with K_h as the size of h-th stratum and like stratified sampling (Sarndal et al. 2003) we can select a sample of size d_h from h-th stratum, which leads

Table 6 Example of POC and POR for a set with 5 units with $R = 2$. According to Poset, there is an order between all the units except b that is incomparable with c, d and e. For these units there are 4 LEs which leads to an unequal size post-stratified initial sample. For these units, 4-th stratum will be empty and instead we will have two units in 3-th stratum. In POC we use mean result of the all LEs and for POR we just select one of them randomly

							POC		POR		
Set	X	Y	LE1	LE2	LE3	LE4	Set	MH	Stratum	LE3	Set	Stratum
e	8	7	e	e	e	b	e	4.75	5	e	e	5
d	4	6	d	d	b	e	d	3.50	3	b	d	3
c	3	5	c	b	d	d	c	2.25	2	d	c	2
b	1	9	b	c	c	c	b	3.50	3	c	b	4
a	0	0	a	a	a	a	a	1.00	1	a	a	1

Table 7 General POC. Please note that $\mathbf{Z}_{\{h\}i} = (X_{\{h\}i}, X_{\{h\}i})$ is a unit that has been fallen into the h-th stratum after $i - 1$ units, according to its mean height MH in respective LEs

Ranks			
1 (1-th Stratum)	2 (2-th Stratum)	$\cdots$	m (m-th Stratum)
$\mathbf{Z}_{\{1\}1}$	$\mathbf{Z}_{\{2\}1}$	$\cdots$	$\mathbf{Z}_{\{m\}1}$
$\mathbf{Z}_{\{1\}2}$	$\mathbf{Z}_{\{2\}2}$	$\cdots$	$\mathbf{Z}_{\{m\}2}$
$\vdots$	$\vdots$	$\ddots$	$\vdots$
$\vdots$	$\vdots$		
		$\cdots$	$\mathbf{Z}_{\{m\}K_m}$
$\mathbf{Z}_{\{1\}K_1}$	$\vdots$		
	$\mathbf{Z}_{\{2\}K_2}$		

to POC (see Table 7). Panahbehagh (2020) showed that we can estimate unbiasedly the elements of $\boldsymbol{\mu}$ by

$$\widehat{\boldsymbol{\mu}}_{\text{POC}} = (\widehat{\mu}_{x.\text{POC}}, \widehat{\mu}_{y.\text{POC}}) = \sum_{h=1}^{m} W_h \bar{\mathbf{Z}}_{\{h\}}, \quad W_h = \frac{K_h}{Km}, \quad \bar{\mathbf{Z}}_{\{h\}} = \frac{1}{d_h} \sum_{i \in s_h} \mathbf{Z}_{\{h\}i}.$$

(2)

where $\mathbf{Z}_{\{h\}i} = (X_{\{h\}i}, Y_{\{h\}i})$ is a unit that has been fallen into the h-th stratum after $i - 1$ units, according to its MH in the respective LEs.

For POC, because $K_h; h = 1, 2, \ldots, m$ are random, with unknown distribution, mean and variance it would be complicated to calculate the variance of $\widehat{\boldsymbol{\mu}}_{\text{POC}}$.

Then it is better to present a design with equal size post stratified-initial sample. For this purpose, POR is presented. In POR, just one of the LEs would be selected to decide about putting the units in the strata. In example of Table 6, assume LE3 is selected, then we will put the units of the set in the strata according to ranks of them in LE3. With repeating this procedure for K sets we will have a post-stratified initial sample with equal size. Now with showing the sample unit in h-th stratum with $\mathbf{Z}_{[h]i} = (X_{[h]i}, Y_{[h]i}); i = 1, 2, \ldots, K$ and get a SRSWOR from h-th stratum of size d_h (an integer smaller than K), say s_h we can estimate the elements of $\boldsymbol{\mu}$ unbiasedly by

$$\widehat{\boldsymbol{\mu}}_{\text{POR}} = (\widehat{\mu}_{x.\text{POR}}, \widehat{\mu}_{y.\text{POR}}) = \frac{1}{m} \sum_{h=1}^{m} \bar{\mathbf{Z}}_{[h]}; \quad \bar{\mathbf{Z}}_{[h]} = \frac{1}{d_h} \sum_{i \in s_h} \mathbf{Z}_{[h]i}, \tag{3}$$

with (setting ψ as x or y)

$$V(\widehat{\mu}_{\psi.\text{POR}}) = \frac{\sigma_\psi^2}{Km} + \frac{1}{m^2} \sum_{h=1}^{m} \frac{1 - \frac{d_h}{K}}{d_h} E_M(\frac{1}{Q} \sum_{q=1}^{Q} S_{[h]q\psi K}^2).$$

where $q = 1, 2, \ldots, Q$ are all the possible combinations of LEs, with the below unbiased estimator of variance (for equal size sampling, $d_h = d$)

$$\widehat{V}(\widehat{\mu}_{\psi.\text{POR}}) = \frac{1}{dm(Km - 1)}[\sum_{h=1}^{m} \sum_{i \in s_{[h]}} (Y_{[h]i} - \widehat{\mu}_{\psi.\text{POR}})^2 + (K - d) \sum_{h=1}^{m} s_{[h]\psi}^2]. \tag{4}$$

where $S_{[h]q\psi K}^2$ and $s_{[h]\psi}^2$ are variance of h-th stratum under q-th combination of LEs and sample variance of h-th stratum for the variable $\psi(= x, y)$ respectively.

3.3.2 Negative Correlation

When the correlations between variables are strongly negative, according to Poset, it is probable that most of the units in a set are incomparable. This can make it meaningless to stratify the sets (note that in this case most of the units will fall in the middle stratum).

For an almost extreme case consider a case with $m = 5$, $R = 2$ and $\rho(X, Y) = -0.95$ in Table 8. As we can see, because of a strong negative correlation between X and Y, all the units are incomparable and then we will have $5! = 120$ possible LEs and all the units will fall in the middle stratum. If this situation happens for all the sets then the design will lead to SRS.

To overcome this problem, if the bivariate correlations between some variables are negative, we can multiple a "-1" to some of them to change the correlations to positive. But if we have more than two variables, sometimes it is not possible to make all the correlations positive. In such cases, it is better to select some more important variables that it is possible to make their correlations positive. We then rank the units using Poset with these new correlations. As we can see in Table 8 with multiple a "-1" to Y, all the units will be comparable and then each of them will fall in a separate stratum. Just please note that, if we decide to multiple "-1" in one of the variables, it should be done for all the selected set and not for some

Table 8 An example of situation in which there is a negative correlation between X and Y with $\rho(X, Y) = -0.95$

Set	X	Y	LE1	LE2	...	LE120	MH	Stratum	Set	X	$-Y$	LE1	MH	Stratum
e	5.0	1.0	e	e	...	a	3	3	e	5.0	-1.0	e	5	5
d	4.0	4.0	d	d	...	b	3	3	d	4.0	-4.0	d	4	4
c	3.0	4.5	c	c	...	c	3	3	c	3.0	-4.5	c	3	3
b	2.0	5.5	b	a	...	d	3	3	b	2.0	-5.5	b	2	2
a	1.0	9.0	a	b	...	e	3	3	a	1.0	-9.0	a	1	1

Table 9 Example involving 3 sets based on POC and POR

Set1	X	Y	MH	Set2	X	Y	MH	Set3	X	Y	MH	Stratum	POC	POR
e	8	7	4.75	q	8	6	4.50	w	9	7	4.80	5	e,w	e,p,w
d	4	6	3.50	p	5	7	4.50	v	6	6	3.60	4	q,p,v	d,q,v
c	3	5	2.25	o	4	5	3.00	u	3	5	2.40	3	d,b,o,r	c,o,u
b	1	9	3.50	g	2	2	2.00	t	1	2	1.20	2	c,g,u	b,g,r
a	0	0	1.00	f	0	0	1.00	r	0	8	3.00	1	a,f,t	a,f,t

of them. In the example of Table 8, we have just one set, and if we have more than one set, we should proceed for all of them like the first set. Also, it is notable that if we use such a procedure, after selecting the final sample, we will use the original data for calculating an unbiased estimator for the population mean vector. For more details see Bruggemann and Carlsen (2011).

3.4 Example to Clarify the Methods and Calculations

Assume we have a population with an interesting two dimensional variable $\mathbf{Z}$ with $\rho(X, Y) \simeq 0.50$ and by $m = 5$ we are going to estimate $\boldsymbol{\mu} = (\mu_x, \mu_y) = (4, 5)$ with a sample of size $n_\bullet = 10$. For this purpose we select 3 sets of size 5 independently. The selected sets with their variables are shown in Table 9. Based on all linear extensions, the MHs of the units are calculated and according to them, the strata are formed for POC. Also based on selecting one of the LEs for each set, the strata are formed for POR. As we can see in Table 9, POC and POR lead to unequal and equal size post-stratified initial sample respectively.

For allocation the sample size to the strata, in POR, we decide to set $n_h = 2, h = 1, 2, \ldots, 5$ and for POC we proceed based on proportional to size of strata allocation, then $n_1 = n_2 = n_4 = \frac{3}{15} \times 10 = 2$, $n_3 = \frac{4}{15} \times 10 \simeq 3$ and finally $n_5 = \frac{2}{15} \times 10 \simeq 1$.

Results for estimating the population means and variances are shown in Tables 10 and 11.

Table 10 Estimating the population means in POC and POR for data of Table 9 based on equations (2) and (3)

POC					POR			
Stratum	Sample	W_h	$\bar{X}_{\{h\}}$	$\bar{Y}_{\{h\}}$	Stratum	Sample	$\bar{X}_{\{h\}}$	$\bar{Y}_{\{h\}}$
5	e	2/15	8.0	7.0	5	p,w	7.0	7.0
4	q,v	3/15	7.0	6.0	4	q,v	7.0	6.0
3	d,b,o	4/15	3.0	6.7	3	c,u	3.0	5.0
2	c,g	3/15	2.5	3.5	2	g,b	1.5	5.5
1	a,t	3/15	0.5	1.0	1	a,t	0.5	1.0
			$\widehat{\mu}_{x.POC} = 3.9$	$\widehat{\mu}_{y.POC} = 4.8$			$\widehat{\mu}_{xPOR} = 3.8$	$\widehat{\mu}_{y.POR} = 4.9$

Table 11 Estimating the variances of the estimators in POR for data of Table 9 based on equation (4)

POR							
Stratum	Sample	$(X_{\{h\}i} - \widehat{\mu}_{x.POR})^2$	$s^2_{x\{h\}}$	$\widehat{V}(\widehat{\mu}_{x.POR})$	$(Y_{\{h\}i} - \widehat{\mu}y.POR^2)^2$	$s^2_{y\{h\}}$	$\widehat{V}(\widehat{\mu}_{y.POR})$
5	p,w	27.04		0.59	4.41		0.51
		1.44	8.0		4.41	0.0	
4	q,v	4.84			1.21		
		17.64	2.0		1.21	0.0	
3	c,u	0.64			0.01		
		0.64	0.0		0.01	0.0	
2	g,b	7.84			16.81		
		3.24	0.5		8.41	24.5	
1	a,t	7.84			8.41		
		14.44	0.5		24.01	2.0	
	Sum	71.2	11.0		44.9	26.5	

4 Conclusion and Discussion

In this chapter, we described a link between sampling strategies and partial order set theory. As we can see, in the case of multivariate variables, it is possible to present efficient strategies based on Poset to consider all the variables in ranking and then estimate the population parameters precisely with reasonable sample size.

Generally, the determination of all linear extensions is computationally a hard and challenging problem. Therefore calculating heights needs themselves sampling techniques as shown by Bubley and Dyer (1999). For such situations, pretty good approximations are presented by Bruggemann et al. (2004) and Bruggemann and Carlsen (2011) and an interesting alternative is shown by Fattore and Arcagni (2018).

Future work (concerning the partial order set concepts):

Although the different concepts to calculate approximatively heights of objects derived from linear extensions, are pretty good, it turns out that sometimes the mathematical simpler concept delivers better results than the more sophisticated one. Therefore it is an important task to develop still better approximations. An alternative is to develop a theoretical framework to decide when which of the two methods is to be preferred.

The set of linear extensions consists of LT elements $(LE_l, l = 1, \ldots, LT)$. Assume that the variables used to describe the objects (or units) are of different importance, then it is clear that linear extensions may have different proximities to the variables (supposed there is a suitable concept of distances). Hence in a general framework, the set of linear extensions should not be seen as a uniform set. Especially if a variable induces a complete order (i.e. an order without ties) then there must be one linear extension, which reproduces this order. This linear extension may play a favorite role. Therefore in the further development of the application of partial order sets in sampling theory, the potential non-uniformity of linear extensions should be conceptually be built in. Linear extensions are images of order-preserving maps, where the order relations found in a Poset are reproduced. The underlying Poset derived from a data matrix reveals more symmetries than expected from the data matrix alone, because of the ordinal interpretation of the data. Hence the MH-values (see Table 6) may have several ties. It can be a useful idea, to develop tie-breaking concepts, to guarantee that within the POC-concept the objects belong as much as possible to different strata.

References

Al-Saleh, M., & Zheng, G. (2002). Estimation of bivariate characteristics using ranked set sampling. *The Australian & New Zealand Journal of Statistics, 44*, 221–232.

Arnold, B. C., Castillo, E., & Sarabia, J. M. (2009). On multivariate order statistics. Application to ranked set sampling. *Computational Statistics and Data Analysis, 53*(12), 4555–4569.

Bouza-Herrera, C., & Al-Omari, A. I. F. (2018). *Ranked set sampling, 65 years improving the accuracy in data gathering*. London/San Diego: Elsevier/Academic.

Bruggemann, R., & Carlsen, L. (2011). An improved estimation of averaged ranks of partial orders. *MATCH Communications in Mathematical and in Computer Chemistry, 65*, 383–414.

Bruggemann, R., Sorensen, P. B., Lerche, D., & Carlsen, L. (2004). Estimation of averaged ranks by a local partial order model. *Journal of Chemical Information and Computer Sciences, 44*, 618–625.

Bubley, R., & Dyer, M. (1999). Faster random generation of linear extensions. *Discrete Mathematics, 201*, 81–88.

Chen, Z., & Shen, L. (2003). Two-layer ranked set sampling with concomitant variables. *The Journal of Statistical Planning and Inference, 115*, 45–57.

Chen, Z., Bai, Z., & Sinha, B. (2004). *Ranked set sampling: Theory and applications*. Lecture notes in statistics. New York: Springer.

Cochran, W. G. (1953). *Sampling techniques*. Oxford: Wiley.

Deville, J. C., & Sarndal, C. E. (1992). Calibration estimators in survey sampling. *Journal of the American Statistical Association, 87*, 376–382.

Deng, L., & Chhikara, R. (1990). On the Ratio and Regression Estimation in Finite Population Sampling. *The American Statistician, 44*(4), 282–284.

Deville, J. C., & Tille, Y. (2004). Efficient balanced sampling: The cube method. *Biometrika, 91*, 893–912.

Dumicic, K. (2011). Representative samples. In M. Lovric (Ed.), *International encyclopedia of statistical science*. Berlin/Heidelberg: Springer.

Fattore, M., & Arcagni, A. (2018). A reduced posetic approach to the measurement of multidimensional ordinal deprivation. *Social Indicators, 136*(3), 1053–1070.

Hajek, J. (1959). Optimum strategy and other problems in probability sampling. *Casopis pro Pestovani Matematiky, 84*, 387–423.

Kruskal, W., & Mosteller, F. (1979a). Representative sampling, I: Non-scientific literature. *International Statistical Review, 47*, 13–24.

Kruskal, W., & Mosteller, F. (1979b). Representative sampling, II: Scientific literature, excluding statistics. *International Statistical Review, 47*, 111–127.

Kruskal, W., & Mosteller, F. (1979c). Representative sampling, III: The current statistical literature. *International Statistical Review, 47*, 245–265.

Kruskal, W., & Mosteller, F. (1980). Representative sampling, IV: The history of the concept in statistics. *International Statistical Review, 48*, 169–195.

Lih-Yuan, D., & Raj, C. (1990). On the ratio and regression estimation in finite population sampling. *The American Statistician, 44*, 282–284.

MacEachern, S. N., Stasny, E. A., & Wolfe, D. A. (2004). Judgement post-stratification with imprecise ranking. *Biometrics, 60*, 207–215.

McIntyre, G. A. (1952). A method of unbiased selective sampling using ranked sets. *Australian Journal of Agricultural Research, 3*, 385–390.

Panahbehagh, B. (2020). Stratified and ranked composite sampling. *Communications in Statistics – Simulation and Computation, 49*(2), 504–515.

Panahbehagh, B., Bruggemann, R., & Salehi, M. (2020). Sampling of multiple variables based on partial order set theory. arXiv:1906.11020v2.

Panahbehagh, B., Bruggemann, R., Parvardeh, A., Salehi, M., & Sabzalian, M. R. (2018). An unbalanced ranked-set sampling method to get more than one sample from each set. *The Journal of Survey Statistics and Methodology, 6*(3), 285–305.

Patil, G. P., Sinha, A. K., & Taillie, C. (1994). Ranked set sampling for multiple characteristics. *International Journal of Ecology and Environmental Sciences, 20*, 94–109.

Rao, J. N. K. (2005). Interplay between sample survey theory and practice: An appraisal. *Survey Methodology, 21*(3), 117–138.

Rao, J. N. K., & Fuller, W. A. (2017). Sample survey theory and methods: Past, present, and future directions. *Survey Methodology, 43*(2), 145–160.

Sarndal, C. E., Swensson, B., & Wretman, J. (2003). *Model assisted survey sampling*. New York: Springer.

Takahasi, K., & Wakimoto, K. (1968). On unbiased estimates of the population mean based on the sample stratified by means of ordering. *Annals of the Institute of Statistical Mathematics, 20*, 1–31.

Tille, Y., & Wilhelm, M. (2017). Probability sampling designs: Principles for choice of design and balancing. *Statistical Science, 32*(2), 176–189.

Wang, Y. G., Chen, Z. H., & Liu, J. (2004). General ranked set sampling with cost considerations. *Biometrics, 60*, 556–561.

Looking for Alternatives? Split-Shots as an Exemplary Case

Lars Carlsen

1 Introduction

In the news we are virtually on a daily basis confronted with now this or that is not good for humans and/or the environment, which obviously calls for alternatives as many of the suspected, or proven harmful substances are actually in their own sense beneficial for their specific purposes. This process may be rather complicated involving a multitude of steps (NAS 2014a, b). The following list is adopted from (NAS 2014a, b):

1. Identify the chemical of concern.
2. Scoping and problem formulation.
3. Identify potential alternatives.
4. Refer cases with limited or no alternatives to research and development.
5. Assess physicochemical properties.
6. Assess human health and ecological hazards, and assess comparative exposure.
7. Integration of information on safer alternatives.
8. Life cycle thinking.
9. Optional assessments: Additional life cycle assessment,
10. Identify acceptable assessments and refer cases with no alternatives to research and development.
11. Compare or rank alternatives.
12. Implement alternatives.
13. Research or de novo design of safer alternatives

L. Carlsen (✉)
Awareness Center, Roskilde, Denmark
e-mail: lc@awarenesscenter.dk

© The Author(s), under exclusive license to Springer Nature Switzerland AG 2021
R. Bruggemann et al. (eds.), *Measuring and Understanding Complex Phenomena*,
https://doi.org/10.1007/978-3-030-59683-5_11

In the present study we focus on step 11: 'compare or rank alternatives', and argue that partial order methodology may be introduced as an advantageous decision support tool, despite the fact that it has been stated that multi-criteria decision analyses (MCDA) "may be useful in some cases, they may be more complicated than required for many assessments" (NAS 2014a, b). The present study applies partial order methodology is applied for the search for suitable alternatives to split-shots as an exemplary case. Data are adopted from the National Academy of Science report on the selection of chemical alternatives (NAS 2014a).

Split-shots have traditionally been produced of lead (Pb) as a relatively soft material that has a good malleability and is corrosion resistant. Further the end product has an excellent availability and is obtainable at a relatively low price. However, Pb is both from an environmental and human health perspective an unwanted compound and thus possible alternatives are searched for. Thus, TURI (2006) reports that

- "Nearly 2,500 metric tons of lead are used each year in the United States to produce fishing sinkers."
- "Many of these sinkers are lost during use. One study found that anglers lost, on average, one sinker every six hours of fishing"
- "Lead sinkers are lethal to waterbirds, such as loons and swans. One study found that the most common cause of death in adult breeding loons was lead toxicity from ingested fishing sinkers"

In the present study 5 alternatives were includes (NAS 2014a, p180; TURI 2006 pp. 3–60 – 3–81), i.e., bismuth (Bi), Ceramic (cer), Steel (ste), Tin (Sn) and Wolfram (W).

The paper discusses the ranking of indicators for the bismuth (Bi), Ceramic (cer), Steel (ste), Tin (Sn) and Tungsten (W) as well as lead (Pb) in order to suggest the optimal alternative to lead split-shots among the 5 other suggested materials applying the concept of average heights (De Loof et al. 2006; Bruggemann and Carlsen 2011). The paper finalizes with a discussion of the ranking probabilities for single alternatives applying the Bubley-Dyer approach to average ranking (Bubley and Dyer 1999) as well as the version 8_3 of the LPOMext module of PyHasse (Bruggemann and Carlsen 2011).

2 Methodology

The present paper describes how selected partial order tools may be applied in the evaluation of alternatives and thus constitutes an advantageous tool in selecting an optimal alternative out of the set, taking several indicators simultaneously into account as an alternative to conventional methods to study multi-indicator systems (MIS) (Bruggemann and Carlsen 2012).

2.1 The Basic Equation of Hasse Diagram Technique

In its basis partial ordering appears pretty simple as the only mathematical relation among the objects is "≤" (Bruggemann and Carlsen 2006a, b; Bruggemann and Patil 2011). As the basis for a comparison of objects, here split-shot alternatives, characterization by the group of indicators is used (vide infra). This series of indicators, r_i, characterizes the single split-shot alternatives. Thus, if one of the alternatives (x) is characterized by the set of indicators $r_i(x)$, i = 1,...,m, where m is the number of indicators, it can be compared to another alternative (y), characterized by the indicators $r_i(y)$. Thus, y < x iff

$$r_i(y) \leq r_i(x) \text{ for all } i = 1, \ldots, m \tag{1}$$

Equation 1 is a very hard and strict requirement for establishing a comparison. It demands that all indicators of x should be better (or at least equal) than those of y. Further, let X be a set of alternatives included in the analysis, i.e., X = {Pb, Bi, cer, ste, Sn, W},[1] x will be ordered higher (better) than y, i.e., x > y, if at least one of the indicator values for x is higher than the corresponding indicator value for y and no indicator for x is lower than the corresponding indicator value for y. On the other hand, if $r_i(x) > r_i(y)$ for some indicator r_i and $r_j(x) < r_j(y)$ for some other indicator r_j, x and y will be called incomparable (notation: x ‖ y) expressing the mathematical contradiction due to conflicting indicator values. A set of mutual incomparable objects is called an antichain. When all indicator values for x are equal to the corresponding indicator values for y, i.e., $r_i(x) = r_i(y)$ for all r_i, the two compared elements will have identical rank and will be considered as equivalent, i.e., x ~ y. The analysis of Eq. 1 can be visualized by a Hasse diagram.

2.2 The Hasse Diagram

The Eq. 1 is the basic for the Hasse diagram technique (HDT) (Bruggemann and Carlsen 2006a, b; Bruggemann and Patil 2011). Hasse diagrams are visual representations of the partial order. In the Hasse diagram comparable objects are connected by a sequence of lines (Bruggemann and Carlsen 2006a, b; Bruggemann and Patil 2011; Bruggemann and Munzer 1993; Bruggemann and Voigt 1995, 2008).

---

[1] Pb: Lead, Bi: Bismut, cer: Ceramics, ste: Steel, Sn: Tin, W: Wolfram.

2.3   The More Elaborate Analyses

In addition to the basic partial ordering tools some more elaborate analyses have been used including average ranks (Bubley and Dyer 1999; De Loof et al. 2006; Bruggemann and Patil 2011) and sensitivity analysis (Bruggemann and Patil 2011; Bruggemann et al. 2014), the latter gives an insight in the relative importance of the included indicators (Bruggemann and Patil 2011; Bruggemann et al. 2014).

The average ranking is expressed as average height from bottom (min. Height = 1) to the top (max height = n, i.e., the maximum number of objects, here n = 6) (Bruggemann and Annoni 2014). The average rank is generated by calculating all linear order preserving sequences (set LE), the "linear extensions of the original partial order. From LE_0 the statistical characterization for each object is obtained. For example the characterization is calculated as the average value an object has, taken all positions of this object within LE_0, the averaged heights. It is clear that this procedure is computationally extremely difficult. Hence, approximations were developed.

For the sentivity analysis (Bruggemann and Patil 2011; Bruggemann et al. 2014), let Q be the set of all indicators, then taken all indicators of Q leads to a partial order, which is called PO_0. The corresponding set of linear extensions is denoted by LE_0. Leaving out one indicator of Q, say r_j, then another partial order results, which is denoted as PO_j.

Both partial orders can be described by an adjacent matrix, say A_0 for PO_0 and A_j for PO_j.

Taken the Euclidian Distance (squared) quantifies the role of indicator qj in PO_0. This is a sensitivity measure for the indicators of set Q, describing the structural changes of the partial order leaving one indicator out. This is not immediately a measure of the sensitivity of the indicators for a ranking, because the ranking is per definition a linear order and here derived over many interim steps.

If a linear order is obtained by all orders in LE_0, the set of linear extensions taken from PO_0, then any PO_j will also lead to a corresponding set LE_j. And this set is the more differing from LE_0 the larger the sensitivity is. Therefore the ranking due to averaged heights is as more affected by indicator r_j as larger its sensitivity is.

For detail information on the single tool the cited literature should be consulted as detailed description is outside the scope of the present paper.

2.4   Software

All partial order analyses were carried out using the PyHasse software (Bruggemann et al. 2014). PyHasse is programmed using the interpreter language Python (version 2.6) (Hetland 2005; Weigend 2006; Ernesti and Kaiser 2008; Langtangen 2008;

Python 2015) Today, the software package contains more than 100 modules and is available upon request from the developer, Dr. R.Bruggemann (brg_home@web.de).

2.5 Indicators

The analysis of possible alternatives to Pb split-shots includes a series of indicators. The indicators are grouped in three main categories, i.e., Technical and Performance Criteria (TPCr), Environmental Criteria (ENCr), Human Health Criteria (HHCr), and Cost criteria (Cost), respectively, where TPCr includes 5 sub-indicators, whereas both ENCr and HHCr includes 3 sub-indicators each and the Cost comprises 2 sub-indicators (NAS 2014a; TURI 2006).

TPCr (Technical and Performance)

dens: Density
hard: Hardness (desirable for "feel" and noise
mall: Malleability (split-shot application)
lowm: Low melting point (desirable for home production)
corr: Corrosion resistant

EnCr (Environmental)

hito: High toxicity to waterfowl (the lower the better)
toaq: Toxic to aquatic species (the lower the better)
dwst: Primary drinking water standards (MCL action level)

HHCr (Human Health)

carc: Carcinogenicity
devt: evelopmental toxicity
ocex: Occupational exposure: REL (8-hour TWA)

Cost

repr: Retail price
avail: Availability of end product

2.6 Data

Data are adopted directly from the National Academy of Science report (NAS 2014a, p180; TURI 2006 pp. 3–60 – 3–81). Here the 5 possible alternatives are compared to Pb in a relative simple way, i.e., better, equal or worse. In the present study we denote these possibilities by 1, 0 and − 1, respectively. In some cases the report (NAS 2014a) states an uncertainty, denoted by "?". This results in two

Table 1 Non-conservative evaluation of the 5 possible alternatives to Pb (uncertainties are regarded as 'equal', i.e., denoted 0 – marked grey)

Criteria	Pb	Bi	cer	ste	Sn	W
dens	0	-1	-1	-1	-1	1
hard	0	1	1	1	0	1
mall	0	-1	-1	-1	0	-1
lowm	0	1	-1	-1	1	-1
corr	0	0	0	-1	0	0
hito	0	1	0	1	1	1
toaq	0	1	0	1	1	1
dwst	0	0	0	1	1	0
carc	0	1	1	1	1	1
devt	0	1	1	1	1	1
ocex	0	0	1	1	1	1
repr	0	-1	-1	0	-1	-1
avail	0	-1	-1	-1	-1	-1

Table 2 Conservative evaluation of the 5 possible alternatives to Pb (uncertainties are regarded as 'worse', i.e., denoted −1 – marked grey)

Criteria	Pb	Bi	cer	ste	Sn	W
dens	0	-1	-1	-1	-1	1
hard	0	1	1	1	0	1
mall	0	-1	-1	-1	0	-1
lowm	0	1	-1	-1	1	-1
corr	0	0	-1	-1	0	0
hito	0	1	-1	1	1	1
toaq	0	1	-1	1	1	1
dwst	0	-1	-1	1	1	-1
carc	0	1	1	1	1	1
devt	0	1	1	1	1	1
ocex	0	-1	1	1	1	1
repr	0	-1	-1	0	-1	-1
avail	0	-1	-1	-1	-1	-1

approaches, i.e., a non-conservative approach whereas in cases of uncertainty an "equal" is assumed and thus denoted by 0 (Table 1) and a conservative approach where an uncertainty denoted by −1, thus assuming a 'worse' (Table 2).

3   Results and Discussion

Prior to the actual analyses it appears appropriate to clarify what an alternatives analysis is and what it is not, which has been stated in the National Academy of Science report (NAS 2014b).

Thus, an alternatives analysis is

- a process for identifying, comparing, and selecting safer alternatives to chemicals of concern.
- intended to facilitate an informed consideration of the advantages and disadvantages of alternatives to a chemical of concern

whereas an alternatives analysis is not

- *a safety assessment*, where the primary goal is to ensure that exposure is below a prescribed standard,
- *a risk assessment*, where risk associated with a given level of exposure is calculated
- *a sustainability assessment*, that considers all aspects of a chemical's life cycle, including energy and material use

3.1 Aggregated Data

Due to the low number of compared elements (6) relative to the total number of indicators (13) a direct partial ordering of the compared elements based on all indicators is not meaningful due to a possible lack of robustness (Sørensen et al. 2000). Even ordering based on the single indicators groups gives only little sense. Thus, some kind of aggregation of the sub-indicators appears appropriate. Hence, the 13 indicators are aggregated into 4 groups, i.e., Technical and Performance Criteria (TPCr), Environmental Criteria (ENCr), Human Health Criteria (HHCr) and Cost criteria (Cost), respectively. The aggregation is carried out by a simple addition of the indicator values for these four groups. Thus,

- TPCr = dens + hard + mall + lowm + corr
- EnCr = hito + toaq + dwst
- HHCr = carc + devt + ocex
- Cost = repr + avail

The resulting data matrices for the non-conservative and conservative approaches are given in Tables 3 and 4, respectively.

Table 3 Aggregated data matrix (non-conservative approach)

Criteria	TPCr	EnCr	HHCr	Cost
Pb	0	0	0	0
Bi	0	2	2	−2
cer	−2	0	3	−2
ste	−3	3	3	−1
Sn	0	3	3	−2
W	0	2	3	−2

Table 4 Aggregated data matrix (conservative approach)

Criteria	TPCr	EnCr	HHCr	Cost
Pb	0	0	0	0
Bi	0	1	1	−2
cer	−3	−3	3	−2
ste	−3	3	3	−1
Sn	0	3	3	−2
W	0	1	3	−2

Fig. 1 Hasse diagram generated based the data in Table 3 using all 4 indicators

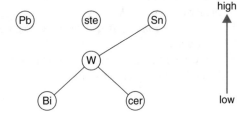

3.2 The Non-conservative Approach

Analyzing the aggregated data for the non-conservative approach using all four indicators leads to the Hasse diagram depicted in Fig. 1. The diagram displays two isolated elements, Pb and ste, which makes an immediate ranking less obvious.

The average ranks of the 6 materials were found to form the following relations: Sn > W > ste = Pb > Bi = cer, the calculated average height being, 5.6 (Sn), 4.2 (W), 3.5 (ste, Pb) and 2.1 Bi, cer), respectively, indicating Sn as the most advantageous alternative to Pb. Hence, Sn and Pb are evaluated equal concerning the technical and performance criteria (TPCr) but Sn appears significantly better than Pb with regard to environment (EnCr) and human health (HHCr) (cf. Tables 3 and 4).

It is obviously of interest to disclose the relative importance of the four indicators. Thus, the relative importances of the indicators TPCr, EnCr, HHCr and Cost were estimated to be 0.333, 0.083, 0.25 and 0.333, respectively. That the technical and performance indicator (TPCr), on a relative scale appears as being of high importance may not be surprising. A similar high importance of the cost indicator (Cost) may not be surprising and these two indicators clearly oust the environmental and human health indicators (EnCr and HHCr). However, this analysis puts a price on environmental and human health. To circumvent this, possible unfair judgment, an analogous analysis was done applying only the three indicators TPCr, EnCr and HHCr.

In Fig. 2 the Hasse diagram generated based on the TPCr, EnCr and HHCr indicators only is depicted. It is immediate seen that a much more clear-cut picture has developed as as all elements are interconnect, which also lead to a clearer picture regarding the average height with Sn > W > ste > Bi > cer > Pb, the corresponding average heights being 6.0 (Sn), 4.8 (W), 3.2 (Bi), 3.0 (ste), 2.4 (cer), 1.6 (Pb), respectively. Thus, excluding the Cost indicator the choice of Sn as the optimal

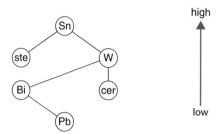

Fig. 2 Hasse diagram generated based the data in Table 3 using TPCr, HHCr and EnCr indicators

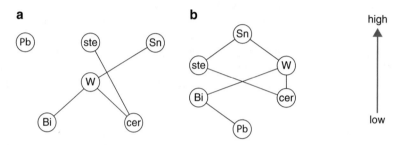

Fig. 3 (**a**) Hasse diagram generated based the data in Table 4 using all 4 indicators. (**b**) Hasse diagram generated based the data in Table 3 using TPCr, HHCr and EnCr indicators

alternative is obvious. It is further interesting to note that in this case all the five suggested alternative materials for split-shots appear to oust Pb.

The dominating influence of the technical and performance indicator is also found in this case, the relative importance of the three indicators TPCr, EnCr and HHCr being equal to 0.5, 0.25, 0.25, respectively.

3.3   The Conservative Approach

A similar set of analyses was carried out for the conservative approach (Table 4), the resulting Hasse diagrams are displayed in Fig. 3.

By comparing Fig.'s 1/3a and 2/3b on can obviously see that the overall trend found for the non-conservative approach is found again here. However, some variations in the relative indicator importance can be noted. Thus, in the case including all four indicators (Fig. 3a) the Cost indicator appears to be the dominating indicator, as the relative importances for TPCr, EnCr, HHCr and Cost were found to 0.273 (TPCr), 0.091 (EnCr), 0.273 (HHCr) and 0.364 (Cost), respectively. Excluding the Cost indicator the TPCr is found again as the dominating indicator as the relative importance for TPCr, EnCr and HHCr are found to 0.4 (TPCr), 0.3 (EnCr) and 0.3 (HHCr), respectively.

The overall picture that Sn appears as the optimal alternative is found again here. Thus, upon inclusion of all four indicators (Fig. 3a) the average ranking was found to be Sn > ste > W > Pb > Bi > cer the average heights being 5.5, 4.3, 4.0, 3.5, 2.0 and 1.7, respectively, whereas excluding the Cost indicator we find Sn > W > ste > Bi > cer > Pb, the average height being 6.0, 4.7, 3.9, 3.1, 1.8 and 1.6, respectively, i.e., the ranking being identical for the conservative and non-conservative approach.

3.4 How Sure Are We on the Ranking?

Looking at the rankings in cases where the Cost indicator is excluded it is clear that Sn turns out as the optimal alternative to Pb split-shots, since Sn is ranked as high as possible (average height = 6.0) as the only maximal element. However, with inclusion of the Cost indicator the picture becomes somewhat more blurred. Thus, decision makers may have a requirement to information concerning how sure we are on the ranking of the alternatives as the ranking presented, based on partial order methodology are average ranking, i.e., the is a finite probability that the single alternatives may take several absolute ranks.

To fulfill such requirements partial order methodology offers several possibilities. In the present study the Bubley-Dyer approach to average ranking (Bubley and Dyer 1999; Bruggemann and Patil 2011) is applied, by which the probabilities for the single elements to have a specific rank are retrieved.

In Table 5 the probabilities for the single alternatives and Pb to have specific ranks are provided. It is seen that although the average ranking placed Sn as the optimal alternative, the probability for Sn to have rank 6 is only 64.7% and to be at rank 5 28.5% while for W to have rank 5 it is at 33.6%. It should here be noted that due to the fact that ste and Pb turn out as elements not comparable to other elements can take all 6 ranks virtually with the same probability around 15–20%.

To further elucidate the of the incomparable alternatives ste, Sn and W to Pb the probabilities for the one element being being ranked higher that another were calculated applying the version 8_3 of the LPOMext module of PyHasse (Bruggemann and Carlsen 2011):

- Sn > Pb: 0.8
- Sn > ste: 0.8
- W > Pb: 0.6
- W > ste: 0.6

Table 5 Probabilities for the six materials to exhibit a specific rank (cf. Fig. 1)

Alt\rank	1	2	3	4	5	6
Pb:	0.158	0.146	0.161	0.173	0.182	0.18
Bi:	0.356	0.371	0.198	0.075	0.0	0.0
cer:	0.335	0.348	0.251	0.066	0.0	0.0
ste:	0.151	0.135	0.162	0.182	0.197	0.173
Sn:	0.0	0.0	0.0	0.068	0.285	0.647
W:	0.0	0.0	0.228	0.436	0.336	0.0

Table 6 Probabilities for the six materials to exhibit a specific rank (cf. Fig. 3a)

	1	2	3	4	5	6
Pb:	0.176	0.14	0.165	0.164	0.188	0.167
Bi:	0.328	0.421	0.21	0.041	0.0	0.0
cer:	0.496	0.376	0.128	0.0	0.0	0.0
ste:	0.0	0.063	0.175	0.293	0.254	0.215
Sn:	0.0	0.0	0.0	0.077	0.305	0.617
W:	0.0	0.0	0.322	0.425	0.253	0.0

A similar analysis was carried out for the conservative approach (Fig. 3a), the resulting calculated probabilities being shown in Table 6. A more or less similar set of probabilities as for the non-conservative approach can be seen. However, one significant difference can be noted due to the fact that ste no longer appear as an isolated element. Thus, the probabilities for ste to have the ranks 4, 5 and 6 are now significantly higher, approx. 30, 25 and 21%, respectively. Simultaneous the probability for W to have rank 5 is reduced to 25.3% and the probability for Sn to have rank 6 is slightly reduced. Pb, as an isolated element can still have all 6 possible ranks with virtually equal probability.

Again the probability relations between Pb and the 3 possible incomparable alternatives (ste, W and Sn) were calculated:

- ste > W: 0.571
- Sn > Pb: 0.8
- Sn > ste: 0.667
- W > Pb: 0.6

4   Conclusions and Outlook

Despite the statement that "MCDA methodsmay be useful in some cases, they may be more complicated than required for many assessments" the present study has shown that partial order methodology is useful in the search for alternatives. Partial order methodology is not specifically complicated and may facilitate assessments.

Initially only the very basics of partial ordering appear necessary as long as a fairly "clear" ordering is obtained, i.e., with a low number of isolated elements. In less clear cases the application of further partial order technics, as here the Bubley-Dyer approach to average ranks and the local partial order approach to mutual probabilities leads to further insights into the ranking, e.g., through the disclosure of probabilities for the single elements to have specific ranks and probability relations between otherwise incomparable elements.

The present study finds Sn (tin) as the optimal alternative and as long as the Cost indicator, including retail price and availability is neglected a pretty clear-cut conclusion. However, apparently the availability and the price of the material play, maybe not surprisingly, a major role that especially in the case of Pb, which is available at rather low prices.

Acknowledgments The author thanks dr. Rainer Bruggemann and dr. Jan W. Owsinski for valuable comments.

References

Bruggemann, R., & Annoni, P. (2014). Average heights in partially ordered sets. *MATCH – Communications in Mathematical and in Computer Chemistry, 71*, 117–142.

Bruggemann, R., & Carlsen, L. (Eds.). (2006a). *Partial order in environmental sciences and chemistry*. Berlin: Springer.

Bruggemann, R., & Carlsen, L. (2006b). Introduction to partial order theory exemplified by the evaluation of sampling sites. In R. Bruggemann & L. Carlsen (Eds.), *Partial order in environmental sciences and chemistry* (pp. 61–110). Berlin: Springer.

Bruggemann, R., & Carlsen, L. (2011). An improved estimation of averaged ranks of partial orders. *MATCH – Communications in Mathematical and in Computer Chemistry, 65*, 383–414.

Bruggemann, R., & Carlsen, L. (2012). Multicriteria decision analyses. Viewing MCDA in terms of both process and aggregation methods: Some thoughts, motivated by the paper of Huang, Keisler and Linkov. *Science of the Total Environment, 425*, 293–295.

Bruggemann, R., & Munzer, B. (1993). A graph-theoretical tool for priority setting of chemicals. *Chemosphere, 27*, 1729–1736.

Bruggemann, R., & Patil, G. P. (2011). *Ranking and prioritization for multi-indicator systems – Introduction*. New York: Springer.

Bruggemann, R., & Voigt, K. (1995). An evaluation of online databases by methods of lattice theory. *Chemosphere, 31*, 3585–3594.

Bruggemann, R., & Voigt, K. (2008). Basic principles of Hasse diagram technique in chemistry. *Combinatorial Chemistry & High Throughput Screening, 11*, 756–769.

Bruggemann, R., Carlsen, L., Voigt, K., & Wieland, R. (2014). PyHasse software for partial order analysis: Scientific background and description of selected modules. In R. Bruggemann, L. Carlsen, & J. Wittmann (Eds.), *Multi-indicator systems and modelling in partial order* (pp. 389–423). Springer: New York.

Bubley, R., & Dyer, M. (1999). Faster random generation of linear extensions. *Discrete Mathematics, 201*, 81–88.

De Loof, K., De Meyer, H., & De Baets, B. (2006). Exploiting the lattice of ideals representation of a poset. *Fundamenta Informaticae, 71*, 309–321.

Ernesti, J., & Kaiser, P. (2008). *Python – Das umfassende Handbuch*. Bonn: Galileo Press.

Hetland, M. L. (2005). *Beginning Python – From novice to professional*. Berkeley: Apress.

Langtangen, H. P. (2008). *Python scripting for computational science*. Berlin: Springer.

NAS. (2014a). A framework to guide selection of chemical alternatives. The National Academies Press at https://abm-website-assets.s3.amazonaws.com/laboratoryequipment.com/s3fs-public/legacyimages/18872_0.pdf

NAS. (2014b). A framework to guide selection of chemical alternatives, report in brief, National Academy of Science, Board on Chemical Sciences and Technology. http://dels.nas.edu/resources/static-assets/materials-based-on-reports/reports-in-brief/Chemical-Alternatives.pdf

Python. (2015). *Python*. https://www.python.org/. Assessed Aug 2018.

Sørensen, P. B., Mogensen, B. B., Carlsen, L., & Thomsen, M. (2000). The influence of partial order ranking from input parameter uncertainty. Definition of a robustness parameter. *Chemosphere, 41*, 595–601.

TURI. (2006). Five chemicals alternatives assessment study, the Massachusetts toxics use reduction Institute, University of Massachusetts Lowell. https://www.turi.org/TURI_Publications/TURI_Methods_Policy_Reports/Five_Chemicals_Alternatives_Assessment_Study._2006

Weigend, M. (2006). *Objektorientierte Programmierung mit Python*. Bonn: mitp-Verlag.

Comparison of Selected Procedures for Generating Activated Carbon with Special Focus on *Miscanthus* Straw as a Sustainable Raw Material

Lars Carlsen and Kamilya Abit

1 Introduction

In a recent paper we reported on the production of activated carbon (AC) from *Miscanthus* straw as a sustainable process to obtain the product as *Miscanthus* straw can be harvested yearly for 15+ years (Abit et al. 2019). The use of *Miscanthus* straw as the basis for AC production is further substantiated since the plant in parallel can be used for soil cleaning as the roots effectively take up heavy metals (Pidlisnyuk et al. 2014; NATO 2017; Nurzhanova et al. 2019; Dias et al. 2007). Since the demand for AC on a global scale is increasing a wide variety of well-documented processes for the production is available cf., e.g. Kundu et al. 2014; Gergova et al. 1994; Bae et al. 2014). Thus, it remains to be analyzed how the process based on *Miscanthus* straw is comparable to already existing processes. The present study compares selected procedures for generated AC with special focus on *Miscanthus* straw applying partial ordering methodology.

1.1 Background: Factors Affecting AC Productions

Some pyrolysis features such as temperature have the most significant effect, as well as retention time, heating rate and nitrogen flow rate.

L. Carlsen (✉)
Awareness Center, Roskilde, Denmark
e-mail: lc@awarenesscenter.dk

K. Abit
al-Farabi Kazakh National University, Almaty, Kazakhstan

© The Author(s), under exclusive license to Springer Nature Switzerland AG 2021
R. Bruggemann et al. (eds.), *Measuring and Understanding Complex Phenomena*,
https://doi.org/10.1007/978-3-030-59683-5_12

In AC production, temperature and activation time play a vital role in influencing the characteristics and properties of the AC produced.

There are processes of physical and chemical activation and simultaneous carbonization and steam/thermal activation. Physical activation includes the stage of carbonization and activation, in which steam and carbon dioxide (CO_2) are the most widely used reagents, significantly affecting the porosity of AC. Thus Gergova et al. (1992) studied the porous structure of activated carbons from agricultural by-products, and Alcaniz-Monge et al. (2012) prepared activated carbon fibres by steam or carbon dioxide activation. The AC generation using chemical activation involves a step in which chemicals, such as potassium hydroxide, phosphoric acid, zinc chloride or other chemicals, can simply be used at room temperature (Menéndez-Díaza and Martín-Gullónb 2006). However, depending on the chemicals used, impurities such as zinc (Zn) and phosphorus (P) can be detected in the final AC product, which at the same time may lead to an increase in AC cost by adding the chemical used and purifying the reaction product from impurities. Such chemical additives for AC activation are potentially harmful to human health and the environment and also significantly increases the cost of the final AC product compared to physical activation, for example, water vapor (cf. Abit et al. 2019) Therefore, physical activation a priori appears as the more appropriate method of activation to prevent environmental pollution and reduce the cost of the final product, while obtaining relatively high surface area of AC.

For commercial purposes, coal activation is usually carried out in a mixture of steam and CO_2 at temperatures above 800 °C. Recently, studies have be reported aiming to optimize the final activation temperature in order to reduce the cost and duration of AC production (cf. Abit et al. 2019). Several studies have been reported that the activation temperature has a large effect on the surface area and the yield of AC (Chowdhury et al. 2011; Wang et al. 2017; Baçaoui et al. 2001). The activation temperature varies from 200 to 1100 °C. However, temperature range from 400 to 500 °C is often used for chemical activation and higher temperatures (800–1000 °C) for physical activation (cf. Abit et al. 2019). It should also be noted that during chemical activation, the processing time of preparing AC is significantly increased due to the long period for creating complete impregnation of raw materials with chemical reagents. According to previously obtained research results, with increasing activation time, the surface area according to the method of Brunauer, Emmett and Teller (BET) (Thommes et al. 2015) gradually increases, while the yield of AC decreases (Baçaoui et al. 2001). This may be due to the volatilization of organic substances from agricultural raw materials.

The yield is an additional indicator that is studied during AC production. The greater the yield of the carbonization-activation reaction, obviously the more productive and cost-effective the technology for AC production will be.

The so-called BET surface area (Thommes et al. 2015) is another important feature, showing the influence of production conditions on the characteristics of the resulting AC. Basically, the surface area of BET increases with increasing activation temperature. This may be due to the advancement of new pores due to the release of volatile matter and the expansion of pores.

Obviously a large surface area is a priori preferable in order to increase the sorption capacity.

Given all the factors affecting the cost of the technology and the final product, as well as the impact on the environment and health, it is necessary to choose the best options for the ratio of the cost of the final product (AC), product yield, sorption properties, and harmfulness of production.

The maximum adsorption capacity of AC, which obviously is of major interest, appears largely to be dependent on the structure of the raw material and the processes of its production.

2 Methods

The study includes 21 different methods for producing activated carbon. The methods comprise various starting material as well as both chemical and physical activation of the initially produced carbon material.

Various materials with high carbon content can be used as raw materials for the production of AC (Ioannidou and Zabaniotou 2007). Some of the widely used starting materials are agro-industrial by-products that are characterized by their renewability, high mechanical strength, low cost, abundance, and low ash content. Hence, various studies report the use of biomass residues from agricultural waste in AC production, such as coconut shells (Laine et al. 1989; Laine and Yunes 1992; Boopathy et al. 2013: Lopez et al. 1996), tropical wood (Hayashi et al. 2000a; Janoš et al. 2009; Phan et al. 2006), jute (Giraldo and Moreno-Piraján 2008), cane sugar bagasse (Foo et al. 2013), walnut shells (Yang and Qiu 2010), Also non-agricultural products, such as phenol-formaldehyde resins (Teng and Wang 2000), bituminous coal (Hsu and Teng 2000), have been reported as sources of AC.

2.1 Data

The data applied in the study have been retrieved from available literature including data from our recent paper on AC from *Miscanthus* straw (Abit et al. 2019).

It should be noted that for some procedures, e.g., that using *Miscanthus* straw both a carbonization process and an activation process are part of the overall AC production, whereas in other procedures only one carbonization/activation process is involved.

To assess the efficiency of the methods for obtaining AC, four specific indicators were selected:

1. The temperature of activation (TempA) that is believed to reflect the energy consumption for the AC production. Thus, the lower the TempA the lower the necessary energy consumption. Note that the negative values of TempA are used

in order to have the same orientation for all indicators, i.e. the higher the better (Bruggemann and Carlsen 2006; Bruggemann and Patil 2011)

2. The eventual yield of AC (Yield)
3. The surface area of the produced AC (Surf) as an indicator for the sorption characteristics of the AC
4. The method of activation (MoA) that can be either physical, i.e., without applying any chemicals (apart from possibly water/steam) or chemical, where specific chemicals are applied as part of the activation process. As itis believed that avoiding using chemicals is preferable. Thus, the indicator values for chemical and physical activation are set to 1 and 2, respectively.

Table 1 presents the characteristics of some studies of various materials used in the production of AC, as a result of which a comparative characteristic was compiled for various methods of obtaining AC.

2.2   Partial Ordering

Partial ordering appears in its basis pretty simple as the only mathematical relation among the elements is "$\leq$" (Bruggemann and Carlsen 2006; Bruggemann and Münzer 1993; Bruggemann and Patil 2011). The basis for a comparison of elements is a group of indicators (vide infra) that characterizes the elements. This series of indicators, r_j, is selected in order to be characteristic for the processes studied, i.e., the preparation of activated carbon. Thus, characterizing one of the methods for AC production (x) by the a set of indicators $r_j(x)$, $j = 1, \ldots, m$, where m is the number of indicators, can be compared to another method (y), characterized by the indicators $r_j(y)$, when

$$r_j(y) \leq r_j(x) \text{ for all } j = 1, \ldots, m \tag{1}$$

Equation 1 is a very hard and strict requirement for establishing a comparison. It demands that all indicators of method x should be better (or at least equal) than those of method y. Further, let X be the group of methods included in the analyses, x will be ordered higher (better) than y, i.e., x > y, if at least one of the indicator values for x is higher than the corresponding indicator value for y and no indicator for x is lower than the corresponding indicator value for y. On the other hand, if $r_j(x) > r_j(y)$ for some indicator j and $r_i(x) < r_i(y)$ for some other indicator i, the methods x and y will be denoted incomparable (notation: x | y) expressing the mathematical contradiction due to conflicting indicator values. A set of mutual incomparable elements is called an antichain. When all indicator values for x are equal to the corresponding indicator values for y, i.e., $r_j(x) = r_j(y)$ for all j, the two methods will have identical rank and will be considered as equivalent, i.e., x ~ y. The analysis of Eq. 1 results in a graph,

Table 1 Indicator values for the 21 method included in the evaluation

№	Method	Method	Indicators			
			1: TempA, °C	2: Yield, %	3: Surf, m²/g	4: MoA
1	M1	Laine et al. 1989	−400	45	1180	1
2	M2a	Lopez et al. 1996	−800	15	620	1
3	M2b	Lopez et al. 1996 Chemical activation with H3PO4 (40%).	−450	40	1450	1
4	M2c	Lopez et al. 1996 Chemical activation with ZnCl2 (25%).	−450	40	1550	1
5	M3a	Laine and Yunes 1992 Physical activation (with CO_2 at 800 °C 3–4 h.)	−800	20	751	1
6	M3b	Laine and Yunes 1992 C3: Chemical activation (with H_3PO_4 30 wt% acid)	−500	30	1360	1
7	M4a	Pastor-Villegas et al.1994 Activation: Air	−750	60	439	2
8	M4b	Pastor-Villegas et al. 1994 Activation: CO_2	−950	60	650	1
9	M4c	Pastor-Villegas et al. 1994 Activation: H_2O	−950	60	759	2
10	M5a	Phan et al. 2006 Activation: CO_2	−950	7	912	1
11	M5b	Phan et al. 2006 Chemical activation: H_3PO_4	−900	33	959	1
12	M6a	Giraldo and Moreno-Piraján 2008 Chemical activation using 50% HNO_3	−900	46	868	1
13	M6b	Giraldo and Moreno-Piraján 2008 Chemical activation using 50% HNO_3	−900	46	967	1
14	M7a	Hayashi et al. 2000a Chemical activation: $ZnCl_2$	−500	60	1000	1
15	M7b	Hayashi et al. 2000a Chemical activation: H_3PO_4	−500	60	700	1
16	M7c	Hayashi et al. 2000a Chemical activation: KOH	−500	65	250	1
17	M8a	Hsu and Teng 2000 Chemical activation: $ZnCl_2$	−600	69	960	1
18	M8b	Hsu and Teng 2000 Chemical activation: $H_3 PO_4$	−600	80	770	1
19	M8c	Hsu and Teng 2000 Chemical activation: KOH	−600	45	1800	1
20	M9	Teng and Wang 2000 Physical activation: 900 °C KOH impregnation	−900	12	2220	1
21	M10	Abit et al. 2019 Activation with H_2O at 800 °C, 60 min.	−800	28	542	2

the so-called Hasse diagram. Hasse diagrams are unique visualizations of the order relations due to Eq. 1.

2.2.1 The Hasse Diagram

The Eq. 1 is the basic for the Hasse diagram technique (HDT) (Bruggemann and Carlsen 2006; Bruggemann and Münzer 1993; Bruggemann and Patil 2011). Hasse diagrams are visual representation of the partial order. In the Hasse diagram comparable elements are connected by a sequence of lines (Bruggemann and Carlsen 2006; Bruggemann and Münzer 1993; Bruggemann and Patil 2011).

By convention, the Hasse diagram originally introduced by Halfon and Reggiani (1986) is drawn with.

- $x \leq y$ locating x below y,
- attempting a symmetric presentation as far as possible and
- by an arrangement of elements in levels that are numbered from the bottom and upwards.
- each element is placed at the highest possible level in the diagram as possible

Two important concepts, in addition to the level structure can directly obtained by inspecting a Hasse diagram:

- Chains are subsets of X, where each element is mutually comparable to the others. Those subsets are denoted as 'completely ordered': Inspecting a Hasse diagram, any sequence of lines upwards or (strictly) downwards is a chain
- Antichains are sets, where each element is mutually incomparable with the others. Hence, levels are subsets of the set of all antichains
- Maximal elements are elements where for a given element x there is no elements y where $x \leq y$
- Minimal elements are elements where for a given element x there is no elements y where $y \leq x$
- If x is at the same time a maximal and a minimal element, then x is called an *isolated element*. Isolated elements are always of interest as they must have a special data structure, which makes them incomparable to any other element of X.

For a detailed explanation see the work by Bruggemann and Patil (2011).

2.3 The More Elaborate Analyses

In addition to the basic partial ordering tools some more elaborate analyses have been used including average ranks (Bubley and Dyer 1999; Bruggemann et al. 2004; De Loof et al. 2006; Bruggemann and Patil 2011, Bruggemann and Carlsen 2011; Bruggemann and Annoni 2014) and sensitivity analysis (Bruggemann and Patil

2011; Bruggemann et al. 2014), the latter gives an insight in the relative importance of the included indicators (Bruggemann and Patil 2011; Bruggemann et al. 2014) and antichain analyses (Bruggemann and Voigt 2011).

2.3.1 Average Ranks

The average ranking is expressed as average height from bottom (min. Height $= 1$) to the top (max height $= n$, i.e., the maximum number of objects, here $n = 21$) (Bruggemann and Annoni 2014). The average rank is generated by calculating all linear order preserving sequences (set LE), the linear extensions of the original partial order. From LE_0 the statistical characterization for each object is obtained. For example the characterization is calculated as the average value an object has, taken all positions of this object within LE_0, the averaged heights. It is clear that this procedure is computationally extremely difficult. Hence, approximations were developed (Bruggemann et al. 2004, Bruggemann and Carlsen 2011).

2.3.2 Sensitivity Analysis

For the sentivity analysis (Bruggemann et al. 2001; Bruggemann and Patil 2011; Bruggemann et al. 2014), let Q be the set of all indicators, then taken all indicators of Q leads to a partial order, which is called PO_0. The corresponding set of linear extensions is denoted by LE_0. Leaving out one indicator of Q, say r_j, then another partial order results, which is denoted as PO_j.

Both partial orders can be described by an adjacent matrix, say A_0 for PO_0 and A_j for PO_j.

Taken the Euclidian Distance (squared) quantifies the role of indicator qj in PO_0. This is a sensitivity measure for the indicators of set Q, describing the structural changes of the partial order leaving one indicator out. This is not immediately a measure of the sensitivity of the indicators for a ranking, because the ranking is per definition a linear order and here derived over many interim steps.

If a linear order is obtained by all orders in LE_0, the set of linear extensions taken from PO_0, then any PO_j will also lead to a corresponding set LE_j. And this set is the more differing from LE_0 the larger the sensitivity is. Therefore the ranking due to averaged heights is as more affected by indicator r_j as larger its sensitivity is.

2.3.3 Indicator Conflicts – Tripartite Graphs

In order visually to display and thus better understand the role of individual indicators for incomparisons, the concept of tripartite graph was introduced by Bruggemann and Voigt (2011). Here an intuitive approach is presented again assuming a case with three indicators. Imagine that Objx has better values (i.e. higher

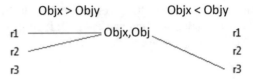

Fig. 1 Example of a tripartite graph

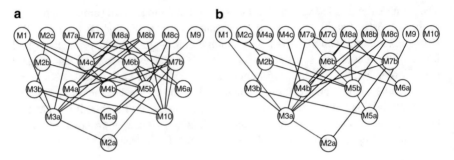

Fig. 2 Hasse diagrams resulting in simultaneous inclusion of A: indicators 1–3 and B: all indicators

values) in comparison to Objy in the first and second indicator, but worse value (i.e. lower value) in the third indicators. This fact can be graphically represented as follows (Fig. 1).

Tripartite graphs may be obtained using the module antichain20_4 from the PyHasse software package.

2.4 Software

All partial order analyses were carried out using the PyHasse software (Bruggemann and Patil 2011; Bruggemann et al. 2014). PyHasse is programmed using the interpreter language Python (version 2.6) (Ernesti and Kaiser 2008; Hetland 2005; Langtangen 2008; Weigend; 2006; Python 2015). Today, the software package contains more than 100 modules and is available upon request from the developer, Dr. R. Bruggemann (brg_home@web.de).

3   Results and Discussion

Two attempts to partially order the 21 different methods included in the parent evaluation were performed: A) disregarding any possible influence of the method of activation, i.e., including only indicators 1–3, and B) including all 4 indicators. In Fig. 2 the resulting Hasse diagrams are shown.

Table 2 Averaged ranking of the 21 methods for producing activated carbon. A: indicators 1–3 and B: all indicators

A			B		
Method	Rkav	Rank	Method	Rkav	Rank
M7a	2.164	1	M7a	2.725	1
M8a	2.968	2	M2c	3.369	2
M2c	3.133	3	M8a	3.927	3
M8b	4	4	M1	5.733	4.5
M1	5.364	5.5	M8c	5.733	4.5
M8c	5.364	5.5	M8b	5.85	6
M2b	6.266	7	M2b	6.738	7
M7b	6.971	8	M6b	7.933	8
M6b	7.533	9	M4c	8	9
M9	8.833	10	M9	8.833	10
M3b	10.032	11	M7b	9.267	11
M7c	11	12	M3b	10.921	12
M4c	12.85	13	M4a	11	14
M6a	16.15	14	M7c	11	14
M5b	16.588	15	M10	11	14
M4a	16.717	16	M6a	16.55	16
M3a	16.786	17	M5b	16.988	17
M4b	17.983	18	M3a	17.408	18
M10	18.913	19	M4b	18.364	19
M2a	19.581	20	M5a	19.881	20
M5a	19.731	21	M2a	19.901	21

It is immediately noted that in case A, i.e. disregarding the actual method of activation the method M10 that is the AC production from *Miscanthus* straw) Abit et al. 2019) is located at the second level (counted from the bottom) covered by M1, M3b, M7b, M8a, M8b, and M8c, respectively, whereas M10 is not covering any other method and is thus a minimal element. Note that the location of M10, which is actually a minimal element, in level 2,is a result of the convention placing the single elements as high as possible in the Hasse diagram (cf. Sect. 2.2.1). In contrast to this it is seen that in case B, i.e. including all four indicators M10 now appears as an isolated element that, again by convention is located at level 5.

Based on these remarks it is obvious that a more precise location of M10 relative to the other 20 methods is not possible only by inspecting the Hasse diagrams in Fig. 2 To get a better insight in the relative ranking of the different method we estimated (applying the module LPOMext8_3 of the PyHasse software; Bruggemann and Carlsen 2011) the so-called averaged rank with 1 as the most optimal method and 21 as the least – based on the included indicators (cf. Table 1). In Table 2 the average ranking of the 21 methods corresponding to the two Hasse diagram shown in Fig. 1 is give. The Rkav values are the calculated ranking values that subsequently is transformed to the '1 – 21' ranking.

In both cases, the production of AC from *Miscanthus* straw (M10) is apparently not one of the best choices. Hence, disregarding the method of activation (Table 2A)

Table 3 Calculated probabilities for method M10 being ranking higher than the other 20 methods

Comparison	P
M10 > M1	0.167
M10 > M2a	0.917
M10 > M2b	0.250
M10 > M2c	0.125
M10 > M3a	0.818
M10 > M3b	0.429
M10 > M4a	0.500
M10 > M4b	0.857
M10 > M4c	0.333
M10 > M5a	0.917
M10 > M5b	0.800
M10 > M6a	0.800
M10 > M6b	0.333
M10 > M7a	0.100
M10 > M7b	0.400
M10 > M7c	0.500
M10 > M8a	0.125
M10 > M8b	0.200
M10 > M8c	0.167
M10 > M9	0.333

Table 4 Relative indicator importance. A: indicators 1–3 and B: all indicators (cf. Table 1)

Indicator	A Relative importance	B Relative importance
TempA	0.147	0.126
Yield	0.412	0.396
Surf	0.441	0.440
MoA		0.101

we find the M10 method at rank 19, whereas applying all four indicators M10 is found at rank 14 (Table 2B). In the latter case it must be remembered that since the M10 is an isolated element the ranking is rather uncertain. To get a further insight in the ranking of M10, the actual probabilities for M10 being ranked higher that the other methods is shown in Table 3.

From Table 3 it is immediate seen that, looking at probabilities higher than 0.5, M10 is indeed located higher than the six methods M2a. M3a, M4b, M5a, M5b, M6a that are all located below M10 in the calculated averaged ranking (Table 2). Hence, these data substantiate the estimated ranking of M10 to 14 is realistic.

The relative importance of the single indicators was studied applying the sensitivity24_5 module of the PyHasse software (Bruggemann et al. 2001; Bruggemann and Patil 2011) (Table 4).

It is immediately noted that in both cases the yield and the surface area are by far the most important indicators with virtually identical importance, whereas the temperature of activation and, in case B the method of activation play only minor roles.

Remaining to be discussed is the reason that M10 in case B appear as an isolated element. This can advantageously be done applying the tripartite tool (sepanalcoloured17_0 module of the PyHasse software; Bruggemann and Voigt 2011). The tripartite graph is a simple visualization of any indicator conflict. Thus, in the present case the element x is the method M10 that is compared to the remaining methods. A line from an indicator to M10 visualized that for M10 this indicator is higher than the same indicator for the method y and vice versa. In Fig. 3 is shown the comparisons of M10 to the remaining 20 method, the comparisons being done level wise.

It immediate seen that several indicator conflicts prevail at all 5 levels. An important factor to be mentioned is the role of the MoA indicator that virtually in all cases are higher for M10 than for the other methods. However, as it is obvious also for the other indicators conflicts prevail at all levels. Thus, as an simple example is level 1, where the indicators MoA and Yield are higher for M10 than for M2a, whereas the indicator Surf is higher for M2a than for M10, thus constituting a conflict and thus an incomparability.

4   Conclusions and Outlook

The importance of AC should obviously not be underestimated, since it satisfies the needs for an adsorbent for the purification of liquid, gas and solid substances. Thus, simple and cost-effective ways to increase the production of AC of high quality are crucial.

In general, we can conclude that the method based on *Miscanthus* straw (M10) obviously is not a specifically efficient way for producing activated carbon. Nevertheless, the following conclusions can be drawn in favor of M10:

1. Activated carbons are prepared using plant material, which is a renewable, fast-growing, perennial plant. This plant annually brings high biomass growth rates (Nsanganwimana et al. 2014). Thus, the method is sustainable.
2. The material shows good sorption capacities to ions of heavy metals and organic substances from the soil (absorption by roots) (Pidlisnyuk et al. 2014; Reddad et al. 2003). In Kazakhstan, work is underway in the frame of an international NATO project to clean Kazakhstan's contaminated soils using this plant (Nurzhanova et al. 2019). Therefore, there is a very high prospect of growing *Miscanthus,* using the root system to clean the soil, and the aboveground part of the biomass converted to activated carbon to clean water bodies. Also, the use of the aboveground part as a cheap raw material for producing AC is economically viable as the disposal of aboveground biomass.
3. Vast unused areas in Kazakhstan can be used for growing *Miscanthus. Miscanthus* material does not require complex pre-treatment, unlike, e, e.g., coconuts

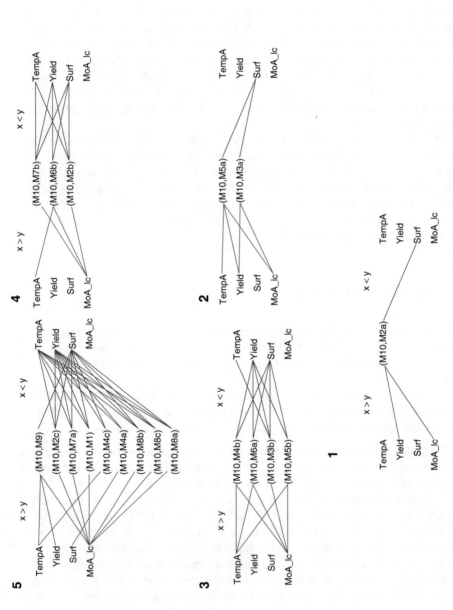

Fig. 3 Tripartite graphs visualizing the indicator conflicts in the comparisons between M10 and the other 20 methods. The comparison is done level wise, the level numbers being giving (compare to Fig. 1)

and apricot kernels. In addition, coconuts are not possible for use in Kazakhstan, since there is no raw material base and as such a relatively expensive raw material.(Antoszczyszyn and Michalska 2016).

4. Although apricot kernels and seeds of other fruit trees a priori look like afdvantageous starting materials thiks material may secrete a large amount of HCN. Further, as a rule, when obtaining AC from the seeds of various fruits, chemical methods are used for activation, which is harmful both for a person working in production and for the environment as a whole.

5. The preparation of AC from traditional raw materials such as bituminous coal, phenol-formaldehyde resins is an economically expensive and environmentally harmful production. This raw material is non-renewable, and its extraction is a laborious and harmful process.

6. The use of wood as a raw material for the production of AC is a laborious process, as it often requires grinding and chemical preparation of the raw material for carbonization. The main task of replacing wood with more accessible material is that a long period is needed to restore and grow large-scale forests, which are cut down annually as a result of the widespread use of wood.

Cost-effective ACs made of new materials are supposed to find its application in various industries, such as processes for cleaning of wastewater, air, and the environment as a whole. Hence, despite the low ranking of the method based on *Miscanthus* straw, this method should obviously not be left out of consideration.

References

Abit, K. E., Carlsen, L., Nurzhanova, A. A., & Nauryzbaev, M. K. (2019). Activated carbons from *Miscanthus* straw for cleaning water bodies in Kazakhstan. *Eurasian Chemico-Technological Journal, 21*, 259–267.

Alcaniz-Monge, J., Perez-Cadenas, M., & Marco-Lozar, J. P. (2012). Removal of harmful volatile organic compounds on activated carbon fibres prepared by steam or carbon dioxide activation. *Adsorption Science and Technology, 30*, 473–482.

Antoszczyszyn, T., & Michalska, A. (2016). The potential risk of environmental contamination by mercury contained in Polish coal mining waste. *Journal of Sustainable Mining, 15*, 191–196.

Baçaoui, A., Yaacoubi, A., Dahbi, A., Bennouna, C., Luu, R. P. T., Maldonado-Hodar, F. J., Rivera-Utrilla, J., & Moreno-Castilla, C. (2001). Optimization of conditions for the preparation of activated carbons from olive-waste cakes. *Carbon, 39*, 425–432.

Bae, W., Kim, J., & Chung, J. (2014). Production of granular activated carbon from food-processing wastes (walnut shells and jujube seeds) and its adsorptive properties. *Journal of the Air & Waste Management Association (1995), 64*, 879–886.

Boopathy, R., Karthikeyan, S., Mandal, A. B., & Sekaran, G. (2013). Adsorption of ammonium ion by coconut shell-activated carbon from aqueous solution: Kinetic, isotherm, and thermodynamic studies. *Environmental Science and Pollution Research, 20*, 533–554.

Bruggemann, R., & Annoni, P. (2014). Average heights in partially ordered sets. *MATCH – Communications in Mathematical and in Computer Chemistry, 71*, 117–142.

Bruggemann, R., & Carlsen, L. (Eds.). (2006). *Partial order in environmental sciences and chemistry*. Berlin: Springer.

Bruggemann, R., & Carlsen, L. (2011). An improved estimation of averaged ranks of partially orders. *MATCH – Communications in Mathematical and in Computer Chemistry, 65*, 383–414.

Bruggemann, R., & Münzer, B. (1993). A graph-theoretical tool for priority setting of chemicals. *Chemosphere, 27*, 1729–1736.

Bruggemann, R., & Patil, G. P. (2011). *Ranking and prioritization for multi-indicator systems – Introduction to partial order applications*. New York: Springer.

Bruggemann, R., & Voigt, K. (2011). A new tool to analyze partially ordered sets - application, ranking of polychlorinated biphenyls and alkanes/alkenes in river main, Germany. *MATCH – Communications in Mathematical and in Computer Chemistry, 66*, 231–251.

Bruggemann, R., Halfon, E., Welzl, G., Voigt, K., & Steinberg, C. (2001). Applying the concept of partially ordered sets on the ranking of near-shore sediments by a battery of tests. *Journal of Chemical Information and Computer Sciences, 41*, 918–925.

Bruggemann, R., Sørensen, P. B., Lerche, D., & Carlsen, L. (2004). Estimation of averaged ranks by a local partial order model. *Journal of Chemical Information and Computer Sciences, 2004*(44), 618–625.

Bruggemann, R., Carlsen, L., Voigt, K., & Wieland, R. (2014). PyHasse software for partial order analysis: Scientific background and description of selected modules. In R. Bruggemann, L. Carlsen, & J. Wittmann (Eds.), *Multi-indicator systems and modelling in partial order* (pp. 389–423). Springer: New York.

Bubley, R., & Dyer, M. (1999). Faster random generation of linear extensions. *Discrete Mathematics, 201*, 81–88.

Chowdhury, Z. Z., Zain, S. M., Khan, R. A., Ahmad, A. A., Islam, M. S., & Arami-niya, A. (2011). Application of central composite design for preparation of Kenaf fiber based activated carbon for adsorption of manganese (II) ion. *The International Journal of Physical Sciences, 6*, 7191–7202.

De Loof, K., De Meyer, H., & De Baets, B. (2006). Exploiting the lattice of ideals representation of a Poset. *Fundamenta Informaticae, 71*, 309–321.

Dias, J. M., Alvim-Ferraz, M. C. M., & Almeida, M. F. (2007). Waste materials for activated carbon preparation and its use in aqueous-phase treatment: A review. *Journal of Environmental Management, 85*, 833–846.

Ernesti, J., & Kaiser, P. (2008). *Python-Das umfassende Handbuch*. Bonn: Galileo Press.

Foo, K. Y., Lee, L. K., & Hameed, B. H. (2013). Preparation of activated carbon from sugarcane bagasse by microwave assisted activation for the remediation of semi-aerobic landfill leachate. *Bioresource Technology, 134*, 166–172.

Gergova, K., Galushko, A., Petrov, N., & Minkova, V. (1992). Investigation of the porous structure of activated carbons prepared by pyrolysis of agricultural by-products in a stream of water-vapor. *Carbon, 30*, 721–727.

Gergova, K., Petrov, N., & Eser, S. (1994). Adsorption properties and microstructure of activated carbons produced from agricultural by-products by steam pyrolysis. *Carbon, 32*, 693–702.

Giraldo, L., & Moreno-Piraján, J. C. (2008). Pb2+ adsorption from aqueous solutions on activated carbons obtained from lignocellulosic residues. *Brazilian Journal of Chemical Engineering, 25*, 143–151.

Halfon, E., & Reggiani, M. G. (1986). On ranking chemicals for environmental hazard. *Environmental Science & Technology, 20*, 1173–1179.

Hayashi, J., Kazehaya, A., Muroyama, K., & Watkinson, A. P. (2000a). Preparation of activated carbon from lignin by chemical activation. *Carbon, 38*, 1873–1878. https://doi.org/10.1016/S0008-6223(00)00027-0.

Hetland, M. L. (2005). *Beginning Python - From novice to professional*. Berkeley: Apress.

Hayashi, J., Kazehaya, A., Muroyama, K., & Watkinson, A. P. (2000b). Preparation of activated carbon from lignin by chemical activation. *Carbon, 38*, 1873–1878. https://doi.org/10.1016/S0008-6223(00)00027-0.

Hsu, L.-Y., & Teng, H. (2000). Influence of different chemical reagents on the preparation of activated carbons from bituminous coal, *Fuel Processing Technology, 64*, 155–166.

Ioannidou, O., & Zabaniotou, A. (2007). Agricultural residues as precursors for activated carbon production: A review. *Renewable and Sustainable Energy Reviews, 1*, 1966–2005.

Janoš, P., Coskun, S., Pilařová, V., & Rejnek, J. (2009). Removal of basic (methylene blue) and acid (Egacid Orange) dyes from waters by sorption on chemically treated wood shavings. *Bioresource Technology, 100*, 1450–1453.

Kundu, A., Redzwan, G., Sahu, J. N., Mukherjee, S., Gupta, B. S., & Hashim, M. A. (2014). Hexavalent chromium adsorption by a novel activated carbon prepared by microwave activation. *Bio Resources, 9*, 1498–1518.

Laine, J., Calafat, A., & Labady, M. (1989). Preparation and characterization of activated carbons from coconut shell impregnated with phosphoric acid. *Carbon, 27*, 191–195.

Laine, J., & Yunes, S. (1992). Effect of the preparation method on the pore size distribution of activated carbon from coconut shell. *Carbon, 30*, 601–604.

Langtangen, H. P. (2008). *Python scripting for computational science*. Berlin: Springer.

Lopez, M., Labady, M., & Laine, J. (1996). Preparation of activated carbon from wood monolith. *Carbon, 34*, 825–827.

Menéndez-Díaza, J. A., & Martín-Gullónb, I. (2006). Types of carbon adsorbents and their production, in: Activated carbon surfaces in environmental remediation, T. Bandosz, ed. *Interface Science and Technology, 7*, 1–47.

NATO. (2017). *New phytotechnology for cleaning contaminated military sites*. Project of NATO G 4687, http://ipbb.kz/eng/2017/01/27/project-g4687-new-phytotechnology-for-cleaning-contaminated-military-sites-is-granted-to-the-institute-as-multy-year-research-project-by-nato-science-for-peace-and-security-program/

Nsanganwimana, F., Pourrut, B., Mench, M., & Douay, F. (2014). Suitability of Miscanthus species for managing inorganic and organic contaminated land and restoring ecosystem services. A review. *Journal of Environmental Management, 143*, 123–134.

Nurzhanova, A., Pidlisnyuk, V., Abit, K., Nurzhanov, C., Kenessov, B., Stefanovsk, T., & Ericson, L. (2019). Comparative assessment of using *Miscanthus × giganteus* for remediation of soils contaminated by heavy metals: A case of military and mining sites. *Environmental Science and Pollution Research, 26*, 13320–13333.

Pastor-Villegas, J., Valenzuela-Calahorro, C., & Gomez-Serrano, V. (1994). Preparation of activated carbon from rockrose char. Influence of activation temperature. *Biomass and Bioenergy, 6*, 453–460.

Phan, N. H., Rio, S., Faur, C., Le Coq, L., Le Cloirec, P., & Nguyen, T. H. (2006). Production of fibrous activated carbons from natural cellulose (jute, coconut) fibers for water treatment applications. *Carbon, 44*, 2569–2577.

Pidlisnyuk, B., Erickson, L., Kharchenko, S., & Stefanovska, T. (2014). Sustainable land management: Growing miscanthus in soils contaminated with heavy metals. *Journal of Environmental Protection, Special Issue in Environmental Remediation, 5*, 723–730.

Python. (2015). *Python*, https://www.python.org/

Reddad, Z., Gerente, C., Andres, Y., & Le Cloirec, P. (2003). Mechanisms of Cr(III) and Cr(VI) removal from aqueous solutions by sugar beet pulp. *Environmental Technology, 24*, 257–264.

Teng, H., & Wang, S.-C. (2000). Preparation of porous carbons from phenol-formaldehyde resins with chemical and physical activation. *Carbon, 38*, 817–824.

Thommes, M., Kaneko, K., Neimark, A. V., Olivier, J. P., Rodriguez-Reinoso, F., Rouquerol, J., & Sing, K. S. W. (2015). Physisorption of gases, with special reference to the evaluation of surface area and pore size distribution (IUPAC Technical Report). *Pure and Applied Chemistry, 87*, 1051–1069.

Yang, J., & Qiu, K. (2010). Preparation of activated carbons from walnut shells via vacuum chemical activation and their application for methylene blue removal. *Chemical Engineering Journal, 165*, 209–217.

Wang, B., Gaob, B., & Fang, J. (2017). Recent advances in engineered biochar productions and applications. *Critical Reviews in Environmental Science and Technology, 47*, 2158–2207.

Weigend, M. (2006). *Objektorientierte Programmierung mit Python*. Bonn: mitp-Verlag.

Uranium Trappers, a Partial Order Study

Nancy Y. Quintero

1 Introduction

Water, soils and subsurface contamination by radionuclides as uranium (U) is a worldwide problem (McCullough et al. 2003); over a long period, U accumulation has led to health risks on humans and to deleterious effects on ecosystems (Prakash et al. 2013). From the chemical point of view, U as well as other actinides, have a pronounced tendency to form complexes with other elements (Markich 2002). For example, one of their important features is that in acidic waters, U (VI) dissolves and forms soluble complexes, facilitating its spread in ground waters (Markich 2002). Although U is a primordial element found in the earth crust (El-Taher et al. 2004), its long half-life, radioactive decay of their daughters and their toxicity, even at low concentrations, have increased its potential as ecological and public health hazards (Committee on Uranium Mining in Virginia; Committee on Earth Resources; National Research Council. Uranium Mining in Virginia: Scientific, Technical, Environmental, Human Health and Safety, and Regulatory Aspects of Uranium Mining and Processing in Virginia. Washington (DC): National Academies Press (US) 2011). Due to these reasons, the final disposal of nuclear wastes has been a matter of concern to the International Atomic Energy Agency (IAEA), scientists and energy companies worldwide.

N. Y. Quintero (✉)
Secretaría de Educación Municipal, Colegio Santos Apóstoles, Cúcuta, Colombia

CHIMA, Mathematical Chemistry Group, Universidad de Pamplona, Pamplona, Colombia

Corporación Colombiana del Saber Científico (Bogotá), SCIO, Bogotá, Colombia
e-mail: yaneth.quintero@udea.edu.co

© The Author(s), under exclusive license to Springer Nature Switzerland AG 2021
R. Bruggemann et al. (eds.), *Measuring and Understanding Complex Phenomena*,
https://doi.org/10.1007/978-3-030-59683-5_13

Contamination of soil and water with U from mining and tailing activities has been extensively treated using conventional methods such as membranes technology, electrolysis, ion exchange, solvent extraction, chemical precipitation and adsorption (Li and Zhang 2012). However, these available treatment technologies are either not effective enough or are very expensive and inadequate especially when treating a large amount of waste waters containing U at low concentration (1 to 100 mg/L, concentrations commonly found in radioactive wastes from) (Li and Zhang 2012; Volesky 2001). These facts have limited the use of these methods at large scale (Li and Zhang 2012). Likewise, these shortcomings led to a search for more eco-friendly approaches like biotechnological methods; these approaches based on the use of biosorbents have emerged in the last decade as one of the most promising cost-effective alternatives (Volesky and Holan 1995). Their advantage relies on the ability of biosorbents, i.e., organisms such as bacteria, fungi or algae, to trap and immobilise the metals by mechanisms such as biosorption (Fourest and Roux 1992). These organisms have been highlighted as potential accumulators having high U uptake capacity.

Once, biosorbents have been used, to avoid spread of radioactive biomass with U adsorbed, it is recommended its suitable treatment for final disposal as in the case of other nuclear wastes (Volesky 2001). The result is helping to the process of ecosystems restoration and to recover the radioactive metals like uranium (McCullough et al. 2003). Compared to conventional physicochemical methods assessed taking into account influencing factors such as pH, presence or absence of carbonates in waste waters, biosorbents are cheaper and attractive because of their low operating cost and high efficiency (Wang and Chen 2009).

Their performance has been assessed taking into account attributes such as pH, metal concentration in solution, biomass concentration, age of biomass, temperature, percentage of metal removal, time requested for removing the metal input and uptake capacity (Yi and Lian 2012). Because this work deals with the biotechnological issue of determining which microorganisms could be better as U trappers in aqueous systems, a comparison is required.

An important point for performing this task was to collect an appropriate methodology. Due to the rapid increase of experimental information coming from many biotechnological studies dealing with U trappers, there are many data in the literature. Specifically, in the biosorption field of U, several studies propose the suitability of biosorbents as U trappers (Wang and Chen 2009); nevertheless, it is not known which are the best U bioaccumulator in aqueous solutions.

Since several attributes affect the biosorption process and they should be simultaneously taken into account, the search of a suitable methodology turns towards multi-criteria decision analysis methods (MCDA) (Lerche et al. 2002). These MCDA approaches do not only collect and extract relevant information from the pair-wise comparisons of attributes but also perform data analysis aiming to screen, to assign priorities and to rank objects (Fattore and Bruggemann 2017). These tasks are also in the focus of the current work, where the comparison of attributes characterising microorganisms should be carried out. In this regard, if the MCDA approach considers a ranking problem, it offers the results in the form of a total or partial ranking, which reflects that no unique ranking methodology could be the best (Fattore and Bruggemann 2017).

In total ranking methods, the order of the objects under study implies to consider judgements and preferences from the decision-makers and to apply weighted sums of single attributes for obtaining composite attributes (Bruggemann and Patil 2011). This procedure adds a certain degree of subjectivity to the ranking, because the weights are not necessarily related to the basic data matrix, but derived from political, ethical and some other grounds. Likewise, judgements of the decision makers can be vague and their preferences as well weights cannot be exactly evaluated with numerical values in practice (Bruggemann and Patil 2011). In contrast, partial order ranking arises as an alternative approach that takes advantage of the use of attributes without including preferences or weights (Bruggemann and Patil 2011).

Due to the disregard of weights in attributes, partial order methods such as the Hasse diagram technique (HDT) are more general and least subjective (Lerche et al. 2002; Bruggemann and Patil 2011). Other feature that increases the objectivity in partial order methods, compared to other MCDA tools, is that the attribute values keep separated without any numerical combination or aggregation (Bruggemann and Patil 2011); this step of aggregation generally can hide valuable information of attributes under study (Bruggemann and Patil 2011). Because of the central concept in partial order for carrying out ranking studies is the comparison without the addition neither subjective preferences nor judgements (Lerche et al. 2002; Bruggemann and Patil 2011), its appropriateness as a MCDA is highlighted. Therefore, the methodology selected for figuring out the biotechnological problem exposed in the goal of this chapter is the partial order theory (POT), specifically the HDT.

The HDT has been a useful approach of POT for decision support, that is very well described in the literature (Bruggemann and Patil 2011). However, a disadvantage is that in many cases, several optimal objects are obtained according to the criteria used to rank objects. Then, the HDT cannot always provide a total ordering of objects, i.e. it is not possible to know which is the best of all, a single object, which is the second best, and so on (Bruggemann and Patil 2011) as in total ranking methods. If decision-makers want to know the best (or worst), this means necessarily a single object, and then, ranking methods are required. Useful approaches to generate rankings based on the HDT have been proposed in the literature; herein two ranking methods were selected: local partial order model (*LPOM0*) (Bruggemann and Carlsen 2011) and extended local partial order model (*LPOMext*) (Bruggemann et al. 2004).

2   Materials and Methods

2.1   Data

The objects to be ranked are 83 microorganisms with potential as U trappers. They are characterised by three attributes according to (Quintero et al. 2017). These attributes are: U uptake capacity, *UC,* percentage of U removal (*%M*) and time requested for removing the U input, *t* (Bruggemann et al. 2004). Table 1 shows the attributes of 83 microorganisms used in the current study (Quintero et al. 2018); this table includes the 38 prokaryotes studied by (Quintero et al. 2017).

2.2   Methodology Selected in the Current Study

Given a finite set of microorganisms it is posible to define partial order relations among them in several ways, whence partial order theory has become a powerful technique in many fields, including the biotechnological one as in the current case (Quintero et al. 2017).

2.2.1   Hasse Diagram Technique

Here, we only give a short description, for more details see (Bruggemann and Patil 2011; Bruggemann and Halfon 1999).

A Hasse diagram (HD) is a visual representation of a partially ordered set (poset) underlying the objects and their attributes. If X is the set gathering the microorganisms with potential as U trappers and each microorganism x is characterised by attributes q, it is possible to order these microorganisms by ordering their atributes (Bruggemann and Patil 2011).

Hence, if it is established that $q_i(x) \leq q_i(y)$, for all i with $x, y \in X$, then $x \preccurlyeq y$. This also holds if at least for one attribute $q_j(x) < q_j(y)$, while for all others $q(x) = q(y)$. The $\preccurlyeq$ relation is a binary relation meeting: $x \preccurlyeq x$; if $x \preccurlyeq y$ and $y \preccurlyeq x \Rightarrow x = y$; and if $x \preccurlyeq y$ and $y \preccurlyeq z \Rightarrow x \preccurlyeq z$; i.e. $\preccurlyeq$ is reflexive, antisymmetric and transitive, which makes, $\preccurlyeq$ an order relation (Trotter 1992).

Any two microorganisms x, y are said to be comparable whenever $x \preccurlyeq y$ or $y \preccurlyeq x$, otherwise they are incomparable (Bruggemann and Patil 2011). The HD shows those comparabilities that cannot be obtained from others, i.e. those that are not obtained by transitivities and it constitutes a map of order relationships for the objects. In the HD, if a sequence of lines is connecting two microorganisms strictly either in the upward or the downward direction, the microorganisms are considered comparable, otherwise, they are incomparable (Bruggemann and Patil 2011).

From the HD, several sets can be derived (Bruggemann and Carlsen 2011; Bruggemann et al. 2004):

Table 1 Attributes of 83 microorganisms used in U removal (Quintero et al. 2018)

Label	Microorganisms	Percentage of U removal [%U]	Uptake capacity (mg U/g biomass dry weight) [UC]	Time requested for removing uranium input (h) [t]	References
1	*Actinomyces flavoviridis* HUT 6147	41	78.06	1	Nakajima and Sakagushi (1986)
2	*Actinomyces levoris HUT 6156*	96.05	45.47	1	Horikoshi et al. (1981)
3	*Anabaena torulosa*	48	56	0.5	Acharya et al. (2012)
4	*Arthrobacter cireus* IAM 1660[a]	2.7	6.66	1	Nakajima and Tsuruta (2004)
5	*Arthrobacter cireus* IAM 12341[a]	17.1	13.6	1	Nakajima and Tsuruta (2004)
6	*Arthrobacter nicotianae* IAM 12342[a]	86.7	68.8	1	Nakajima and Tsuruta (2004)
7	*Arthrobacter* sp. US-10	>90	23.54	1	Nakajima and Tsuruta (2004)
8	*Arthrobacter simplex* IAM 1660	30.63	58.31	1	Tsuruta (2007)
9	*Citrobacter freudii* IAM 12471[a]	21.3	16.9	1	Nakajima and Tsuruta (2004)
10	*Citrobacter* N14	>90	91	17	Kulkarni et al. (2013)
11	*Bacillus badius* IAM 11059	71.86	31.12	1	Horikoshi et al. (1981)
12	*Bacillus cereus* AHU 1030	89.62	30.7	1	Horikoshi et al. (1981)
13	*Bacillus cereus* AHU 1355	87.84	30.28	1	Horikoshi et al. (1981)
14	*Bacillus cereus* AHU 1356	87.39	27.92	1	Horikoshi et al. (1981)
15	*Bacillus cereus* IAM 1656	20.13	38.32	1	Nakajima and Sakagushi (1986)
16	*Bacillus stearother-mophilus* IAM 11062	71.70	32.72	1	Horikoshi et al. (1981)

(continued)

Table 1 (continued)

Label	Microorganisms	Percentage of U removal [%U]	Uptake capacity (mg U/g biomass dry weight) [UC]	Time requested for removing uranium input (h) [t]	References
17	*Bacillus subtilis* AHU 1219	70.94	36.33	1	Horikoshi et al. (1981)
18	*Bacillus subtilis* AHU 1390	74.31	52.81	1	Horikoshi et al. (1981)
19	*Bacillus subtilis* IAM 11062	97.91	48.94	1	Horikoshi et al. (1981)
20	*Bacillus thuringiensis* IAM 11064	46.37	37.04	1	Horikoshi et al. (1981)
21	*Bacillus licheniformis* IAM111054[a]	57.9	45.9	1	Nakajima and Tsuruta (2004)
22	*Bacillus licheniformis* ATCC 14580[a]	90.9	85	1	Yi and Yao (2012)
23	*Bacillus megaterium* IAM1166[a]	47.7	37.8	1	Nakajima and Tsuruta (2004)
24	*Bacillus mucilaginosus* ACCC 10012[a]	87.5	172	1	Yi and Lian (2012)
25	*Bacillus* sp. US-9	>90	23.72	1	Tsuruta (2004)
26	*Bacillus subtilis* IAM1026[a]	66	52.4	1	Nakajima and Tsuruta (2004)
27	*Brevibacterium helvolum* IAM 1637	6.38	12.14	1	Nakajima and Sakagushi (1986)
28	*Corynebacterium equi* IAM1038 (*Rhodococcus equi*)[a]	27	21.4	1	Nakajima and Tsuruta (2004)
29	*Corynebacterium glutamicum* IAM 12435[a]	7.5	5.9	1	Nakajima and Tsuruta (2004)
30	*Deinococcus* -PhoN	>90	214	8	Kulkarni et al. (2013)

(continued)

Table 1 (continued)

Label	Microorganisms	Percentage of U removal [%U]	Uptake capacity (mg U/g biomass dry weight) [UC]	Time requested for removing uranium input (h) [t]	References
31	*Deinococcus proteolyticus* IAM 12141	14.38	27.37	1	Nakajima and Sakagushi (1986)
32	*Deinococcus radiodurans* DrPhoN	85	202.3	6	Appukuttan et al. (2011)
33	*Enterobacter aerogenes* IAM 1183	20.75	39.51	1	Nakajima and Sakagushi (1986)
34	*Erwinia herbicola IAM 1562*	16.25	30.94	1	Nakajima and Sakagushi (1986)
35	*Escherichia coli* IAM 1268[a]	4.5	17.61	1	Nakajima and Tsuruta (2004)
36	*Escherichia coli* AHU 1520	94.71	22.59	1	Horikoshi et al. (1981)
37	*Micrococcus luteus* IAM 1056[a]	48.9	38.8	1	Nakajima and Tsuruta (2004)
38	*Micrococcus varians* IAM 13594[a]	4.5	3.57	1	Nakajima and Tsuruta (2004)
39	*Micromonospora chalcea* KCCA 0124	27.75	52.84	1	Choudary and Sar (2011)
40	*Myxococcus xanthus* C.E.C.T. 422	4.8	4.8	1	González-Muñoz et al. (1997)
41	*Nocardia erythropolis* IAM 1399[a]	64.5	51.2	1	Nakajima and Tsuruta (2004)
42	*Pseudomonas fluorescens* IAM 12022[a]	26.7	21.18	1	Nakajima and Tsuruta (2004)
43	*Pseudomonas aeruginosa* IAM 1054[a]	39.9	31.65	1	Nakajima and Tsuruta (2004)
44	*Pseudomonas aeruginosa* J007	99	275	6	Choudary and Sar (2011)

(continued)

Table 1 (continued)

Label	Microorganisms	Percentage of U removal [%U]	Uptake capacity (mg U/g biomass dry weight) [UC]	Time requested for removing uranium input (h) [t]	References
45	*Pseudomonas aeruginosa* IAM 1095	35	66.64	1	Nakajima and Sakagushi (1986)
46	*Pseudomonas radiola* IAM 12098	7.5	14.28	1	Nakajima and Sakagushi (1986)
47	*Pseudomonas saccharophilia* IAM 1504	47.75	87.11	1	Nakajima and Sakagushi (1986)
48	*Pseudomonas* sp. EPS-5028	>90	55	1	Marqués et al. (1991)
49	*Pseudomonas stutzeri* IAM 12097[a]	44.1	34.99	1	Nakajima and Tsuruta (2004)
50	*Pseudomonas* MGF-48	86	174	0.083	Malekzadeh et al. (2002)
51	*Serratia marcescens* IAM 1022	19.75	37.6	1	Nakajima and Sakagushi (1986)
52	*Streptomyces albidoflavus* HUT 6129	89.91	48.38	1	Horikoshi et al. (1981)
53	*Streptomyces albidus* HUT 6129	88.44	40.03	1	Horikoshi et al. (1981)
54	*Streptomyces albosporeus* HUT 6130	79.85	38.84	1	Horikoshi et al. (1981)
55	*Streptomyces albus* HUT 6132	92.32	27.49	1	Horikoshi et al. (1981)
56	*Streptomyces albus* HUT 6047	45.88	v87.35	1	Nakajima and Sakagushi (1986)
57	*Streptomyces albogriseolus* HUT 6054	59.11	46.89	1	Nakajima and Tsuruta (2002)
58	*Streptomyces antibioticus* HUT 6137	84.66	55.69	1	Horikoshi et al. (1981)

(continued)

Table 1 (continued)

Label	Microorganisms	Percentage of U removal [%U]	Uptake capacity (mg U/g biomass dry weight) [UC]	Time requested for removing uranium input (h) [t]	References
59	*Streptomyces chartreusis* HUT 6140	92.26	38.18	1	Horikoshi et al. (1981)
60	*Streptomyces cinereoruber* HUT 6142	24.25	46.17	1	Horikoshi et al. (1981)
61	*Streptomyces echinatus* HUT 6090	42.88	81.63	1	Nakajima and Sakagushi (1986)
62	*Streptomyces flavoviridis HUT 6147*	77.70	61.64	1	Nakajima and Tsuruta (2002)
63	*Streptomyces fradie HUT 6054*	48.90	38.79	1	Nakajima and Sakagushi (1986)
64	*Streptomyces griseoflavus* HUT 6153	50.70	40.22	1	Nakajima and Sakagushi (1986)
65	*Streptomyces griseolus* HUT 6099	31.88	60.69	1	Nakajima and Sakagushi (1986)
66	*Streptomyces hiroshimensis* HUT 6033	22.50	17.85	1	Nakajima and Sakagushi (1986)
67	*Streptomyces levoris*[a]	57	90.44	1	Tsuruta (2004)
68	*Streptomyces lilacinofulvus* HUT 6210	9.13	17.37	1	Nakajima and Sakagushi (1986)
69	*Streptomyces novaecae-sareae* HUT 6158	71.73	45.55	1	Horikoshi et al. (1981)
70	*Streptomyces obiraceus HUT 6061*	39.63	75.45	1	Nakajima and Sakagushi (1986)
71	*Streptomyces olivaceus HUT 6061*	66	52.36	1	Nakajima and Tsuruta (2002)
72	*Streptomyces* sp.	60	214.2	0.33	Golab et al. (1991)
73	*Synechococcus elongatus* BDU 75042	72	53.5	1	Acharya et al. (2009)

(continued)

Table 1 (continued)

Label	Microorganisms	Percentage of U removal [%U]	Uptake capacity (mg U/g biomass dry weight) [UC]	Time requested for removing uranium input (h) [t]	References
74	*Thiobacillus novellus* IAM12110[a]	11.1	11.66	1	Nakajima and Tsuruta (2004)
75	*Sphingomonas* sp. *Strain* BSAR-1	>90	306	6–7	Kulkarni et al. (2013)
76	*Paenibacillus* sp. JG-TB8	32	77.1	1	Reitz et al. (2014)
77	*Sulfolobus acidocaldarius* DMS 639[a]	9.24	17	1	Reitz et al. (2010)
78	*Streptomyces violaceus* HUT 6164[a]	80.24	28.27	1	Horikoshi et al. (1981)
79	*Streptomyces viridochromogenes* HUT 6031[a]	38	72.35	1	Nakajima and Sakagushi (1986)
80	*Streptomyces viridochromogenes* HUT 6166[a]	83.68	23.35	1	Horikoshi et al. (1981)
81	*Streptomyces viridochromogenes* HUT 6167[a]	99.43	20.19	1	Horikoshi et al. (1981)
82	*Thiobacillus novellus* IFO 12443[a]	25.13	47.84	1	Nakajima and Tsuruta (2004)
83	*Zoogloea ramigera* IAM 12136[a]	37.35	71.88	1	Nakajima and Sakagushi (1986)

[a]$\%U$ values were calculated from data given by authors (see supplementary data in Quintero et al. 2017)

$F(x)$, the up set: $F(x) = \{y \in X : y \succcurlyeq x\}$. i.e. $F(x)$ gathers those objects that are comparable to x and are located above x.

$O(x)$, the down set: $O(x) = \{y \in X : y \preccurlyeq x\}$. $O(x)$ gathers those comparable objects to x, which are depicted below x.

$P(x)$, the predecessors set: $P(x) = F(x) - \{x\}$.

$S(x)$, the successors set: $S(x) = O(x) - \{x\}$

$I(x)$, the set of incomparable with x: $I(x) = \{y \in X : x \parallel y\}$ $I(x)$. When for the objects x, y it is valid that $q(x) \not\preceq q(y)$ and $q(x) \not\succeq q(y)$, then x and y are incomparable $(x \parallel y)$.

$C(x)$,the set of comparable with x that is given by: $C(x) := \{y \in X : x \preceq y \; or \; y \preceq x\}$. When for the microorganisms x, y it is valid that $q(x) \preceq q(y)$ or $q(x) \succeq q(y)$, then x and y are comparable $(x \perp y)$.

The objects located at the top of the HD that have no predecessors in a poset are called maximal objects and which have no successors are called minimal objects (Bruggemann and Patil 2011). In PyHasse software, minimal objects are located near the bottom of the drawing plane.

A special case arises when a HD contains two or more pieces called components. A component of a poset $(X, \preceq)$ is a local poset $(C(X_i), \preceq)$ of $(X, \preceq)$ (Patil and Taillie 2004).

For $y \in X$,given a poset $(X, \preceq)$, if $X_i \subseteq X$, then $C(X_i) = \{y : y \perp x, x \in X_i\}$. Components partition the poset into disjoint subsets such that, if x is an arbitrary member of one component and y is an arbitrary member of a different component, then $x \parallel y$ (Bruggemann and Patil 2011).

A poset $(X, \preceq)$ is called a weak order if $\preceq$ is transitive and meets linearity. Hence, the difference between weak and total order is that the former is not antisymmetric but the latter is Quintero et al. (Quintero et al. 2018).

Likewise, a ranking of X is a two-step procedure where (1) a weak order is found for X and (2) an ordinal (rank) is assigned to each object of X (Quintero et al. 2018). To assess $\preceq$, the attributes need to be rightly oriented in such a way that, for example, high values indicate similar ranking aims (common monotonicity) (Bruggemann and Patil 2011). Further conventions to draw a Hasse diagram are found in (Bruggemann and Patil 2011; Bruggemann and Halfon 1999).

2.2.2 Ranking Methodologies

Applying HDT on a set under study, a linear order or ranking of objects may not be directly found; one way to overcome this issue is to determine the so-called average heights from the concept of linear extensions (De Loof et al. 2011): A linear order derived from a poset, preserving all its order relations is a linear extension (Bruggemann and Patil 2011); the sequence of objects due to a linear extension is described by their values of height; the object at the bottom of a linear extension has height $= 1$, the next, height $= 2$, and so on (Bruggemann and Patil 2011).

The number of linear extensions suggests how complex is the poset and how many pairs of objects are incomparable. The calculation of average heights is of interest because, once estimated, a weak order (tied ranks are not excluded) can be derived (Bruggemann and Annoni 2014). Therefore, average heights are often called average ranks (*rkavs*). But, the direct calculation of these average heights by counting the heights of objects in each linear extension is most often computationally intractable (Brightwell and Winkler 1991); the reason is because the number of linear extensions in a poset grows with the factorial of the number of ways of ordering incomparable objects in a poset (Bruggemann et al. 2014). Then, other methods or approximations are needed (Bruggemann and Annoni 2014). In this study, two approximations for calculating average ranks of each object in study have been applied, i.e. local partial order models (*LPOM*) (Bruggemann and Carlsen 2011; Bruggemann et al. 2004).

2.3 Software

The calculations described in this chapter are performed using the software PyHasse (Bruggemann et al. 2014).

3 Application of Partial Order Theory to Data Matrix in Study

3.1 Orientation of Attributes

In the current study, it was set up that the orientation follows the criterion that large attribute values indicate better recovery of U, than lower one (Quintero et al. 2017). Hence, from the original data, %*M* and *UC* were already rightly oriented, while *t* required reorientation. The reason was that the larger the time for U trapping the more undesirable for ranking by biotechnological reasons (Quintero et al. 2017).

At this last point, it is more efficient an organism trapping U in short time compared to another spending a larger time (Quintero et al. 2017). In this regard, in the set *X* gathering 83 microorganisms with potential as U trapping, it was needed that organisms requiring less time had high values of the reoriented attribute (Quintero et al. 2018). Then, *t* was multiplied by −1 and the maximum *t* value added to obtain positive values of the reoriented time, *eff* in this study (Quintero et al. 2018) (Quintero et al. 2017).

3.2 Hasse Diagram

Table 1 (Quintero et al. 2018) shows the presence of values not clearly defined for %*U* and *eff* for some microorganisms, namely 7, 10, 25, 30, 48 and 75. These microorganisms have intervals in attributes %*U* and *eff*. With the aim of studying microorganisms having interval attributes, in (Quintero et al. 2017) it was devised a set of hypothetical microorganisms. Likewise, it was explored how different interval values affected their order relationships through the HDT. From Quintero et al.

(2017), it was shown that the arbitrary selection of attributes within intervals led to drawing different conclusions as if other arbitrary selection within the interval were made. According to (Quintero et al. 2017), in the current study, microorganisms having %$U>$ 90 and 0 $\leq eff \leq$ 1 were set up to %$U=$ 90.5 and $eff=$ 1. The corresponding HD is shown in Fig. 1 (Quintero et al. 2018).

At the top of the HD shown in Fig. 1 (Quintero et al. 2018), eight maximal objects (microorganisms) are found; the numbers in parenthesis represent the labels in the HD:

Bacillus subtilis IAM 11062 (19)
Bacillus licheniformis ATCC 14580 (22)
Bacillus mucilaginosus ACCC 10012 (24)
Pseudomonas aeruginosa J007 (44)
Pseudomonas MGF-48 (50)
Streptomyces sp. (72)
Sphingomonas sp. BSAR-1 (75) and,
Streptomyces viridochromogenes HUT 6167 (81).

Likewise, four minimal objects (microorganisms) are shown at the bottom of HD (Fig. 1); they are:

Arthrobacter cireus IAM 1660 (4)
Citrobacter N14 (10)
Deinococcus radiodurans DrPhoN (32)
Micrococcus varians IAM 13594 (38)

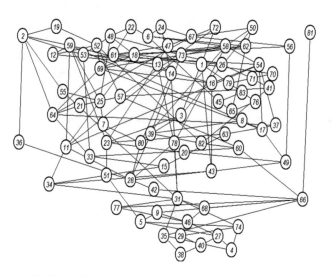

Fig. 1 Hasse Diagram of 83 microorganisms with potential for U trapping. Left side: component with 78 microorganisms performing the U biosorption process in a time ≤1 h; right side: component with 5 microorganisms characterised by performing the U removal in a long time (> 6 h). Reprinted from Quintero et al. (Quintero et al. 2018)

A special comment arises here, related to two minimal microorganisms labelled 10 and 32; these species are genetically modified (Kulkarni et al. 2013; Appukuttan et al. 2011). As expected from this biotechnological feature, they might have a good U trapping potential due to their engineered cellular systems. However, their positions as minimal microorganisms show that further studies are required for improving their U uptake abilities in aqueous solution (Quintero et al. 2017).

3.3   Average Ranks Obtained from the Application of Ranking Methodologies

LPOM0 and *LPOMext* ranking methods were applied to the microorganisms and their attributes; the results are summarised in Table 2 (reprinted from Quintero et al. 2018). Rank 1 means sought microorganisms and higher ranks less preferred microorganisms. Note that information for the last ranks is not available, for there are microorganisms having the same rank due to ties. These cases are highlighted with an asterisk (Quintero et al. 2017, 2018). Likewise, in the columns showing the average ranks for each ranking method, it is also included the label of each microorganism in parenthesis.

Average ranks calculated by *LPOM0* and *LPOMext* can be pretty different; as a token of that, in Table 2 (Quintero et al. 2018) is shown that microorganisms labelled 1, 9, 13, 25, 39, 83 and other ones differ in average ranks values calculated by these two methods. In *LPOM0*, the number of all microorganisms under study (X), the number of successors of x, ($|S(x)|$) and the number of incomparable objects of x, $I(x)$, are considered (Bruggemann and Carlsen 2011). Furthermore, in *LPOM0* microorganisms y incomparable with a certain microorganism x are supposed isolated (Bruggemann and Carlsen 2011). As seen, it is a simple method given its mathematical structure that ignores connections between some objects in the poset (Bruggemann et al. 2004). In contrast, *LPOMext* does consider connections among microorganisms and for each microorganism $y \in I(x)$, it is checked how many positions are accessible above and below x (Bruggemann et al. 2004).

4   Results and Discussion

4.1   Mathematical Interpretation from the HD and Average Ranks

In Fig. 1 taken from Quintero et al. (2018) the existence of two components shows that the microorganisms located in each of them are incomparable regarding the organisms located on the other component. This incomparability results from the existence of attributes with very high values that do not appear in organisms of

Table 2 Average ranks calculated by *LPOM0*, and *LPOMext* for 83 microorganisms with potential for U trapping

Rank	Microorganism label		Average ranks, *rkavs* (microorganism label)	
	LPOM0	*LPOMext*	*LPOM0*	*LPOMext*
1	22	22	82.73(22)	82.38(22)
2	24	24	82.64(24)	81.99(24)
3	50	50	82.62(50)	81.85(50)
4	19	19	82.35(19)	81.33(19)
5	72	72	82.21(72)	80.51(72)
6	48	48	80.89(48)	79.79(48)
7	2	2	80.27(2)	77.92(2)
8	6	6	79.24(6)	77.71(6)
9	81	67	78.40(81)	74.61(67)
10	59	52	77.00(59)	73.62(52)
11	52	62	76.85(52)	73.50(62)
12	67	58	76.70(67)	73.41(58)
13	3	59	75.87(3)	71.50 59)
14	58	47	75.60(58)	70.41(47)
15	62	56	75.43(62)	69.82(56)
16	47	44,75[a]	74.00(47)	66.9(44,75)
17	56	73	73.50(56)	66.85 73)
18	55	18	73.04(55)	66.37 (18)
19	53	53	71.71(53)	66.00(53)
20	36	3	70.74(36)	64.40(3)
21	73	61	70.00(73)	63.76(61)
22	18	55	69.70(18)	61.15(55)
23	61	1	67.61(61)	60.50 (1)
24	44,75[a]	26	67.20(44,75)	60.17(26)
25	1	71	65.56(1)	57.07 (71)
26	26	70	64.91(26)	55.75(70)
27	12,71[a]	12	63.00(12,71)	54.26(12)
28	70	54	62.46(70)	54.16(54)
29	41	69	61.09(41)	54.05(69)
30	54	41	60.90(54)	53.96(41)
31	76	79	60.67(76)	52.04 (79)
32	25	81	60.48(25)	52.02(81)
33	79	76	60.31(79)	50.06(76)
34	69	57	60.00(69)	48.81(57)
35	83	83	58.15(83)	48.34 (83)
36	13	13,25[a]	57.93(13)	45.72(13,25)
37	7	21	57.12(7)	45.28(21)
38	57	36	54.78(57)	44.71(36)
39	45	45	54.60(45)	44.25(45)
40	21	64	52.27(21)	40.96(64)
41	14	65	52.14(14)	39.64(65)

(continued)

Table 2 (continued)

Rank	Microorganism label	Average ranks, *rkavs* (microorganism label)		
42	65	37	50.00(65)	37.02(37)
43	8	7	48.00(8)	36.92(7)
44	64	8	47.25(64)	36.53(8)
45	78	14	46.06(78)	35.24(14)
46	17	17	45.40(17)	34.87(17)
47	11	63	44.47(11)	34.58(63)
48	16	30	44.33(16)	33.87(30)
49	37	16	42.86(37)	32.80(16)
50	39	39	42.00(39)	32.61(39)
51	63	11	41.14(63)	31.59(11)
52	80	23	39.53(80)	30.80(23)
53	23	78	36.24(23)	29.33(78)
54	20,30,82[a]	82	33.60(20,30,82)	28.71(82)
55	60	20	31.29(60)	27.84(20)
56	49	60	30.69(49)	26.07(60)
57	43	49	27.49 43)	24.95(49)
58	33	80	24.89(33)	24.47(80)
59	15	33	22.10(15)	22.90(33)
60	32	43	21.00(32)	22.33(43)
61	51	32	19.93(51)	21.50 (32)
62	28	15	17.75(28)	20.30 15)
63	10	51	16.80(10)	17.99(51)
64	42	28	16.563(42)	17.61(28)
65	34	10	15.75(34)	16.93(10)
66	66	42	14.76(66)	16.15(42)
67	31	34	13.39(31)	15.50(34)
68	9	66	10.65(9)	14.60(66)
69	5	31	8.055(5)	13.47(31)
70	68,77[a]	9	7.946(68,77)	12.44(9)
71	46	68,77[a]	6.632(46)	10.10 (68,77)
72	74	5	5.676(74)	10.01(5)
73	27	46	4.421(27)	7.461(46)
74	35	35	3.600(35)	7.440(35)
75	29	74	3.316(29)	7.338(74)
76	40	27	2.182(40)	5.380(27)
77	4	29	1.105(4)	4.503(29)
78	38	40	1.077(38)	2.730(40)
79	–	4	–	2.149(4)
80	–	38	–	1.314(38)
81	–	–	–	–
82	–	–	–	–
83	–	–	–	–

Reprinted from Quintero et al. (2018)
[a]Microorganisms having the same average rank, therefore equal rank (Quintero et al. 2018)

the other component. Microorganisms isolated in the component on the right (five microorganisms) are characterised by performing the biosorption process in a very long time greater than 6 hours, whereas the microorganisms in the component on the left, trap U in 1 h or less than 1 h as in the case of bacteria 3, 50 and 72 whose U removal time is equal to 0.5, 0.083 and 0.33 h, respectively.

In the HD shown in Fig. 1 (Quintero et al. 2018), the maximal microorganisms are located at the top (19, 22, 24, 44, 50, 72, 81) whereas minimal microorganisms (4, 10, 32, 38) are located at the bottom; maximal and minimal ones correspond to the best and the least suitable microorganisms for U removal respectively. Figure 1 (Quintero et al. 2018) also shows that by the simultaneous consideration of %*U*, *UC* and *eff* it is not possible to find a single best microorganism in the set under study. Instead, several microorganisms are found as maximal ones. A top eight microorganisms are found applying *LPOM*-methods (Quintero et al. 2018); both ranking methodologies are coincident in the first eight positions, they are from the first to the eighth:

Bacillus licheniformis ATCC 14580 (22)
Bacillus mucilaginosus ACCC 10012 (24)
Pseudomonas MGF-48(50)
Bacillus subtilis IAM 11062 (19)
Streptomyces sp. (72)
Pseudomonas sp. EPS-5028 (48)
Actinomyces levoris HUT 6156 (2)
Arthrobacter nicotianae IAM 12342 (6)

Numbers in parenthesis represent the label in the HD (Fig. 1) (Quintero et al. 2018).

Because of the presence of some ties, i.e., same values in average ranks for some microorganisms, the last ranks are not available in Table 2 (Quintero et al. 2018). These results show that the microorganism 22 (*Bacillus licheniformis* ATCC 14580) is the best U trapper. It belongs to the maximal microorganisms (Fig. 1). If the maximal microorganisms in Fig. 1 are inspected, five out of eight are considered as well as good U trappers according to the top eight obtained from the application of order theoretical methods; these five microorganisms are 19, 22, 24, 50 and 72 (Quintero et al. 2018). The top three microorganisms in the current study also coincide with the results found in the study by (Quintero et al. 2017).

4.2   Biological Interpretation

From the taxonomical point of view, the top three microorganisms in the group under study are bacteria classified in two genera, *Bacillus* and *Pseudomonas*, that include microorganisms traditionally used as potent U biosorbents. However not all bacteria belonging to the same genus are good U trappers from aqueous systems (Quintero et al. 2017).

Bacteria are generally divided into two main groups, Gram-positive and Gram-negative, based on their Gram stain retention property; regarding the top eight of the set of 83 microorganisms herein studied, ranks 1, 2, 4, 5, 7 and 8 correspond to Gram-positive and ranks 3 and 6 to Gram-negative bacteria, i.e. there is no correlation between U biosorption and Gram stain classification for these microorganisms (Quintero et al. 2017; Happel 1996). However, there are other studies showing that Gram-positives bacteria absorb more U than negative ones [Quintero and Restrepo, 2017], for the first have 40% more binding sites for trapping metal ions than the second ones (Borrok et al. 2005). In relation to the particular top eight microorganisms in Table 2, the number of binding-sites for U uptake has not yet been determined (Quintero et al. 2017).

According to the top eight shown in Table 2 (Quintero et al. 2018), different microorganisms belonging to the same genus (*Bacillus* for microorganisms 1, 2 and 3 and *Pseudomonas* for microorganisms 3 and 6) have good U uptake capacities. However, this ability is not joined to the genus or species; in the current study, there are populated genera by different number of bacteria and each bacterium is characterised by pretty different *UC*; some examples of these genera and the number of bacteria studied belonging to these genera, respectively are *Bacillus* (16); *Pseudomonas* (9) and *Streptomyces* (25).

Likewise, strains belonging to the same species do not trap U in similar amounts, although they share genetic similarity in the DNA (Quintero et al. 2017; Happel 1996); according to the literature, it is still not clear why these strains do not trap U in quite similar amounts (Happel 1996). Likewise, different conditions used in the harvesting from bacterial cultures could lead to different densities of binding sites at the cell wall level.

Moreover, it has been proposed that bacterial growth phase and biomass pre-treatment (including washing conditions and presence of inhibitors (Premuzic et al. 1991) can exert some influence, i.e., if cells are in stationary or growing stage rather than resting stage or if they have been washed with acidic solutions that could increase their ability for trapping metal ions (Premuzic et al. 1991). Although there is no such information for the microorganisms here studied, *Pseudomonas* sp. EPS-5028 was exposed to inhibitors and its *UC* was increased (Marqués et al. 1991).

In addition, the nutrient composition in the growth media may influence the bacterial growth and lead to variability in the number of functional groups onto the cell walls (Quintero et al. 2017); additionally it has been proposed that extracellular polymeric substances (EPS), i.e., macromolecules situated outside of cell walls could enhance the *UC*, in some strains of bacteria (Ates 2015). In this study, the top eight microorganisms were cultured in different nutrient solutions, being difficult to assess the influence of EPS in their high efficiency as U trappers (Quintero et al. 2017). However, the two first bacteria in Table 2 (Quintero et al. 2018), *B. licheniformis* and *B. mucilaginosus* secrete EPS or capsules made of extracellular polysaccharides (Yi and Lian 2012; Yi and Yao 2012). Still, additional work at the molecular level is needed to understand the chemical properties responsible for the specificity of the cell wall U interactions found in bacteria (Borrok et al. 2005).

5 Conclusions

From the biotechnological point of view, it is important to remark that the results of the application of ranking methodologies and the relations among attributes are based on the set of 83 microorganisms collected and on the kind of attributes considered. The presence of two components in the HD evidences the incomparability among the microorganisms: In the rightward component, 5 microorganisms are characterised by high values in UC and time for removing U input ($t \geq 6$ h, i.e. low eff), whereas 78 microorganisms in the leftward component evidenced relative low values in UC and their time for the U removal is ≤ 1 h (i.e., high eff).

From the application of order theoretical methods, the ordering of 83 microorganisms is shown, being the best one *Bacillus licheniformis* ATCC 14580 [Quintero et al. 2017; Quintero et al. 2018]. Although the two ranking approaches yielded similar results, some ties between the two rankings are found. These ties can be gathered in equivalence classes as follows: four in *LPMO*, i.e. [44,75], [12,71], [20,30,82] and [68,77] and three equivalence classes in the case of LPOMext, namely [44,75],[13,25] and [68,77].

Regarding partial order, the results could be eminently broadened, if the density of active sites for U trapping in cell walls of microorganisms could be included as an attribute in the data matrix. However, until now this information is not available.

From the biological point of view, it has been found that a high uranium uptake capacity associated to the microorganisms studied does not depend upon the taxonomic hierarchy, e.g., species, genera or phyla, even strains in the same species show different ability for U biosorption (Quintero et al. 2017; Happel 1996). Differences in U uptake capacities may be joined to specific genetic traits of each microorganism expressed in its chemical composition and in the number of binding sites for trapping U cations in the cell walls (Borrok et al. 2005).

As an outlook, the following biotechnological perspectives can be desirable:

- The suitability of the U trappers proposed in the current study could be further tested in studies combining batch experiments at the wet lab.
- Although the extrapolation of knowledge from the laboratory scale to industrial application has been a slow process, if an industrial application of the optimal microorganisms here reported is sought for, other features need to be studied, e.g. economical ones. This would allow obtaining an estimation of the overall cost of the sorbent and biosorption process to treat large volumes of waste waters with low U concentrations in nuclear facilities.
- In addition to the aforementioned suggestions, a step towards the practical application is to study the adsorption equilibria supplemented with adsorption kinetics. Due to the economic importance of recovering U from the biomasses, it is recommended to carry out studies focused on the separation of this actinide from the biomass. Likewise, it is proposed to assess the reuse of the microorganisms through adsorption-desorption cycles for improving the benefit-cost relation inherent to the process at industrial scale.

Finally, we hope this chapter, besides serving as a preliminary step towards the potential application of microorganisms for removing U input in aqueous systems, shows the methodological advantages of partial order in bioremediation, a promising biotechnological field with many data coming from experiments dealing with U trappers.

References

Acharya, C., Joseph, D., & Apte, S. K. (2009). Uranium sequestration by a marine cyanobacterium *Synechococcus elongatus* strain BDU/75042. *Bioresource Technology, 100*, 2176–2181.

Acharya, C., Chandwadkar, P., & Apte, S. K. (2012). Interaction of uranium with a filamentous, heterocystous, nitrogen-fixing cyanobacterium, *Anabaena torulosa*. *Bioresources Technology, 116*, 290–294.

Appukuttan, D., Seetharam, C., Padma, N., Rao, A. S., & Apte, S. K. (2011). PhoN-expressing, lyophilized, recombinant *Deinococcus radiodurans* cells for uranium bioprecipitation. *Journal of Biotechnology, 154*, 285–290.

Ates, O. (2015). 2015. Systems biology of microbial exopolysaccharides production. *Frontiers in Bioengineering and Biotechnology, 3*, 1–16.

Borrok, D., Turner, B. F., & Fein, J. B. (2005). A universal surface complexation framework for modelling proton binding onto bacterial surfaces in geologic settings. *American Journal of Science, 305*, 826–853.

Brightwell, G. & Winkler, P. (1991). Counting linear extensions is #P-complete. Proceeding STOC'91 Proceedings of the twenty-third annual ACM symposium on Theory of computing, New Orleans, May 05–08, 1991, 175–181.

Bruggemann, R., & Annoni, P. (2014). Average heights in partially ordered sets. *MATCH Communications in Mathematical and in Computer Chemistry, 71*, 101–126.

Bruggemann, R., & Carlsen, L. (2011). An improved estimation of averaged ranks of partial orders. *MATCH – Communications in Mathematical and in Computer Chemistry, 65*, 383–414.

Bruggemann, R., & Halfon, E. (1999). Introduction to the general principles of the partial order ranking theory. In *Order theoretical tools in environmental sciences. Proceedings of the second workshop*, Sørensen P. B., Carlsen L., Mogensen B.B., Bruggemann R., Luther B., Pudenz S., Simon U et al. Roskilde, Denmark. NERI Technical Report No. 318. Ministry of Environment and Energy National Environmental Research Institute, pp. 1–172.

Bruggemann, R., & Patil, G. (2011). *Ranking and prioritization for multi-indicator systems, Introduction to partial orders applications*. Environmental and Ecological Statistics. Germany: Springer.

Bruggemann, R., Sørensen, P. B., Lerche, D., & Carlsen, L. (2004). Estimation of averaged ranks by a local partial order model. *Journal of Chemical Information and Modelling, 44*, 618–625.

Bruggemann, R., Carlsen, L., Voigt, K, & Wieland, R. (2014). PyHasse software for partial order analysis: Scientific background and description of selected modules. In: Bruggemann R., Carlsen L., & Wittmann J. *Multi-indicator systems and modelling in partial order*. Part V, Germany: Springer.

Choudary, S., & Sar, P. (2011). Uranium biomineralization by metal resistant *Pseudomonas aeruginosa* strain isolated from contaminated mine waste. *Journal of Hazardous Materials, 186*, 336–343.

Committee on Uranium Mining in Virginia; Committee on Earth Resources; National Research Council. Uranium Mining in Virginia: Scientific, Technical, Environmental, Human Health and Safety, and Regulatory Aspects of Uranium Mining and Processing in Virginia. Washington (DC): National Academies Press (US). (2011, December 19). 5, Potential human health effects of uranium mining, processing, and reclamation. Available from: https://www.ncbi.nlm.nih.gov/books/NBK201047/. Accessed 06 Dec 2017

De Loof, K., De Baets, B., & De Meyer, H. (2011). Approximation of average ranks in posets. *MATCH – Communications in Mathematical and in Computer Chemistry., 66*, 219–229.

El-Taher, A., Nossair, A., Azzam, A. H. Kratz, K. L., & Abdel-Halim, A. S. (2004). Determination of traces of uranium and thorium concentration in some Egyptian environmental matrices by instrumental neutron activation analysis. In *Proceedings of the Environmental Physics*, Minya, Egypt. 24–28 February, 2004, 141–149

Fattore, M., & Bruggemann, R. (2017). *Partial order concepts in applied sciences* (Preface) (pp. v–vi). Berlin: Springer.

Fourest, E., & Roux, J.-C. (1992, June). Heavy metal biosorption by fungal mycelial by products: Mechanisms and influence of pH. *Applied Microbiology and Biotechnology, 37*(3), 399–403.

Golab, Z., Orlowska, B., & Smith, R. W. (1991). Biosorption of lead and uranium by *Streptomyces* sp. *Water Air Soil Pollution, 60*, 99–106.

González-Muñoz, M. T., Merroun, M. L., Ben, O. N., & Arias, J. M. (1997). Biosorption of uranium by *Myxococcus xanthus*. *International Biodeterioration &. Biodegradation, 40*, 107–114.

Happel, A. (1996). Evaluation of actinide biosorption by microorganisms. Glenn T. Seaborg Institute for Transactinium Sciences Lawrence Livermore National Laboratory. UCRL-ID-124223, Livermore.

Horikoshi, T., Nakajima, A., & Sakaguchi, T. (1981). Studies on the accumulation of heavy metals elements in biological systems. XIX. Accumulation of uranium by microorganisms. *European Journal of Applied Microbiology and Biotechnology, 12*, 90–96.

Kulkarni, S., Ballal, A., & Apte, S. K. (2013). Bioprecipitation of uranium from alkaline waste solutions using recombinant *Deinococcus radiodurans*. *Journal of Hazardous Materials, 262*, 853–861.

Lerche, D., Bruggemann, R., Sørensen, P., & Nielsen, O. J. (2002, September). A comparison of partial order techniques with three methods of multi-criteria analysis for ranking of chemical subatances. *Journal of Chemical Information and Computer Sciences., 42*(5), 1086–1098.

Li, J., & Zhang, Y. (2012). Remediation technology for the uranium contaminated environment: A review. *Procedia Environmental Sciences, 13*, 1609–1615.

Malekzadeh, F., Farazmand, A., Ghafourian, H., Shahamat, M., Levin, M., & Colwell, R. R. (2002). Uranium accumulation by a bacterium isolated from electroplating effluent. *World Journal of Microbiology and Biotechnology, 18*, 295–302.

Markich, S. J. (March 15, 2002). Uranium speciation and bioavailability in aquatic systems: An overview. Mini-Review. *The Scientific World Journal., 2*, 707–729.

Marqués, A. M., Roca, X., Dolores, S.-P. M., Carmen, F. M., & Congregado, F. (1991). Uranium accumulation by *Pseudomonas* sp. EPS-5028. *Applied Microbiology and Biotechnology, 35*(3), 406–410.

McCullough, J., Hazen, T., & Benson, S. (2003). *Bioremediation of metals and radionuclides: What it is and how it works* (2nd ed.) Natural and Accelerated Bioremediation Research (NABIR) Program. Bioremediation of metals and radionuclides, 2003. Available from: https://cloudfront.escholarship.org/dist/prd/content/qt7md2589q/qt7md2589q.pdf. Accessed 06 Oct 2017

Nakajima, A., & Sakagushi, T. (1986). Selective accumulation of heavy metals by microorganisms. *Applied Microbiology and Biotechnology, 24*, 59–64.

Nakajima, A., & Tsuruta, T. (2002). Competitive biosorption of thorium and uranium by actinomycetes. *Journal of Nuclear Science and Technology, 3*, 528–531.

Nakajima, A., & Tsuruta, T. (2004). Competitive biosorption of thorium and uranium by *Micrococcus luteus*. *Journal of Radioanalytical and Nuclear Chemistry, 260*, 13–18.

Patil, G. P., & Taillie, C. (2004). Multiple indicators, partially ordered sets, and linear extensions: Multicriterion ranking and Priorization. *Environmental and Ecological Statistics, 11*, 119–228.

Prakash, D., Gabani, P., Chandel, A. K., Ronen, Z., & Singh, O. (2013, July). Bioremediation: A genuine technology to remediate radionuclides from the environment. *Microbial Biotechnology, 6*(4), 349–360.

Premuzic, E. T., Lin, M., Zhu, H. L., & Gremme, A. M. (1991). Selectivity in metal uptake by stationary phase microbial populations. *Archives of Environmental Contamination and Toxicology, 20*, 234–240.

Quintero N. Y., & Restrepo, G. (2017). Formal Concept Analysis Applications in Chemistry: From radionuclides and Molecular structure to toxicity and diagnosis. *In: Partial Order in Applied Sciences*, Fattore, M.; Brüggemann, R., Eds.; Springer: Berlin, Germany, pp. 207–217.

Quintero, N. Y., Bruggemann, R., & Restrepo, G. (2017, April). Ranking of 38 prokaryotes according to their uranium uptake capacity in aqueous solutions: An approach from order theory through the Hasse diagram technique. *Toxicological and Environmental Chemistry, Mathematical Approaches to Environmental Chemistry, 99*(7–8), 1242–1269.

Quintero, N. Y., Bruggemann, R., & Restrepo, G. (2018). Mapping Posets into low dimensional spaces: The case of U trappers. *MATCH – Communications in Mathematical and in Computer Chemistry., 80*, 793–820.

Reitz, T., Merroun, M., Rossberg, A., & Selenska-Pobell, S. (2010). Interactions of *Sulfolobus acidocaldarius* with uranium. *Radiochimica Acta, 98*, 249–257.

Reitz, T., Rossberg, A., Barkleit, A., Selenska-Pobell, S., & Merroun, M. L. (2014). Decrease of U(VI) immobilization capability of the facultative anaerobic strain *Paenibacillus* sp. JG-TB8 under anoxic conditions due to strongly reduced phosphatase activity. *PLoS One, 9*, e102447.

Trotter, W. T. (1992). *Combinatorics and partially ordered sets, dimension theory*. Baltimore: The Johns Hopkins University Press.

Tsuruta, T. (2004). Adsorption of uranium from acidic solution by microbes and effect of thorium on uranium adsorption by *Streptomyces levoris*. *Journal of Bioscience and Bioengineering, 97*, 275–277.

Tsuruta, T. (2007). Removal and recovery of uranium using microorganisms isolated from north American uranium deposits. *American Journal of Environmental Science, 3*, 60–66.

Volesky, B. (2001, February). Detoxification of metal-bearing effluents: Biosorption for the next century. *Hydrometallurgy, 59*, 203–216.

Volesky, B., & Holan, J. R. (1995, May–June). Biosorption of heavy metals. *Biotechnology Progress, 11*(3), 235–250.

Wang, J., & Chen, C. (2009, March). Biosorbents for heavy metals removal and their future. *Biotechnology Advances, 27*(2), 195–226.

Yi, Z., & Lian, B. (2012). Adsorption of U (VI) by *Bacillus mucilaginosus*. *Journal of Radioanalytical and Nuclear Chemistry, 293*, 321–329.

Yi, Z., & Yao, J. (2012). Kinetic and equilibrium study of uranium (VI) adsorption by *Bacillus licheniformis*. *Journal of Radioanalytical and Nuclear Chemistry, 293*(2), 907–914.

Part III
Indicators in Social Sciences

There Is No Such Thing as a Free Lunch! Who Is Paying for Our Happiness?

Lars Carlsen

1 Introduction

In a report published by the Danish Ministry of Environment (HRI 2012) it is stated that "*it is no longer possible to imagine a future where the pursuit of happiness is not somehow connected to sustainability. As the human species continues its quest for happiness and well-being, more emphasis must be placed on sustainability and the interaction between sustainability and happiness*" and further "*there is a growing awareness of how sustainability and happiness can go hand-in-hand*". However, the term happiness is not uniquely defined and a somewhat broad definition could be "the experience of joy, contentment, or positive well-being, combined with a sense that one's life is good, meaningful, and worthwhile" (Lyubomirsky 2008). A more well-defined and structured index for happiness has been reported based on seven indicators (HI 2016, 2017, 2018):

1. GDP per capita is in terms of Purchasing Power Parity (GPD)
2. Social support (or having someone to count on in times of trouble) (SocSup)
3. The time series of healthy life expectancy at birth (LifeExp)
4. Freedom to make life choices (FreeCho)
5. Generosity (Gener)
6. Perceptions of corruption (PerCor)
7. The country's own perception of doing better or worse than the hypothetical country Dystopia (Dys)

In a recent study, comprising the 157 countries included in the World Happiness Index study (HI 2016) these indicators were analyzed (Carlsen 2018) applying

L. Carlsen (✉)
Awareness Center, Roskilde, Denmark
e-mail: lc@awarenesscenter.dk

© The Author(s), under exclusive license to Springer Nature Switzerland AG 2021
R. Bruggemann et al. (eds.), *Measuring and Understanding Complex Phenomena*,
https://doi.org/10.1007/978-3-030-59683-5_14

partial ordering techniques, disclosing, among other features that on an average basis the following 10 countries were found as the happiest countries: Iceland, Australia, Switzerland, Norway, New Zealand, Denmark, the Netherlands, Finland and Austria, whereas the bottom of the list displays Madagascar, Congo (Brazzaville), Egypt, Benin, Chad, Gabon, Burundi, Angola, Armenia and Yemen as the least happy countries, results that is somewhat different from the original study where the index is generated by a simple arithmetic aggregation of the 7 indicator values (HI 2016, 2017, 2018).

In today's world nothing is free, so the obvious question that arises is now: who is paying for our happiness? To some extent a study of the Happy Planet Index, which is focused on sustainable wellbeing for all and is based on 4 indicators, i.e., experienced wellbeing (EWB), life expectancy (LEX), inequality of outcomes (IoO) and the ecological footprint (EFP) (Jeffrey et al. 2016) may give some answers.

The present study focus on answering the above question by partial order analyses of the World Happiness Index and the Happy Planet index in parallel.

2 Methodology

The present paper describes how selected partial order tools may be applied in the evaluation of a series of countries taking several indicators simultaneously into account as an alternative to conventional methods to study MIS (Bruggemann and Carlsen 2012).

2.1 The Basic Equation of Partial Ordering

In its basis partial ordering appears pretty simple as the only mathematical relation among the objects is "$\leq$" (Bruggemann and Carlsen 2006a, b Bruggemann and Patil 2011). The basis for a comparison of objects, here countries, characterized by the subset of indicators describing their performance in relation a) to happiness as well as b) to the planetary 'happiness' (vide infra). This series of indicators, r_j, characterizes the single countries. Thus, characterizing one country (x) by a set of indicators $r_j(x)$, $j = 1,...,m$, where m is the number of indicators, can be compared to another country (y), characterized by the indicators $r_j(y)$, when

$$r_j(y) \leq r_j(x) \text{ for all } j = 1, \ldots, m \tag{1}$$

Equation 1 is a very hard and strict requirement for establishing a comparison. It demands that all indicators of x should be better (or at least equal) than those of y. Further, let X be the subset of countries included in the analyses, x will be ordered higher (better) than y, i.e., $x > y$, if at least one of the indicator values for x is higher than the corresponding indicator value for y and no indicator for x is lower than the corresponding indicator value for y. On the other hand, if $r_j(x) > r_j(y)$ for some indicator j and $r_i(x) < r_i(y)$ for some other indicator i, x and y will be called incomparable (notation: $x \parallel y$) expressing the mathematical contradiction due to conflicting indicator values. A set of mutual incomparable objects is called an antichain. When all indicator values for x are equal to the corresponding indicator values for y, i.e., $r_j(x) = r_j(y)$ for all j, the two objects/nations will have identical rank and will be considered as equivalent, i.e., $x \sim y$. The analysis of Equation 1 results in a graph, the Hasse diagram. Hasse diagrams are unique visualizations of the order relations due to Equation 1.

2.2 The Hasse Diagram

The Eq. 1 is the basic for the Hasse diagram technique (HDT) (Bruggemann and Carlsen 2006a, b; Bruggemann and Patil 2011). Hasse diagrams are visual representation of the partial order. In the Hasse diagram comparable objects are connected by a sequence of lines (Bruggemann and Carlsen 2006a, b; Bruggemann and Patil 2011; Bruggemann and Münzer 1993; Bruggemann and Voigt 1995, 2008).

2.3 The More Elaborate Analyses

In addition to the basic partial ordering tools some more elaborate analyses have been used including average ranks (Bruggemann and Annoni 2014; Morton et al. 2009; De Loof et al. 2006; Lerche et al. 2003; Bruggemann et al. 2004; Bruggemann and Carlsen 2011) and sensitivity analysis (Bruggemann and Patil 2011; Bruggemann et al. 2014), the latter gives an insight in the relative importance of the included indicators (Bruggemann and Patil 2011; Bruggemann et al. 2014).

The average ranking is expressed as average height from bottom (min. Height = 1) to the top (max height = n, i.e., the maximum number of objects) (Bruggemann and Annoni 2014). The average rank is generated by calculating all linear order preserving sequences (set LE), the "linear extensions of the original partial order. From LE_0 the statistical characterization for each object is obtained. For example the characterization is calculated as the average value an object has, taken all positions of this object within LE_0, the averaged heights. It is clear that this procedure is computationally extremely difficult. Hence, approximations were developed.

For the sensitivity analysis (Bruggemann and Patil 2011; Bruggemann et al. 2014), let Q be the set of all indicators, then taken all indicators of Q leads to a partial order, which is called PO_0. The corresponding set of linear extensions is denoted by LE_0. Leaving out one indicator of Q, say r_j, then another partial order results, which is denoted as PO_j.

Both partial orders can be described by an adjacent matrix, say A_0 for PO_0 and A_j for PO_j.

Taken the Euclidian Distance (squared) quantifies the role of indicator qj in PO_0. This is a sensitivity measure for the indicators of set Q, describing the structural changes of the partial order leaving one indicator out. This is not immediately a measure of the sensitivity of the indicators for a ranking, because the ranking is per definition a linear order and here derived over many interim steps.

If a linear order is obtained by all orders in LE_0, the set of linear extensions taken from PO_0, then any PO_j will also lead to a corresponding set LE_j. And this set is the more differing from LE_0 the larger the sensitivity is. Therefore the ranking due to averaged heights is as more affected by indicator r_j as larger its sensitivity is.

For detail information on the single tool the cited literature should be consulted as a detailed description is outside the scope of the present paper.

2.4 Software

All partial order analyses were carried out using the PyHasse software (Bruggemann et al. 2014). PyHasse is programmed using the interpreter language Python (version 2.6) (Ernesti and Kaiser 2008; Hetland 2005; Langtangen 2008; Weigend 2006; Python 2015) Today, the software package contains more than 100 modules and is available upon request from the developer, Dr. R.Bruggemann (brg_home@web.de).

2.5 Indicators

The seven indicators applied in the World Happiness Index (HI 2016, 2017, 2018) has been stated above in the introduction.

As mentioned in the introduction, the Happy Planet Index (HPI), focussing on sustainable wellbeing is based experienced wellbeing (EWB), life expectancy (LEX), inequality of outcomes (IoO) and the ecological footprint (EFP), the latter being expressed in global hectares per capital. One global hectare is the world's annual amount of biological production for human use and human waste assimilation, per hectare of biologically productive land and fisheries. An approximate formula for calculating HPI is given by

$$HPI \approx \frac{LEX * EWB * IoO}{EFP} \tag{2}$$

The eventual calculation of HPI uses a somewhat more elaborate formula applying 'some technical adjustments are made to ensure that no single component dominates the overall score' (Jeffrey et al. 2016), where inequality adjusted values of LEX and EWB are used and some scaling constants are incorporated.

$$HPI = \frac{0.452 * ((EWB_{IA} - 0.158) * LEX_{IA} + 3.951)}{(EFP + 2.067)} \tag{3}$$

The subscript IA denotes that the EWB and LEX indicators have been 'inequality adjusted' for economic inequalities in the countries. For details Jeffrey et al. (2016) and nef (2016) should consulted.

It should be noted that in order to achieve a sensible ranking picture it is mandatory that all indicators included have the same orientation, e.g., the larger the better. Thus, in the case of the HPI the EFP indicator will be multiplied by -1 in order to guarantee co-monotony with the EWB and LEX indicators.

2.6 Data

The data used for the analysis can be found in the appropriate cited reports (HI 2016; Carlsen 2018; Jeffrey et al. 2016). The full set of indicators and the complete set of countries (approx. 150) have been used for the calculations.

3 Results and Discussion

3.1 The World Happiness Index

Let us initially look at what makes us happy. Here we take the onset in the Word Happiness Index (HI 2016, 2017, 2018). As mentioned in the introduction this index is calculated by a simple arithmetic aggregation of the 7 indicators mentioned above. Obviously, such an aggregation of data may lead to more or less strange results due to compensation effects (Munda 2008), roughly speaking adding apples and oranges getting bananas. Hence, in a recently paper (Carlsen 2018) the happiness index was revisited applying partial order methodology, among other things to disclose the relative importance of the seven indicators. In Fig. 1 the relative importance of the seven indicators are depicted as calculated applying the sensitivity module sensitivity23_1 of the PyHasse software package (Bruggemann and Patil 2011; Bruggemann et al. 2014) on the 2016 happiness index data (HI 2016).

The result summarized in Fig. 1 has in details been discussed by Carlsen (2018), a discussion that shall not be reproduced here. However, it is worthwhile to mention just 3 specific indicators, i.e., GPd, Gener and Dys, respectively.

First it can be noted that in an overall evaluation of happiness money, here expressed as the gross domestic product or more precisely as the purchasing power

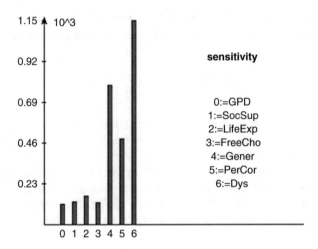

Fig. 1 Relative importance of the seven indicators used to generate the 2016 World Happiness Index (HI 2016)

parity (PPP), apparently plays only a minor role, actually displaying the lowest importance of the seven indicators. This is in agreement with the old myth that 'money can't buy you happiness'. Second it is, in the context interesting to look at the second most important indicator is generosity (Gener). Hence, if the GDP indicator is a measure of receiving/having it is immediately clear that to helping others and to give is a much more important factor for our happiness as pointed out in Acts 20:35 "It is more blessed to give than to receive"(KJBO 2016; see also McConnell 2010).

Third, it is immediately seen the Dys indicator appears as the most important factor in our perception of happiness. The Dys indicator reveals the single country's own, obviously subjective perception of doing better or worse than the hypothetical country Dystopia, a country where it, roughly speaking, couldn't be worse (HI 2016, 2017, 2018; Carlsen 2018). This dominance of the Dys indicator is not surprising. It has been nice expressed by Fyodor Dostoevsky: "The greatest happiness is to know the source of unhappiness"(Brainyquote 2001). In Table 1 the top-10 countries based on average ranking are shown. The numbers in parentheses after the single countries refer to the placement based on the HI for the years 2016–2018 (HI 2016, 2017, 2018; Carlsen 2018).

It can be noted (Table 1) that apart from a single case (Austria in 2016) the Top-10 countries based on an average ranking including all seven indicators fits reasonable well with the original HI. However, it also puts a question mark to the annual discussion in Danish news media that we are no longer the most happy people in the world (2017 and 2018) since Denmark based on the average ranking never was.

A short video presentation highlighting the main finding of the study can be found at https://www.researchsquare.com/article/rs-113102/v1.

Table 1 Top-10 countries based on average ranking of the seven Hi indexes for 2016–2018. The number in parenthesis refer to the placement based on the HI for the years

	2016	2017	2018
1	Canada (6)	Switzerland (4)	Switzerland (5)
2	Iceland (3)	Iceland (3)	Norway (2)
3	Australia (9)	Norway (1)	Iceland (4)
4	Switzerland (2)	Canada (7)	Canada (7)
5	Norway (4)	Denmark (2)	Finland (1)
6	New Zealand (8)	New Zealand (8)	Australia (10)
7	Denmark (1)	Netherlands (6)	Denmark (3)
8	Netherlands (7)	Australia (10)	Netherlands (6)
9	Finland (5)	Sweden (9)	New Zealand (8)
10	Austria (12)	Finland (5)	Sweden (9)

Table 2 Ecological footprint, inequality-adjusted life expectancy and wellbeing for the top-10 countries by the Happy Planet index

HPI Rank	Country	Footprint (gha/capita)	Inequality-adjusted life expectancy	Inequality-adjusted wellbeing
1	Costa Rica	2.84	72.62	6.79
2	Mexico	2.89	66.31	6.83
3	Colombia	1.87	63.10	5.72
4	Vanuatu	1.86	60.32	5.94
5	Vietnam	1.65	64.79	5.22
6	Panama	2.79	68.33	6.32
7	Nicaragua	1.39	63.44	4.76
8	Bangladesh	0.72	56.62	4.27
9	Thailand	2.66	66.35	5.98
10	Ecuador	2.17	64.09	5.52

3.2 The Happy Planet Index

Turning to the Happy Planet Index (HPI) a quite different picture develops. Let us first look at the top-10 and bottom-10 countries based on the HPI (Eq. 3).

In the top-10 countries Bangladesh is surprisingly found in the top-10, i.e., at rank 8 (Table 2). However, looking at the details (Table 2) the answer is found. Thus, although the Inequality-adjusted life expectancy (56.62) as well as the inequality-adjusted wellbeing indicators (4.27) are found relatively low also the ecological footprint for Bangladesh is extremely low, i.e. 0.72, which obviously let to the high ranking (cf. Eq. 3).

Turning to the bottom-10 countries based on HPI (Table 3) again some surprising results are seen. In general these countries have rather low Inequality-adjusted life expectancy and the inequality-adjusted wellbeing indicators which in combination with low ecological footprint (cf. Eq. 3) lead to the low rank. However, 3 countries appearing on this list (Table 5) are surprising, especially with regards to the

Table 3 Ecological footprint, inequality-adjusted life expectancy and wellbeing for the bottom-10 countries by the Happy Planet index

HPI Rank	Country	Footprint (gha/capita)	Inequality-adjusted Life expectancy	Inequality-adjusted wellbeing
131	Burundi	0.80	33.01	3.03
132	Swaziland	2.01	31.81	4.44
133	Sierra Leone	1.24	28.18	3.98
134	Turkmenistan	5.47	48.33	5.12
135	Cote d'Ivoire	1.27	30.64	3.51
136	Mongolia	6.08	56.87	4.61
137	Benin	1.41	37.27	2.82
138	Togo	1.13	39.64	2.42
139	Luxembourg	15.82	78.97	6.70
140	Chad	1.46	27.32	3.67

ecological footprint. Thus, Turkmenistan (5.47), Mongolia (6.08) and, virtually out of scale Luxembourg (15.82). In the case of Luxembourg it is worthwhile to mention that one reason for the extreme ecological footprint may be sought for in the fact that the country is rather small (2.6 km^2 x 1000) and dominated by the city Luxembourg. Hence, Luxembourg as a country may be regarded as urban area with a population density of 231 people per square kilometer (World Bank 2017) in contrast to the other much larger countries like, e.g., Mongolia with an area od 1564.1 km^2 x 1000 and a pollution density of 2 people per square kilometer (World Bank 2017) For these countries obviously a somewhat higher values for the Inequality-adjusted life expectancy and the inequality-adjusted wellbeing indicators cannot compensate for the high ecological footprint.

The data presented in Tables 2 and 3 and the associated discussion point at the importance of the ecological footprint (EFP). This is confirmed by looking at the relative importance of the 3 indicators, EFP, LEX and WB (Fig. 2).

Not surprisingly an average ranking differ here significantly from the simple HPI ranking based on Eq. 3. In Tables 4 and 5 the top-10 and bottom-10 countries based on an average ranking applying the 3 HPI indicators (see Sect. 2.5) is shown. The original HPI calculated based on Eq. 3 is given in addition to the ecological footprint for the single countries. For comparison the result of the average ranking for the 10 countries based on the seven Hi indicators are shown. Denmark and Luxembourg are further included (Table 4) for comparison to the HI.

Immediately (Tables 4 and 5) is it noted that significant variations in the average HPI ranking compared to the average HI ranking prevail.

Looking at the ecological footprint as a key factor to the HPI it appears interesting to elucidate the variation in the average HPI ranking with a changed EFP. Using Luxembourg as a spectacular example it is found that a reduction of the Luxembourg EFP by 10 gha/capita moves the country from place 103 to place 39.

Fig. 2 Relative importance of the three indicators used to generate the 2016 Happy Planet Index (Jeffrey et al. 2016)

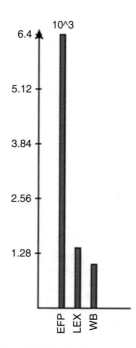

Table 4 Top-10 countries plus Denmark and Luxembourg based on average ranking of the HPI indicators

Rkav	Country	HPI	EFP	HI (Rkav)
1	Bangladesh	8	0.72	84
2	Costa Rica	1	2.84	20
3	Pakistan	63	0.79	47
4	Norway	12	4.98	5
5	Spain	15	3.67	32
6	Colombia	3	1.87	56
7	Tajikistan	25	0.91	54
8	Philippines	20	1.1	51
9	Vietnam	5	1.65	100
10	Nicaragua	7	1.39	23
50	Denmark	32	5.51	7
103	Luxembourg	139	15.82	21

For comparison the original HPI and the ecological footprint are given in addition to the average ranking of the same countries applying the HI indicators. All 2016 data

3.3 Including the Financial Aspect

Now, with reference to the HI, it might be of interest to including the financial aspect. Thus, adding the Purchasing Power Parity (PPP) as a fourth indicator, PPP

Table 5 Bottom-10 countries based on average ranking of the HPI indicators

Rkav	Country	HPI	EFP	HI (Rkav)
131	Gabon	120	2.02	153
132	Trinidad and Tobago	130	7.92	90
133	Benin	137	1.41	151
134	Estonia	118	6.86	87
135	South Africa	128	3.31	136
136	Djibouti	127	2.19	na
137	Latvia	121	6.29	82
138	Botswana	126	3.83	130
139	Turkmenistan	134	5.47	49
140	Mongolia	136	6.08	99

For comparison the original HPI and the ecological footprint are given in addition to the average ranking of the same countries applying the HI indicators. All 2016 data

Fig. 3 Relative importance of the three original indicators used to generate the 2016 Happy Planet Index plus the Purchasing Power Parity (Jeffrey et al. 2016)

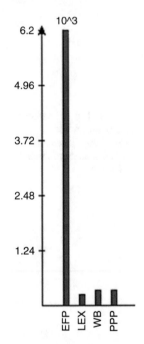

compares different countries' currencies through a "basket of goods" approach. In Fig. 3 the relative indicator importance is visualized.

In excellent agreement with the HI it is seen that again the financial aspect plays a very minor role. However, not surprisingly inclusion of the PPP indicator does make some changes to the average HPI ranking both in the top-10 (Table 6) and the bottom-10 (Table 7). Of the more significant changes Norway, Denmark and Luxembourg can be mentioned (Table 6) where Norway climbs to the top rank, whereas Denmark climbs by 17 places and Luxembourg from 103 to 73,

Table 6 Top-10 countries plus Denmark and Luxembourg based on average ranking of the original three

HPI (Rkav)	Country	HPI	EFP	HI (Rkav)
1	Norway	12	4.98	5
2	Spain	15	3.67	32
3	Colombia	3	1.87	56
4	Pakistan	36	0.79	47
5	Philippines	20	1.1	51
6	Uruguay	14	2.91	33
7	Bangladesh	8	0.72	84
8	Palestine	22	1.19	138
9	Netherlands	18	5.28	8
10	Costa Rica	1	2.84	20
37	Denmark	32	5.51	7
73	Luxembourg	139	15.82	21

HPI indicators plus the PPP indicator. For comparison the original HPI and the ecological footprint are given in addition to the average ranking of the same countries applying the HI indicators. All 2016 data

Table 7 Bottom-10 countries plus Denmark and Luxembourg based on average ranking of the original three

Rkav	Country	HPI	EFP	HI (Rkav)
131	Benin	137	1.41	151
132	Trinidad and Tobago	130	7.92	90
133	Mauritania	117	2.54	125
134	Guinea	129	1.41	97
135	Niger	122	1.56	127
136	Djibouti	127	2.19	na
137	Estonia	118	6.86	87
138	Latvia	121	6.29	82
139	Turkmenistan	134	5.47	49
140	Mongolia	136	6.08	99

HPI indicators plus the PPP indicator. For comparison the original HPI and the ecological footprint are given in addition to the average ranking of the same countries applying the HI indicators. All 2016 data

Table 8 Comparison between the four indicators for Norway and Luxembourg

HPI	Country	EFP	LEX_{IA}	EWB_{IA}	PPP
12	Norway	4.98	78.60	7.42	101,564
139	Luxembourg	15.82	78.97	6.70	105,447

in agreement with the relative high PPP for these countries. Hence, the PPP for Denmark, Norway and Luxembourg in 2016 were 57,636, 101,564 and 105,447 thousand USD, respectively (Jeffrey et al. 2016). For comparison the PPP for Bangladesh in 2016 was only 859 thousand USD (Jeffrey et al. 2016).

A direct comparison between Norway and Luxembourg is and exemplary case to illustrate the effects of the different indicators (Table 8).

The original HPI rank for the two countries are clearly having 3 positive contributions, i.e., LEX, EWB and PPP, respectively, and one significant negative contribution, i.e., the EFP. Assuming the latter for Luxembourg to be changed by 10 gha/capita the country changes its average HPI ranking from 73 (Table 6) to 24 again supporting the assumption the EFP is the main controlling factor.

4 Conclusions and Outlook

It has been revealed that the most important sub-indicator for our happiness as expressed by the analysis of the World Happiness Index appears to be the 'Dystopia' indicator, which is a rather subjective measurement that fits quite nicely with the Lyubomirsky definition of happiness (Lyubomirsky 2008) as "the experience of joy, contentment, or positive well-being, combined with a sense that one's life is good, meaningful, and worthwhile" as well the Dostoevsky quote:" The greatest happiness is to know the source of unhappiness "(Brainyquote 2001). On the other hand it was found that the gross domestic product per capita in terms of purchasing power parity plays only an inferior role. This latter finding is found again looking at the Happy Planet index. Hence, introducing the GDP expressed as the Purchasing Power Parities (PPP) again discloses the minor role of financial wealth as a factor for sustainability in terms of happiness.

It has been demonstrated that the original ranking based on HPI is significantly different from that based on HI and a posetic based data analysis of the HPI dataset leaves no doubt that the culprit in this respect unequivocally is the ecological footprint, which point directly to the Sustainability Development Goal No. 12, i.e., Responsible consumption and production (SDG 2018). Of less importance for the average HPI ranking is inequality adjusted life expectancy and wellbeing that both increase the HPI. Here reference to Sustainability Development Goal No. 3, i.e., Good health and well-being and No. 10, i.e., Reduced inequalities, appears (SDG 2018) appropriate.

One serious question apparently remains: *Who is paying for our happiness*? The answer appears rather simple as it point to us. Hence, apparently through our (non-sustainable) exploitation of nature we let our planet pay for our happiness! This answer unequivocally leads to a further question: *Are we ready for a change*? The more optimistic answer is a maybe, as there might still be time. Let the words by Frederika Stahl (2015) from 'The world to come' close this:

I breathe you in
Soon you'll be gone
Look at the mess you're in
See what we've done

The more pessimistic, also expressed by Frederika Stahl is:

I breathe you in
Kiss you one last goodbye
We knew that we could save you
But never really tried

References

Brainyquote. (2001). https://www.brainyquote.com/quotes/fyodor_dostoevsky_154347

Bruggemann, R., & Annoni, P. (2014). Average heights in partially ordered sets. *MATCH – Communications in Mathematical and in Computer Chemistry, 71*, 117–142.

Bruggemann, R., & Carlsen, L. (Eds.). (2006a). *Partial order in environmental sciences and chemistry*. Berlin: Springer.

Bruggemann, R., & Carlsen, L. (2006b). Introduction to partial order theory exemplified by the evaluation of sampling sites. In R. Bruggemann & L. Carlsen (Eds.), *Partial order in environmental sciences and chemistry* (pp. 61–110). Berlin: Springer.

Bruggemann, R., & Carlsen, L. (2011). An improved estimation of averaged ranks of partially orders. *MATCH – Communications in Mathematical and in Computer Chemistry, 65*, 383–414.

Bruggemann, R., & Carlsen, L. (2012). Multicriteria decision analyses. *Viewing MCDA in terms of both process and aggregation methods: some thoughts, motivated by the paper of Huang, Keisler and Linkov Sci. Total Environ. 425*, 293–295.

Bruggemann, R., & Münzer, B. (1993). A graph-theoretical tool for priority setting of chemicals. *Chemosphere, 27*, 1729–1736.

Bruggemann, R., & Patil, G. P. (2011). *Ranking and prioritization for multi-indicator systems – Introduction*. New York: Springer.

Bruggemann, R., & Voigt, K. (1995). An evaluation of online databases by methods of lattice theory. *Chemosphere, 31*, 3585–3594.

Bruggemann, R., & Voigt, K. (2008). Basic principles of Hasse diagram technique in chemistry. *Combinatorial Chemistry & High Throughput Screening, 11*, 756–769.

Bruggemann, R., Sørensen, P. B., Lerche, D., & Carlsen, L. (2004). Estimation of averaged ranks by a local partial order model. *Journal of Chemical Information and Computer Sciences, 44*, 618–625.

Bruggemann, R., Carlsen, L., Voigt, K., & Wieland, R. (2014). PyHasse software for partial order analysis: Scientific background and description of selected modules. In R. Bruggemann, L. Carlsen, & J. Wittmann (Eds.), *Multi-indicator systems and modelling in partial order* (pp. 389–423). Springer: New York.

Carlsen, L. (2018). Happiness as a sustainability factor. the world happiness index. A Posetic based data analysis. *Sustainability Science, 13*, 549–571. https://doi.org/10.1007/s11625-017-0482-9.

De Loof, K., De Meyer, H., & De Baets, B. (2006). Exploiting the lattice of ideals representation of a poset. *Fundamenta Informaticae, 71*, 309–321.

Ernesti, J., & Kaiser, P. (2008). *Python – Das umfassende Handbuch*. Bonn: Galileo Press.

Hetland, M. L. (2005). *Beginning Python – From Novice to professional*. Berkeley: Apress.

HI. (2016). World Happiness Report 2016. Helliwell, J., Layard, R. and Sachs, J., Eds.; http://worldhappiness.report/ed/2016/

HI. (2017). World Happiness Report 2017. Helliwell, J., Layard, R., & Sachs, J., Eds.; http://worldhappiness.report/ed/2017/

HI. (2018). World Happiness Report 2018. Helliwell, J., Layard, R., & Sachs, J., Eds.; http://worldhappiness.report/ed/2018/

HRI. (2012). *Sustainable happiness*. Danish Ministry of Environment: Why waste prevention may lead to an increase in quality of life. http://mst.dk/media/130530/141203-sustainable-happiness.pdf.

Jeffrey, K., Wheatley, H., & Abdallah, S. (2016). *The happy planet index: 2016. A global index of sustainable well-being*. https://static1.squarespace.com/static/5735c421e321402778ee0ce9/t/57e0052d440243730fdf03f3/1474299185121/Briefing+paper+-+HPI+2016.pdf

KJBO. (2016). King James Bible. The preserved and living word of god, acts 20:35.; http://www.kingjamesbibleonline.org/Acts-Chapter-20/

Langtangen, H. P. (2008). *Python scripting for computational science*. Berlin: Springer.

Lerche, D., Sørensen, P. B., & Bruggemann, B. (2003). Improved estimation of the ranking probabilities in partial orders using random linear extensions by approximation of the mutual ranking probability. *Journal of Chemical Information and Computer Sciences, 43*, 1471–1480.

Lyubomirsky, S. (2008). *The how of Happiness. A new approach to getting the life you want*. Penguin Press, New York.; https://www.amazon.com/How-Happiness-Approach-Getting-Life/dp/0143114956

McConnell, A. R. (2010). Giving really is better than receiving. Does giving to others (vs. the self) promote happiness? Psychology Today; https://www.psychologytoday.com/blog/the-social-self/201012/giving-really-is-better-receiving

Morton, J., Pachter, L., Shiu, A., Sturmfels, B., & Wienand, O. (2009). Convex Rank Tests and Semigraphoids. *SIAM Journal on Discrete Mathematics, 23*, 1117–1134.

Munda, G. (2008). *Social multi-criteria evaluation for a sustainable economy* (Operation) (p. 227). Heidelberg/New York: Springer. https://doi.org/10.1007/978-3-540-73703-2.

Nef. (2016). *Happy planet index 2016*. Methods paper. https://static1.squarespace.com/static/5735c421e321402778ee0ce9/t/578cc52b2994ca114a67d81c/1468843308642/Methods+paper_2016.pdf. Accessed Feb 2019

Python. (2015). *Python*. https://www.python.org/. Assessed Aug 2018

SDG. (2018). *Sustainability development goals*. United Nations, Division for Sustainable Development Goals, https://sustainabledevelopment.un.org/?menu=1300

Stahl, F. (2015). 'The world to come' from the album tomorrow. https://genius.com/Fredrika-stahl-the-world-to-come-lyrics

Weigend, M. (2006). *Objektorientierte Programmierung mit Python*. Bonn: mitp-Verlag.

World Bank. (2017). *WV. World development indicators: Size of the economy*. http://wdi.worldbank.org/table/WV.1

Posetic Tools in the Social Sciences: A Tutorial Exposition

Marco Fattore and Alberto Arcagni

1 Why Partial Orders in the Social Sciences?

Why is partial order theory of interest in the social sciences? Simply because many socio-economic problems are naturally conceptualized and formalized in terms of *order relations* and must then be addressed in ordinal terms, i.e. by using concepts and tools from the theory of partial orders (Davey and Priestley 2002). A typical example is the evaluation of social traits, like deprivation or well-being, when statistical units (e.g. individuals or households) are scored against multidimensional systems of ordinal attributes (i.e. data systems with many variables or indicators measured on a set of statistical units). If the observed achievement profiles of the units have so-called conflicting scores (e.g. unit *a* scores better than unit *b* on a dimension and scores worse on another), and this is quite often the case, data can be ordered only partially, producing a partially ordered set (poset). What kind of information can be extracted out of such a data structure and how? Here the theory of order relations comes into play and provides the proper analytical toolbox.

Remark Interestingly, partial orders are useful also when numerical data systems are to be addressed and one does not want to, or cannot, mix variables through aggregated procedures, like those leading to composite indicators. In this respect, one can argue whether using ordinal scales and partially ordered structures is intrinsic to the phenomena under study or whether it depends upon the perspective taken by the researcher. In any case, the data structure and the tools adopted in

M. Fattore (✉)
Department of Statistics and Quantitative Methods, University of Milano-Bicocca, Milan, Italy
e-mail: marco.fattore@unimib.it

A. Arcagni
Department MEMOTEF, Sapienza University of Rome, Rome, Italy
e-mail: alberto.arcagni@uniroma1.it

© The Author(s), under exclusive license to Springer Nature Switzerland AG 2021
R. Bruggemann et al. (eds.), *Measuring and Understanding Complex Phenomena*,
https://doi.org/10.1007/978-3-030-59683-5_15

any statistical analysis must be as faithful and consistent as possible with the phenomenon of interest and, in many situations, posets are the appropriate choice.

Although they have not attracted much statistical research yet, partial orders are ubiquitous in socio-economics and make their appearance whenever multi-criteria decision problems based on multi-indicator systems (MISes) are to be addressed (so, for example, partial orders are also applied in environmental chemistry or ecology, just to mention two further scientific research disciplines). To give the flavor of possible applications of partial order theory to concrete problems in the socio-economic field, we list a few cases that can be found in literature.

1. **Refugees' relocation in the EU** (Carlsen 2017). The study builds a ranking of European countries, for refugees' relocation, based on a multidimensional system of indicators reflecting territorial refugee absorption capacity.
2. **Temporal analysis of the Fragile/Failed State Index** (Carlsen and Bruggemann 2014, 2017). Here, poset theory is used to investigate the Fragile and the Failed State Index, by considering explicitly the multidimensional configurations of scores on the attributes/variables used in the assessment process, with the aim to identify temporal trends and anomalous units.
3. **European opinions on services** (Annoni and Bruggemann 2009). This study investigates Eurobarometer data, on social and political aspects of citizens' daily life, through purely "ordinal" computations, clustering European countries in terms of satisfaction/dissatisfaction for the prices of different goods/services.p
4. **Comparison of fiscal policies**. These studies (Bachtrögler et al. 2016; Badinger and Reuter 2015) analyze, from a multidimensional point of view, the properties of the fiscal frameworks and the fiscal rules of 81 countries, building a ranking of fiscal stringency.
5. **Immigrants' deprivation and vulnerability** (Arcagni et al. 2019). Here, the deprivation and the economic vulnerability of immigrants, in the Italian region of Lombardy, are evaluated. The individual deprivation profiles of a sample of units are multidimensionally compared to some reference profiles, getting overall deprivation measures for the entire population of immigrants and for its main subgroups.
6. **Ranking Emergency Departments**. In di Bella et al. (2018), partial order theory is used to build a ranking of Emergency Departments in the Italian Liguria region, assessed against a multi-dimensional system of attributes, addressing a very relevant issue, given the increasing importance of emergency services in regional healthcare systems.
7. **Life satisfaction**. In Caperna and Boccuzzo (2018), partial order theory is applied to the study of life satisfaction in Italy, adapting posetic analytical tools to a big data setting and revealing significant differences in satisfaction degrees among regions, between genders and among levels of formal education.

8. **Supporting employees with intellectual disabilities**. Partial order theory has been recently used to support the definition of the requirements, for the development of an assistive control panel and display interface to enable employees with intellectual disabilities to extend their range of tasks and to increase their level of responsibility (Fuhrmann et al. 2018).

All in all, what these examples show is that the concepts and the tools from partial order theory allow for "classical" problems (like ranking or evaluation...) to be consistently addressed in an ordinal setting, paving the way to deeper, more reliable and more effective representations of complex socio-economic traits. This should be enough to acknowledge the key role of partial order theory in supporting decision-making processes in an increasingly complex world (for a more general discussion, see Fattore and Maggino 2014).

2   A Few Technical Notes

In this section, we fix the terminology and collect some essential concepts of partial order theory, also briefly touching upon available software resources, for practical applications. More details and the mathematical proofs can be found in cited references.

2.1   Basic Definitions

A partially ordered set (or a *poset*) $\pi = (X, \unlhd_\pi)$ is a set X endowed with a partial order relation $\unlhd_\pi$, i.e. with a *reflexive, antisymmetic* and *transitive* binary relation (Davey and Priestley 2002; Schröder 2016). Two elements x_i and x_j of the poset are called *comparable*, if either $x_i \unlhd_\pi x_j$ or $x_j \unlhd_\pi x_i$, otherwise they are called *incomparable* (written $x_i \|_\pi x_j$). A poset where any two elements are comparable is called a *linear order* or a *complete order* or a *total order*. A subset of a poset is called a *chain* if any two of its elements are comparable: at the opposite, it is called an *antichain*, if any two of its elements are incomparable. The *upset* of an element $x_i \in \pi$, written $x_i \uparrow$, is the set of elements dominating it: $x_i \uparrow = \{x \in \pi : x_i \unlhd_\pi x\}$; analogously, the *downset* of $x_i \in \pi$, written $x_i \downarrow$, is the set of elements dominated by x_i, i.e. $x_i \downarrow = \{x \in \pi : x \unlhd_\pi x_i\}$. Given $x_i, x_j \in \pi$, x_j is said to *cover* x_i (written $x_i \prec_\pi x_j$) if $x_i \unlhd_\pi x_j$ and there is no other element $x_h \in \pi$ ($x_h \neq x_i, x_j$) such that $x_i \unlhd_\pi x_h \unlhd_\pi x_j$. If the poset has a finite number n of elements, the cover relation determines the partial order relation, since $x_i \unlhd_\pi x_j$ holds if and only if there exists a sequence of elements $x_0, x_1, \ldots, x_k$, such that $x_i = x_0 \prec_\pi x_1 \prec_\pi \ldots \prec_\pi x_k = x_j$. In the following, we consider only finite posets.

2.2 Matrix Representations of Posets

The simplest way to represent algebraically finite partial order relations is by means of the $n \times n$ *incidence* matrix Z, defined as $Z_{ij} = 1$ if $x_i \trianglelefteq_\pi x_j$ and $Z_{ij} = 0$ otherwise. Alternatively, one can define the $n \times n$ *cover* matrix G, whose entries are given by $G_{ij} = 1$ if $x_i \prec_\pi x_j$ and $G_{ij} = 0$ otherwise. Since the cover relation and the partial order relation determine each other, matrices Z and G can be obtained one from the other, by simple algebraic formulas (Patil and Taillie 2004). Matrices Z and G are extremely useful in practical computations and provide easy ways to inspect and investigate relevant features of the underlying poset.

2.3 Graphical Representation of Posets

When the number of elements is small enough, posets can be depicted graphically, in various ways (Neggers and Kim 1998). The most useful and widely adopted one is by means of Hasse diagrams, which are directed acyclic graphs, reproducing the cover relation. In a Hasse diagram, poset elements are represented by *vertices*, or *nodes*; if $x_i \trianglelefteq_\pi x_j$, then node corresponding to x_j is placed higher than node corresponding to x_i and, if $x_i \prec_\pi x_j$, an edge is inserted between them. By transitivity, one can then recover all of the comparabilities of the input poset. Many examples of Hasse diagrams will be shown in the rest of the chapter (and across the entire book).

2.4 Linear Extensions

Given two partially ordered sets $\pi = (X, \trianglelefteq_\pi)$ and $\sigma = (X, \trianglelefteq_\sigma)$ on the same set X, we say that σ is an extension of π, if it is obtained from the latter by turning some incomparabilities into comparabilities, i.e. if $x_i \trianglelefteq_\pi x_j$ in π implies $x_i \trianglelefteq_\sigma x_j$ in σ. If σ is an extension of π and also a linear order, then it is called a *linear extension* of π. The linear extensions of π are all the possible orderings of elements of π which are compatible with the partial order relation $\trianglelefteq_\pi$, i.e. that do not switch or eliminate any comparability of π. A simple, but fundamental, theorem in partial order theory states that any finite poset π is uniquely identified by the set $\Omega(\pi)$ of its linear extensions. As discussed later in the chapter, this is a key property for applications of poset theory to data analysis.

2.5 Mutual Ranking Probabilities

If $x_i \trianglelefteq_\pi x_j$ in poset π, then it is also $x_i \trianglelefteq_\lambda x_j$ in each linear extension λ of π; on the contrary, if $x_i \|_\pi x_j$, i.e. if the two elements are incomparable in π, then in some linear extension λ we have $x_i \trianglelefteq_\lambda x_j$ and in some other linear extension ρ we have $x_j \trianglelefteq_\rho x_i$. The fraction p_{ij} of linear extensions $\lambda \in \Omega(\pi)$ where $x_i \trianglelefteq_\lambda x_j$ is called the *mutual ranking probability* (MRP) of x_j over x_i (De Loof 2009; De Loof et al. 2006, 2008); informally, the MRP expresses the "degree of dominance" of x_j over x_i. MRPs are usually arranged into the $n \times n$ *mutual ranking probability matrix* M, whose entry ij is p_{ij}. As it will be shown in subsequent paragraphs, M plays a key role in practical applications.

2.6 Software Resources

Currently, there are two main software resources to perform statistical analysis on partially ordered data. The PyHasse suite (Koppatz and Bruggemann 2017), developed in Python, is available at https://pyhasse.org/. It provides a huge number of modules and procedures for various statistical analyses. A system of functions for poset manipulation and socio-economic analysis on partially ordered data in the R environment is provided in the package parsec (Arcagni 2017). Some other useful routines, mainly to compute mutual ranking probability matrices, are also available in the package netrankr (Schoch 2017).

3 Posetic Tools in Socio-economics

In this section, we present some of the main posetic tools, for the analysis of ordinal MISes and partially ordered data in socio-economics. We organize the exposition around some reference topics, which are of main interest for social scientists. For each tool, we give the main formulas, summarize its properties and provide a short example of its use on real data.

3.1 Scoring and Ranking Partially Ordered Data

Perhaps the main problem in the study of multi-dimensional MISes and partially ordered data is that of ranking statistical units, based on their score profiles or on their "degree of dominance" with respect to other units. Usually, this requires computing a non-negative *score* function $s(\cdot)$ (i.e. a function assigning to each profile a non-negative real number) and then ranking units based on it. In order to be consistent with the input partial order relation, the score function $s(\cdot)$ is required to be *strictly order preserving*, i.e. such that $x \lhd y$ (which means $x \unlhd y$ and $x \neq y$) in the input poset implies $s(x) < s(y)$; thus, the ranking problem reduces to the definition of "reasonable" strictly order-preserving maps on posets.

In the daily practice of socio-economic statistics, it is quite typical to compute the score function, by coding ordinal scores as numbers and by applying tools from classical data analysis, or even by computing simple averages. This approach is definitely inconsistent, for two main reasons: first, since ordinal scores cannot be treated as cardinals, unless forcing the nature of the data; second, since partially ordered data need not be obtained from MISes, so that no attribute scores even exist (for example, one could partially order products or services based on personal taste, with no explicit reference to any underlying quality dimensions). In a posetic setting, however, all the information useful for scoring and ranking is comprised in the structure of the partial order relation adopted to describe the data and must be extracted out of it. The issue thus becomes how score functions can be computed directly over partial orders.

Currently there are two main algorithms, to score units and to extract rankings out of a partially ordered set, namely the *average height* algorithm (Bruggemann and Patil 2011) and the *dominance eigenvector* algorithm (Fattore et al. 2019); both draw upon mutual ranking probabilities, which carry information on the relative dominance of pairs of poset elements.

3.1.1 Average Height

Given a finite poset π, the average height $avh(x_i)$ of an element $x_i \in \pi$ is defined as the arithmetic mean of the heights of x_i in the linear extensions of π,

$$avh(x_i) = \frac{1}{|\Omega(\pi)|} \sum_{\lambda \in \Omega(\pi)} h_\lambda(x_i), \tag{1}$$

where the height $h_\lambda(x_i)$ of x_i in λ is given by *1 + the number of elements below x_i in λ*. It can be shown (De Loof 2009) that the average height is linked to mutual ranking probabilities by the following simple relation:

$$avh(x_i) = \sum_{j=1}^{n} p_{ji}. \tag{2}$$

Once the average height is computed for each element of the poset, a ranking is obtained by ordering elements in a decreasing way. By construction, the average height is a strictly order preserving map.

3.1.2 Dominance Eigenvector

This scoring procedure is based on the Singular Value Decomposition (Meyer 2000) of the mutual ranking probability matrix M, associated to the input poset π. M is a non-negative matrix and so, by the Perron-Frobenius (Meyer 2000), its first right singular vector $\boldsymbol{v} = (v_1, \ldots, v_n)$, i.e. the eigenvector[1] of $M^T M$ relative to the greatest eigenvalue, has strictly positive components. In addition, it can be proven that such components are such that $x_i \lhd x_j$ in π implies $v_i < v_j$. Therefore, the score function $s(\cdot)$

$$s : \pi \mapsto R^+$$

$$: x_i \to v_i$$

is strictly order preserving and can be used to rank the elements of the input poset. Vector $\boldsymbol{v}$ is a linear combination of the rows of M, so the score associated to element x is a weighted average of the probabilities that x dominates poset elements. By the properties of the Singular Value Decomposition, vector $\boldsymbol{v}$ has the optimal property to provide the best uni-dimensional approximation to the mutual ranking probability matrix; more precisely, it turns out that the rows of the rank-one matrix $\hat{M}$ which best approximates matrix M in the Euclidean norm (here called, Frobenius norm) are proportional to $\boldsymbol{v}$.

Both the average rank and the dominance eigenvector preserve linear orders, i.e. if the input poset is a linear order λ, then the final ranking is λ itself. This natural property is not shared by other scoring functions proposed in the literature (Todeschini et al. 2015; Saaty and Hu 1998), which extract eigenvectors directly from the mutual ranking probability M and not from matrix $M^T M$. Notice also that, in general, scoring functions can produce ties, whenever different profiles occupy "equivalent" positions in the input poset.

3.1.3 Real Example

Table 1 reports the values of three economic indicators for EU-28 member states (year 2017). These indicators are numerical, but refer to different features of national economies; combining them into a composite index can be misleading and so we keep them separate and represent the dataset as a poset, whose Hasse diagram is drawn in Fig. 1. Quite naturally, country j dominates country i in the diagram, if

---

[1]Let A be a square matrix; a vector x is called *eigenvector* of A relative to the *eigenvalue* a if it holds $Ax = ax$, where a is a real number. Eigenvectors, when they exist, provide deep information on the structure of the input matrix and are often used in multivariate statistics, to produce optimal data synthesis.

Table 1 Economic indicators for EU-28 member states: GDP per inhabitant (Purchasing Power Standards); deficit/surplus (% of GDP); gross debt (% of GDP). (Source: Eurostat 2017)

State		GDP per inhabitant	Deficit/surplus	Gross debt
AT	Austria	38,100	−0.8	78.3
BE	Belgium	35,000	−0.9	103.4
BU	Bulgaria	14,800	1.1	25.6
CY	Cyprus	25,400	1.8	96.1
CZ	Czechia	26,900	1.5	34.7
DE	Germany	37,100	1.0	63.9
DK	Denmark	38,400	1.1	36.1
EE	Estonia	23,600	−0.4	8.7
ES	Spain	27,600	−3.1	98.1
FI	Finland	32,700	−0.7	61.3
FR	France	31,200	−2.7	98.5
GR	Greece	20,200	0.8	176.1
HR	Croatia	18,500	0.9	77.5
HU	Hungary	20,300	−2.2	73.3
IE	Ireland	54,300	−0.2	68.4
IT	Italy	28,900	−2.4	131.2
LT	Lithuania	23,500	0.5	39.4
LU	Luxembourg	75,900	1.4	23.0
LV	Latvia	20,000	−0.6	40.0
MT	Malta	29,300	3.5	50.9
NL	Netherlands	38,400	1.2	57.0
PO	Poland	20,900	−1.4	50.6
PT	Portugal	23,000	−3.0	124.8
RO	Romania	18,800	−2.9	35.1
SE	Sweden	36,300	1.6	40.8
SI	Slovenia	25,500	0.1	74.1
SK	Slovakia	22,900	−0.8	50.9
UK	United Kingdom	31,700	−1.8	87.4

the former has no economic indicators worse than the latter and at least one of them which is better (the sign of gross debt has been reversed, so as to have concordant indicators).

In view of the computation of the average height and the dominance eigenvector, we preliminarily compute the matrix of mutual ranking probability of the country poset. To this goal, we must (i) import the data, (ii) define the partial order relation on the set of countries, (iii) compute the incidence matrix of the resulting poset and (iv) finally get the MRP matrix. The R code, employing the package parsec (Arcagni 2017), is reported below.

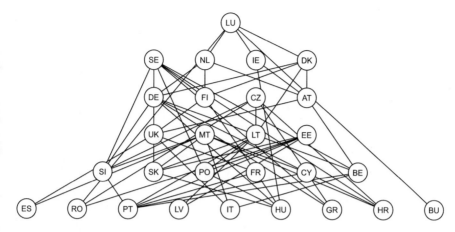

Fig. 1 Hasse diagram for the data reported in Table 1

```
library(parsec)

% data <- data.frame(
%   "GDP per inhabitant" = c(38100, 35000, ... ),
%   "deficit/surplus" = c(-0.8, -0.9, ...),
%   "gross debt" = c(-78.3, -103.4, ...)
% )
% rowames(data) <- c("AT", "BE", ...)
% The above instructions are commented,
% since data are to be completed according to Table 1

X <- rownames(data)
r <- function(x, y) all(data[x,] <= data[y,])
r <- Vectorize(r)
Z <- outer(X, X, FUN = r)
dimnames(Z) <- list(X, X)

% Function Vectorize() produces a wrapper of r
% so as to pass elements one-by-one in the
computation of Z

Z <- validate.partialorder.incidence(Z) % checking
     whether Z
% actually represents a partial order relation

M <- MRP(Z) % MRP matrix computation
```

(function MRP depends upon package netrankr Schoch (2017) that provides different approaches to evaluate the MRP matrix, namely *exact*, *sampled* and

Fig. 2 Average height and
dominance eigenvector
scores, for the data reported
in Table 1 (the vector of
average heights has been
normalized to have euclidean
norm equal to 1, as the
dominance eigenvector)

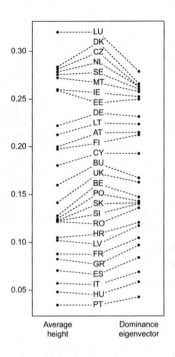

approximated; by default, MRP employs the *exact* approach, which is generally
suitable for small posets). Finally, the average height and the dominance eigenvector
are computed out of M (see Fig. 2).

```
avr_height <- colSums(M)
eigenvector <- abs(svd(M)$v[,1])
```

3.2   Evaluation and Comparison to Multidimensional Benchmarks

A fundamental problem in socio-economics is the evaluation and the measurement
of multidimensional phenomena, like poverty, quality-of-life, well-being, but also
literacy, freedom, sustainability and many others more.... Usually, when the input is
a MIS, evaluation is performed by aggregative procedures, where the input attributes
are combined together into a composite indicator Joint Research Centre-European
Commission et al. (2008). This approach is questionable from many points of view
and, in particular, it proves scarcely consistent and scarcely effective when truly
multidimensional social traits are to be evaluated. The prototypical example is that
of *multidimensional poverty* where, overcoming the GDP-based perspective on the
societal wealth, one tries to consider jointly different aspects of quality-of-life and
well-being, beyond income or consumption (e.g. access to services; employment

status; ownership of goods...). In such cases, there is no common unit of measure among attributes and it is often forcing to look for one. Moreover, poverty data are often, if not always, of an ordinal kind, so that aggregative procedures are technically unfeasible.

Poset theory offers a natural solution to this kind of problems, allowing for multidimensional comparisons among profiles (i.e. set of achievements on attributes) to be performed, without any preliminary attribute aggregation. This not only solves a technical issue, but is also more consistent with the nature of the evaluation problem itself; it is more natural to assess the level of poverty (to take the same example, as above) of a subject by comparing his/her achievements to one or more poverty benchmark profiles, rather than pretending to compress them into an "absolute" deprivation score, to be compared with some numerical threshold. Moreover, poverty is a multifaceted trait that may assume many different shapes; this is easily accounted for in a poset setting, where different incomparable profiles can be chosen as structurally different poverty benchmarks, while aggregative/compensative procedures are forced to identify just a single threshold level. In other words, the posetic approach to evaluation is much more "complexity preserving" than the composite indicator one, being capable to extract information directly from the multidimensional comparison system (i.e. from the partial order relation) and not from a "compressed" unidimensional reduction of the input MIS.

As detailed in Fattore (2016), the posetic evaluation of multidimensional ordinal traits can be performed as follows:

1. Take the set of all possible profiles generated by the attributes of the input MIS and structure them as a poset π.
2. Identify one or more reference profiles, to be used as benchmarks in the evaluation process. These benchmarks represent "one or more alternative reference forms of deprivation", identified based on socio-economic considerations. As such, they must constitute an antichain τ, otherwise they would identify different levels of deprivation, rather than its "border". This antichain is the multidimensional ordinal analogue of the threshold level in aggregative procedures.
3. Given the antichain τ, poset elements can be partitioned into three disjoint subsets: U, comprising elements of π that are ordered above all of the elements of τ; D, comprising τ itself and all of the poset elements that are ordered below *at least* one element of τ and I, comprising poset elements which neither belong to U nor to D. Referring again to the poverty example, U comprises non-deprived profiles, D comprises deprived profiles (consistently with the interpretation of τ as a poverty threshold) and I comprises profiles which are "partly" (in a fuzzy sense) deprived.
4. Two evaluation functions can then be computed. The first, called *identification function* and written $idn(\cdot)$, measures to which degree a poset element belongs to D (in our reference example, this means measuring to what extent a profile can be classified as poor). This function takes value 0 on U, value 1 on D and values in $(0, 1)$ on I. The second function, called *severity function* and written $svr(\cdot)$, computes the "intensity" of the trait, for elements belonging to D or to I. The

identification function $idn(x)$ of a profile $x \in \pi$ is computed as the fraction of linear extensions where x is ordered below at least one element of τ. The severity function $svr(x)$ is computed as the average of the distances of x from the least element above the threshold in each linear extension of π, normalized to $[0, 1]$ (for details and formulas, see Fattore 2016).

5. Each statistical unit in the dataset then inherits the identification and severity scores of the profile it shares.
6. Finally, synthetic indicators of various kind can be computed at population level, starting from the individual scores.

Remark At first, the above procedure can seem somehow "artificial" and deserves a short comment on the "naturality" of the posetic approach. The sequence of steps 1–6 can be seen as a *composition of actions*, each of which is "natural", i.e. inherently consistent with its input (e.g., we structure input data as a poset, we apply algorithms that are consistent with the poset structure...). Assuming that "the composition of consistent actions is overall consistent", we can consider the described procedure as a "natural" way to approach the evaluation of multidimensional deprivation or similar socio-economic traits.

The procedure just outlined can be tuned to different contexts and modified according to the needs of the study, as done in Arcagni et al. (2019), where a "welfare" threshold has been added to the poverty one, to better assess and compare populations' deprivation.

3.2.1 Real Example

To exemplify the above procedure, here we report part of the analysis developed in Arcagni et al. (2019), about deprivation of migrants in Lombardy (Italy). Data come from the 2014 ORIM[2] Survey, which involved 4000 subjects, from different countries of origin, and pertain to migrants' socio-economic conditions.

Among other social traits, the cited paper focuses on migrants' *social fragility*, described through a MIS comprising the following four ordinal attributes:

- *Working dynamics*, coded on a 6-degrees scale: 1 – Persistent unemployment; 2 – Run into unemployment; 3 – Worsening condition 4 – Non-active status; 5 – Improving condition; 6 – Stable condition.
- *Legal status*, coded on a 2-degrees scale: 1 – Illegal; 2 – Legal.
- *Dependent family in country of origin*, coded on a 2-degrees scale: 1 – Yes; 2 – No.
- *One-income family*, coded on a 2-degrees scale: 1 – Yes; 2 – No.

---

[2]ORIM stands for "Osservatorio regionale per l'integrazione e la multietnicità" ("Regional observatory on integration and multiethnicity").

Table 2 Weight distribution on the fragility profiles

Profile	Weight	Profile	Weight	Profile	Weight
1111	0.66	5121	0.00	3212	161.45
2111	0.66	6121	0.25	4212	85.82
3111	2.75	1221	79.33	5212	49.32
4111	0.00	2221	27.98	6212	676.55
5111	0.00	3221	56.37	1122	12.59
6111	7.51	4221	145.62	2122	5.98
1211	15.48	5221	19.61	3122	19.56
2211	21.57	6221	223.55	4122	0.69
3211	59.36	1112	14.23	5122	5.80
4211	45.67	2112	4.22	6122	4.99
5211	19.55	3112	5.58	1222	178.97
6211	250.44	4112	1.59	2222	104.29
1121	1.01	5112	2.72	3222	181.55
2121	0.00	6112	13.55	4222	402.49
3121	7.87	1212	57.92	5222	94.38
4121	0.00	2212	51.45	6222	732.79

To evaluate migrants' fragility, we first compute the set of all of the profiles generated by the above attributes (see Table 2):

```
library(parsec)

prf <- var2prof(varmod = list(
   "Working dynamics" = 1:6,
   "Legal status" = 1:2,
   "Dependent family in country of origin" = 1:2,
   "One-income family" = 1:2
))
```

Next, we set the fragility threshold to the pair of profiles {3122, 3221} and pass the profiles to the evaluation function to compute the identification and severity scores (the function, in an automatic way, builds a poset according to the "natural" criterion that profile j is less fragile than profile i if the former has no attribute worse than the latter and at least one which is better):

```
res <- evaluation(
   prf,
   threshold = c("3122", "3221"),
   weights = data
)
```

where object data is a vector assigning to each profile the weights reported in Table 2 obtained as sums of sample weights. The procedure produces a list of results comprising, among other information, vectors res$idn_f and res$svr_abs, i.e. the identification and severity functions, respectively. The following code produces their frequency and cumulative distributions, depicted in Fig. 3.

```
ord <- order(res$idn_f)
plot(
  res$idn_f[ord], res$prof_w[ord],
  type = "h",
  xlab = "Identification",
  ylab = "Frequency"
)
plot(
  res$idn_f[ord], cumsum(res$prof_w[ord]),
  type = "s",
  xlab = "Identification",
  ylab = "Cumulative frequency"
)

ord <- order(res$svr_abs)
plot(
  res$svr_abs[ord], res$prof_w[ord],
  type = "h",
  xlab = "Severity",
  ylab = "Frequency"
)
plot(
  res$svr_abs[ord],
  cumsum(res$prof_w[ord]),
  type = "s",
  xlab = "Severity",
  ylab = "Cumulative frequency"
)
```

3.3 Comparing Populations Over Posets

When two or more populations must be compared on a MIS, the easiest approach is to score each statistical unit on the latent trait underlying the indicators and then to compare the resulting distributions, either by using some synthetic indicator (typically in the class of power means), or in terms of stochastic dominance. In any case, some information get lost, when multidimensional data are "compressed" to unidimensional scores, prior to the comparison. A more efficient and powerful approach is to compare the distributions *directly over the poset generated by the indicator system*, extending stochastic dominance to partially ordered structures. In its essence, the idea is quite simple. Let π be the poset generated by the MIS

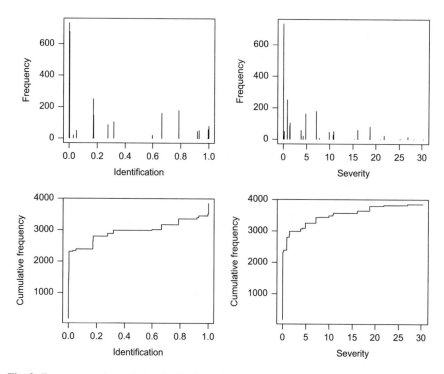

Fig. 3 Frequency and cumulative distributions of the identification and severity functions

and let $\Omega(\pi)$ be the set of its linear extensions. On each linear extension of π, the distributions can be compared, by using the standard first-order dominance criterion[3] (Fattore and Arcagni 2018). This way, to each pair of populations i and j, on each linear extension λ of π, a stochastic dominance degree Δ_{ij}^{λ} is associated, measuring to what extent population j stochastically dominates population i, on λ. Averaging such degrees over $\Omega(\pi)$, one gets an overall *fuzzy first-order dominance* degree Δ_{ij} between all pairs of distributions (Fattore and Arcagni 2018). The resulting matrix Δ comprises all of the information on pairwise dominance among the populations; although its entries are not mutual ranking probabilities, a final population ranking can be obtained, by using the dominance eigenvector approach described previously.

---

[3]Given two cumulative distributions $F(t)$ and $G(t)$, defined on the same totally ordered set T (which can be either continuous or discrete), we say that G first-order dominates F, if $G(t) \leq F(t)$, $\forall t \in T$. As described in Fattore and Arcagni (2018), the notion of stochastic dominance can be made fuzzy, computing a degree of dominance in [0, 1].

3.3.1 Real Example

We report here a simple example from the original paper (Fattore and Arcagni 2018), pertaining to the comparison of populations' health. Data are taken from The National Health Interview Survey 2010, held in Denmark. Following Hussain et al. (2016), four binary dimensions of health are considered[4] and scored in binary terms (1 – No health problem; 0 – Health problem) on five populations distinguished by the educational attainment of the respondents ($D_1 = Basic$, $D_2 = Vocational$, $D_3 = Shorthigher$, $D_4 = Mediumhigher$, $D_5 = Longhigher$). Table 3 reports the relative frequency distributions of the populations, on the set of profiles over the four health dimensions. Such distributions are compared as described above, getting the matrix of pairwise fuzzy dominance degrees, reported in Table 4. Synthetic dominance scores extracted by the dominance eigenvector are reported in the last column of the same table. As it can be seen, the health level increases with the educational attainment of individuals; in particular, there are noticeable score gaps (negative and positive, respectively), between the least and the most trained populations, with respect to the subgroups with middle educational level. The R code producing the results shown in tables is reported below.

Table 3 Frequency distributions on the health profiles, for levels of educational attainment (data have been expressed as relative frequencies)

Profile	Basic	Vocat.	Short	Medium	Long
0000	0.0912	0.0254	0.0189	0.0220	0.0096
1000	0.0415	0.0369	0.0512	0.0238	0.0065
0100	0.0092	0.0016	0.0038	0.0062	0.0011
1100	0.0930	0.0810	0.0814	0.0669	0.0528
0010	0.0384	0.0248	0.0224	0.0159	0.0032
1010	0.0790	0.0553	0.0452	0.0407	0.0091
0110	0.0217	0.0052	0.0049	0.0045	0.0000
1110	0.2337	0.2294	0.1995	0.1709	0.1232
0001	0.0144	0.0164	0.0100	0.0060	0.0103
1001	0.0405	0.0337	0.0603	0.0517	0.0476
0101	0.0022	0.0014	0.0039	0.0028	0.0023
1101	0.0679	0.1162	0.1358	0.1826	0.2067
0011	0.0183	0.0094	0.0000	0.0041	0.0095
1011	0.0555	0.0554	0.0408	0.0485	0.0638
0111	0.0038	0.0045	0.0040	0.0032	0.0074
1111	0.1896	0.3034	0.3180	0.3501	0.4468

---

[4]Dimension 1: Subjective and self-reported health. Dimension 2: Pain or discomfort in shoulder, back, arms, legs…; headaches; sleeping problems, depression, anxiety…Dimension 3: Asthma, allergy; migraine; diabetes; hypertension; chronic bronchitis. Dimension 4: Tobacco use; excessive alcohol consumption, obesity; unhealthy life style…

Table 4 Matrix Δ of pairwise fuzzy dominance degrees between distributions of Table 3 (element Δ_{ij} is the degree of dominance of distribution j over distribution i); the last column reports the final score from the dominance eigenvector

Profile	Basic	Vocat.	Short	Medium	Long	Score
Basic	1.00	0.67	0.67	0.70	0.76	0.375
Vocat.	0.47	1.00	0.59	0.62	0.70	0.441
Short	0.47	0.58	1.00	0.62	0.70	0.443
Medium	0.44	0.56	0.56	1.00	0.68	0.462
Long	0.38	0.50	0.51	0.55	1.00	0.505

```
library(parsec)
prf <- var2prof(varmod = list(
    dim1 = 0:1,
    dim2 = 0:1,
    dim3 = 0:1,
    dim4 = 0:1
))
res <- FFOD(profiles = prf, distributions = data)
```

(data is an object of class data.frame replicating Table 3). The output of the above code is an object of class FODposet which comprises matrix Δ (extracted by res$delta) and various other results on pairwise dominance degrees. By calling

```
abs(svd(res$delta)$v[,1])
```

the Singular Value Decomposition of Δ is finally computed, getting the dominance eigenvector.

3.4 Synthetic Indicators Over Posets and the Measurement of Inequality

A major problem, in socio-economic statistics, is the computation of synthetic indicators for frequency distributions defined over ordinal multidimensional MISes. This issue combines together two sources of complexity, namely multidimensionality and "ordinality". The first poses non-trivial conceptual problems. For example, when extending inequality measures from the unidimensional case, it is necessary to state unambiguously what is meant by a "more unequal" distribution in a multidimensional setting, an issue that is not trivial to solve, given the increased number of degrees of freedom in the shape of multidimensional distributions. On the other hand, dealing with ordinal attributes adds technical difficulties since, as previously discussed, mathematical tools designed for cardinal variables cannot be employed. Again, partial order theory provides the right conceptual and formal setting, to overcome both these issues.

The formal development of the theory of synthetic indicators over posets is not trivial (details can be found in Fattore 2017), but the basic idea is quite simple and relies, again, on the equivalence between finite posets and their set of linear extensions. Let π be an n-element poset generated by a MIS and let $\boldsymbol{p} = (p_1, \ldots, p_n)$ be a relative frequency distribution over it (so that p_i is the fraction of statistical units associated to the i-th element of π). Suppose, to set the stage, that we want to measure how unequally $\boldsymbol{p}$ is distributed on π, by means of a non-negative synthetic indicator $G(\boldsymbol{p}, \pi)$. Notice that $G(\cdot, \cdot)$ not only depends upon the frequency vector $\boldsymbol{p}$, but also upon the underlying order structure, i.e. upon poset π. To see why this dependence is essential, just consider two posets on three elements $\{a, b, c\}$, $\pi_1 = a \lhd b \lhd c$ (a chain) and $\pi_2 = a||b||c$ (an antichain), and let $p_a = 0.5$, $p_b = 0$, $p_c = 0.5$ be the frequency distribution defined on both of them. Intuitively, in the chain case inequality is higher, since elements a and c are at the "vertical extremes" of the poset, while in the antichain case there is no such "vertical" dimension. Keeping fixed the frequency distribution $\boldsymbol{p}$, indicator $G(\cdot, \cdot)$ is thus required to depend upon the underlying ordinal relation and, in particular, not to decrease as the poset gets extended. In addition, since all finite posets can be reconstructed by their linear extensions, the value of $G(\cdot, \cdot)$ on π is also required to be a function of the degree of inequality of $\boldsymbol{p}$ on the linear extensions of π; as derived in Fattore (2017), such a function must belong to the class of power means. In summary, the inequality degree of a frequency distribution on a poset π is computed as some power mean of its inequality degrees over the linear extensions of π. But linear extensions are complete orders and, on them, inequality can be measured by using any of the ordinal unidimensional indicators available in literature Maggino and Fattore (2019). This way, the measurement of inequality on a complex ordered structure gets reduced to an aggregation of inequality indicators over simple linear orders. Needless to say, this very same approach can be applied to many other kinds of synthetic indicators, for which unidimensional ordinal counterparts are available.

3.4.1 Real Example

We apply the procedure outlined above to the measurement of inequality of childhood poverty in the Democratic Republic of Congo (DRC). Data are taken from Table 3 of Nanivazo (2015), here replicated in Table 5, and assess poverty in terms of four binary attributes (1 – Deprivation; 0 – No deprivation):

1. *Sanitation deprivation* – Children with no access to any kind of improved latrines or toilets.
2. *Water deprivation* – Children with only access to surface water for drinking or for whom the nearest source of water is more than a 15 min walking distance from their dwellings.
3. *Shelter deprivation* – Children living in dwellings with more than five people per room or with no flooring material (e.g., a mud floor).

Table 5 Frequency distributions on deprivation profiles for the Democratic Republic of Congo and its regions: *DRC* democratic republic of congo, *KSS* kinshasa, *BCO* bas-congo, *BDD* bandundu, *ETR* equateur, *KOC* kasai-occidental, *KOT* kasai-oriental, *KTG* katanga, *MNM* maniema, *NKV* north-kivu, *ORT* orientale, *SKV* south-kivu. (Source: Nanivazo Nanivazo 2015)

Profile	DRC	KSS	BCO	BDD	ETR	ORT	NKV	MNM	SKV	KTG	KOT	KOC
0000	0.31	0.03	0.31	0.34	0.57	0.47	0.40	0.19	0.22	0.22	0.17	0.36
0001	0.15	0.03	0.24	0.40	0.16	0.08	0.07	0.11	0.10	0.06	0.07	0.22
0010	0.03	0.02	0.01	0.11	0.01	0.01	0.01	0.01	0.02	0.01	0.01	0.02
0011	0.01	0.03	0.02	0.02	0.01	0.00	0.00	0.00	0.00	0.00	0.00	0.01
0100	0.14	0.01	0.06	0.07	0.13	0.25	0.08	0.35	0.16	0.10	0.24	0.16
0101	0.06	0.01	0.10	0.02	0.05	0.04	0.04	0.17	0.05	0.07	0.11	0.11
0110	0.01	0.02	0.03	0.00	0.00	0.00	0.02	0.01	0.03	0.00	0.04	0.01
0111	0.01	0.02	0.02	0.00	0.01	0.00	0.02	0.02	0.03	0.00	0.00	0.01
1000	0.04	0.04	0.01	0.02	0.01	0.03	0.15	0.03	0.11	0.06	0.02	0.03
1001	0.03	0.06	0.02	0.01	0.00	0.01	0.10	0.03	0.03	0.06	0.03	0.06
1010	0.02	0.06	0.01	0.00	0.00	0.02	0.03	0.01	0.05	0.02	0.02	0.00
1011	0.02	0.12	0.08	0.00	0.00	0.00	0.01	0.00	0.01	0.02	0.01	0.01
1100	0.04	0.04	0.00	0.00	0.01	0.04	0.01	0.03	0.04	0.10	0.08	0.00
1101	0.04	0.09	0.02	0.00	0.00	0.03	0.01	0.02	0.04	0.08	0.12	0.00
1110	0.03	0.12	0.02	0.00	0.01	0.01	0.03	0.00	0.04	0.07	0.05	0.00
1111	0.06	0.31	0.06	0.00	0.00	0.01	0.01	0.00	0.07	0.13	0.03	0.00

4. *Health deprivation* – Children for whom the nearest health service provider is more than a 15 min walking distance from their dwellings.

The deprivation poset π is composed of 16 binary profiles (see first column of Table 5), partially ordered as depicted in Fig. 4.

As unidimensional inequality index (normalized to [0,1]), we adopt the so-called *Leti index* (Leti 1983)

$$I(\boldsymbol{p}, \lambda) = \frac{4}{n-1} \sum_{x_i \trianglelefteq_\lambda x_j} p_i p_j \left[h_\lambda(x_i) - h_\lambda(x_j) \right]$$

where $\lambda = (X, \trianglelefteq_\lambda)$ is a linear extension of the poset π and $h_\lambda(x_s)$ is the height of x_s in λ. Inequality over π is then computed as the average of $I(\boldsymbol{p}, \lambda)$ over the set of linear extensions of π:

$$I(\boldsymbol{p}, \pi) = \frac{1}{|\Omega(\pi)|} \sum_{\lambda \in \Omega(\pi)} I(\boldsymbol{p}, \lambda).$$

The computation of the above index for DRC and each of its regions can be performed using the function `evaluation` in package `parsec`:

```
library(parsec)
prf <- var2prof(varmod = list(
```

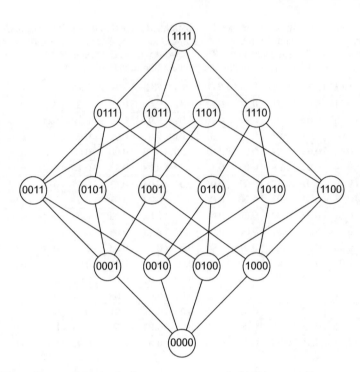

Fig. 4 Hasse diagram of the deprivation poset π composed of 16 binary profiles

```
    Water = 0:1,
    Sanitation = 0:1,
    Shelter = 0:1,
    Health = 0:1
))
res <- sapply(data, function(x)
    evaluation(prf,
        threshold = "0000",
        weights = x,
        inequality = TRUE
    )$inequality
)
```

(data is a data.frame replicating Table 5). In order to run function evaluation, a threshold must be provided; since we are not interested in the identification or severity functions, this can be set to any profile. The final inequality measures (normalized to [0,1]) are shown in Table 6.

Table 6 Inequality measures for DRC and its regions

DRC	KSS	BCO	BDD	ETR	ORT	NKV	MNM	SKV	KTG	KOT	KOC
0.6848	0.6484	0.7199	0.2751	0.3254	0.4423	0.5660	0.4769	0.7172	0.8229	0.6961	0.4417

4 Future Research and Perspectives

In the previous sections, we have concisely presented the main motivations why partial order theory is of interest for the statistical analysis of socio-economic data, providing an overview of the main posetic tools currently available for practical applications. Surely this is just a rough summary of what can be achieved using posetic algorithms and much more details and examples can be found in the references cited along the text. As a matter of fact, however, the use of posets in socio-economic analysis is still at its early stage and several research paths are open and require further investigation. There are at least four main fields of development, to be carried on. The first is, in a sense, "cultural": as socio-economic phenomena get increasingly complex and nuanced, social scientists must change the way they represent and measure them, accepting that complexity is irreducible and that it must be dealt with and accounted for. Posets, which have both a "vertical" dimension (comparability) and a "horizontal" dimension (incomparability), naturally account for both the "intensity" of multi-dimensional ordinal socio-economic traits and for their intrinsic "variability" and can represent them in a way that fits better to their structure. This, however, requires some mind-changing and the acquisition of a new language. This point of view is brilliantly exposed by Sen (Sen (1992), pages 48–49): "[…] if an underlying idea has an essential ambiguity, a *precise* formulation of that idea must try to *capture* that ambiguity rather than lose it. Indeed, the nature of interpersonal comparisons of well-being as well as the task of inequality evaluation as a discipline may admit incompleteness as a regular part of the respective exercises. An approach that can rank the well-being of every person against that of every other in a straightforward way, or one that can compare inequalities without any room for ambiguity or incompleteness, may well be at odds with the nature of these ideas. Both well-being and inequality are broad and partly opaque concepts. Trying to reflect them in the form of totally complete and clear-cut orderings can do less than justice to the nature of these concepts." The second field concerns the development of new statistical procedures, to address data analysis problems not solved yet, in an ordinal setting. Examples of open issues are the development of algorithms for cluster analysis and for dimensionality and complexity reduction on partially ordered structures, as well as for employing partial orders in inferential procedures, like linear or logistic regression. The third area pertains to the development of complete and efficient software resources, so as to spread the use of posets in applied statistics and make it more effective, on larger datasets. Finally, the fourth development area is the integration of posetic concepts and tools into older statistical frameworks, in particular in the field of evaluation (of personal and social traits, like literacy or poverty, but also of social processes and systems, like the healthcare, the educational or the welfare systems), where the "posetic perspective" may well be a key factor to support governance and decision-making.

References

Annoni, P., & Bruggemann, R. (2009). Exploring partial order of European countries. *Social Indicators Research, 92*(3), 471.

Arcagni, A. (2017). PARSEC: An R package for partial orders in socio-economics. In M. Fattore & R. Bruggemann (Eds.), *Partial order concepts in applied sciences* (pp. 275–289). Cham: Springer.

Arcagni, A., di Belgiojoso, E. B., Fattore, M., & Rimoldi S. M. L. (2019). Multidimensional analysis of deprivation and fragility patterns of migrants in lombardy, using partially ordered sets and self-organizing maps. *Social Indicators Research, 141*, 551–579.

Bachtrögler, J., Badinger, H., de Clairfontaine, A. F., & Reuter, W. H. (2016). Summarizing data using partially ordered set theory: An application to fiscal frameworks in 97 countries. *Statistical Journal of the IAOS, 32*(3), 383–402.

Badinger, H., & Reuter, W. H. (2015). Measurement of fiscal rules: Introducing the application of partially ordered set (poset) theory. *Journal of Macroeconomics, 43*, 108–123.

Bruggemann, R., & Patil, G. P. (2011). *Ranking and prioritization for multi-indicator systems: Introduction to partial order applications*. New York: Springer Science & Business Media.

Caperna, G., & Boccuzzo, G. (2018). Use of poset theory with big datasets: A new proposal applied to the analysis of life satisfaction in italy. *Social Indicators Research, 136*(3), 1071–1088.

Carlsen, L. (2017). An alternative view on distribution keys for the possible relocation of refugees in the european union. *Social Indicators Research, 130*(3), 1147–1163.

Carlsen, L., & Bruggemann, R. (2014). The 'failed state index' offers more than just a simple ranking. *Social Indicators Research, 115*(1), 525–530.

Carlsen, L., & Bruggemann, R. (2017). Fragile state index: Trends and developments. A partial order data analysis. *Social Indicators Research, 133*(1), 1–14.

Davey, B. A., & Priestley, H. A. (2002). *Introduction to lattices and order*. Cambridge: Cambridge University Press.

De Loof, K. (2009). *Efficient computation of rank probabilities in posets*. Ph.D. thesis, Ghent University.

De Loof, K., De Meyer, H., & De Baets, B. (2006). Exploiting the lattice of ideals representation of a poset. *Fundamenta Informaticae, 71*(2–3), 309–321.

De Loof, K., De Baets, B., & De Meyer, H. (2008). Properties of mutual rank probabilities in partially ordered sets. In *Multicriteria ordering and ranking: Partial orders, ambiguities and applied issues* (pp. 145–165). Warsaw: Systems Research Institute, Polish Academy of Sciences.

di Bella, E., Gandullia, L., Leporatti, L., Montefiori, M., & Orcamo, P. (2018). Ranking and prioritization of emergency departments based on multi-indicator systems. *Social Indicators Research, 136*(3), 1089–1107.

Fattore, M. (2016). Partially ordered sets and the measurement of multidimensional ordinal deprivation. *Social Indicators Research, 128*(2), 835–858.

Fattore, M. (2017). Functionals and synthetic indicators over finite posets. In M. Fattore & R. Bruggemann (Eds.), *Partial order concepts in applied sciences* (pp. 71–86). Cham: Springer.

Fattore, M., & Arcagni, A. (2018). F-FOD: Fuzzy first order dominance analysis and populations ranking over ordinal multi-indicator systems. *Social Indicators Research*, 1–29. First online.

Fattore, M., & Maggino, F. (2014). Partial orders in socio-economics: A practical challenge for poset theorists or a cultural challenge for social scientists? In R. Bruggemann, L. Carlsen, & J. Wittmann (Eds.), *Multi-indicator systems and modelling in partial order* (pp. 197–214). New York: Springer.

Fattore, M., Arcagni, A., & Maggino, F. (2019, Forthcoming). Optimal scoring of partially ordered data, with an application to the ranking of smart cities. In *Smart statistics for smart applications – SIS* 2019 conference, Milan.

Fuhrmann, F., Scholl, M., & Bruggemann, R. (2018). How can the empowerment of employees with intellectual disabilities be supported? *Social Indicators Research, 136*(3), 1269–1285.

Hussain, M. A., Jørgensen, M. M., & Østerdal, L. P. (2016). Refining population health comparisons: A multidimensional first order dominance approach. *Social Indicators Research, 129*(2), 739–759.

Joint Research Centre-European Commission, et al. (2008). *Handbook on constructing composite indicators: Methodology and user guide*. Paris: OECD Publishing.

Koppatz, P., & Bruggemann, R. (2017). Pyhasse and cloud computing. In M. Fattore & R. Bruggemann (Eds.), *Partial order concepts in applied sciences* (pp. 291–300). Cham: Springer.

Leti, G. (1983). *Statistica descrittiva*. Bologna: Il Mulino.

Maggino, F., & Fattore, M. (2019). Social polarization. *Wiley statsRef: Statistics reference online* (pp. 1–4).

Meyer C. D. (2000). *Matrix analysis and applied linear algebra*. Philadelphia: SIAM.

Nanivazo, M. (2015). First order dominance analysis: Child wellbeing in the democratic republic of congo. *Social Indicators Research, 122*(1), 235–255.

Neggers, J., & Kim, H. S. (1998). *Basic posets*. Singapore: World Scientific.

Patil, G. P., & Taillie, C. (2004). Multiple indicators, partially ordered sets, and linear extensions: Multi-criterion ranking and prioritization. *Environmental and Ecological Statistics, 11*(2), 199–228.

Saaty, T. L., & Hu, G. (1998). Ranking by eigenvector versus other methods in the analytic hierarchy process. *Applied Mathematics Letters, 11*(4), 121–125.

Schoch, D. (2017). *netrankr: An R package to analyze partial rankings in networks*.

Schröder, B. S. W. (2016). *Ordered sets: An introduction with connections from combinatorics to topology*. Cham: Springer.

Sen, A. (1992). *Inequality reexamined*. Oxford: Clarendon Press.

Todeschini, R., Grisoni, F., & Nembri, S. (2015). Weighted power–weakness ratio for multi-criteria decision making. *Chemometrics and Intelligent Laboratory Systems, 146*, 329–336.

Assessing Subjective Well-being in Wide Populations. A Posetic Approach to Micro-data Analysis

Leonardo Salvatore Alaimo and Paola Conigliaro

1 Measuring Subjective Well-being: A Multidimensional Issue[1]

The growing attention to the issues of quality of life and well-being in social studies and in political analysis led to the use of an increasingly wide range of subjective indicators. The so-called "Stiglitz Report" (Stiglitz et al. 2009) asserts the need to analyze the subjective condition using indicators able to detect both aspects of conscious evaluation and emotional states. Further studies added a third dimension, the eudaimonic one. This three-dimensional classification was explained in particular in the OECD guidelines on subjective well-being (OECD 2013), which collect and systematize the most accredited theoretical production in this field.

The definition of measures to assess subjective well-being (hereinafter: SWB) involves many disciplines, different approaches and a very articulated range of models, indicators and inquiring tools. This topic naturally belongs to the field

---

[1] Although this paper should be considered the result of the common work of the two authors, L. S. Alaimo has mainly written Sections "Exploratory Data Analysis", "Application of the partial ordering methodology" and "Conclusions"; P. Conigliaro has mainly written Sections "Measuring subjective well-being: a multidimensional issue", "The issue of subjective indicators synthesis" and "Data".

L. S. Alaimo
Department of Social Sciences and Economics, Sapienza University of Rome, Rome, Italy

Italian National Institute of Statistics, Istat, Rome, Italy
e-mail: leonardo.alaimo@istat.it

P. Conigliaro (✉)
Italian National Institute of Statistics, Istat, Rome, Italy
e-mail: paola.conigliaro@istat.it

R. Bruggemann et al. (eds.), *Measuring and Understanding Complex Phenomena*,
https://doi.org/10.1007/978-3-030-59683-5_16

of psychology in general, and in particular to positive psychology (Seligman and Csikszentmihalyi 2000). Indeed, the latter identifies two macro-dimensions of SWB: (1) the hedonic one (related to pleasure, personal satisfaction, positive emotions and sensations); (2) the eudaimonic one (related to the realization of the self and the development of authentic human nature). The statistical surveys in order to measure subjective aspects of quality of life often use tools borrowed from psychometry, psychology, medicine, social psychology or sociology. These usually assume the form of scales (verbal, numerical, continuous)[2]. They can detect conditions, perceptions, relational aspects, attitudes, behaviours.

The OECD guidelines (OECD 2013) identify three main dimensions in the assessment of SWB. According to these guidelines, none of the three dimensions should be neglected in analyzing SWB.

The first one is the *evaluative (or cognitive) dimension*, concerning the conscious assessment of satisfaction for life or for some specific aspects of it. Individuals assess their satisfaction through a process of implicit comparison with expectations and with a condition deemed suitable or desirable. As A. C. Michalos (1985) argues, the standards of reference include comparisons with one's own needs, past conditions, aspirations, and the results achieved by other people considered to be nearer and more significant. The "measurement" of satisfaction for life has a consolidated tradition in social surveys and uses tools that easily allow statistical processing. They usually consist on questions such as: "... how satisfied are you with your life on the whole?". The answers consist in a numerical or verbal scale. According to some authors, a 7-level scale should be more precise and reliable, but many surveys adopt a scale [0–10], supposed to be more familiar to the respondents, especially in those countries like Italy that use this scale of assessment at school (Macrì 2017). However, even if such references are easy to perceive, interpretative distortion can still occur. Some studies have shown that people tend to avoid extreme responses, especially if the scales have a wide range, and they focus on medium-high responses, if there are no central responses.

Some reports adopt satisfaction for life as a measure of happiness. But in most cases, this condition is considered an expression of the *hedonic dimension* or as an *emotional state ("affect"* in the OECD guidelines). The definition of the emotional dimension of SWB has significant interpretative discrepancies, because the concept presents aspects of polyvalence and polysemy. Analyzing the data we must always carefully observe the structure of the questions and answers. Furthermore, an important aspect concerns whether we attribute to the emotional dimension the role of determinant or effect of well-being. In the first case, the researchers believe that the emotional state is mainly caused by individual temperament. In the second case, the emotional state is considered as an expression of SWB and assumes the role of its indicator. However, it may happen to meet studies that treat the same object with both functions (e.g. see Eurostat 2015 and 2016).

---

[2] *e.g.* the attitude scales (Likert, Thurstone), social distance scale (Bogardus), semantic differential scales (Osgood).

In order to assess emotional state, social studies often adopt tools – internationally used and tested in epidemiology – measuring mental health or psychological unease. One of the most used is the battery of questions on mental health (hereinafter: MHI) adopted in the SF-36[3] questionnaire. This tool considers two positive and three negative emotional states and the question is how long a person has experienced each of the five emotions during the previous 4 weeks. The answers are articulated in six possible items, ranging from *"never having experienced that state"* to *"being always in that state"*. Ware et al. (1993) summarize the measure of mental health adopting the average score between the five answers. This way to elaborate data follows two assumptions (even though implicitly): (1) the modalities are quantitative and continuous; (2) the value of synthesis admits compensation. Many studies (clinical and epidemiological, but also social statistics) base their analysis on these assumptions. This model implies the idea that the experience of a negative emotion can be subtracted from the experience of a positive one, generating a balanced measure of the respondent's emotional state. This presupposes that positive and negative emotions lay on a single continuum, e.g. a latent variable corresponding to the emotional state (or mental health). Anyway, this statement is neither obvious nor incontrovertible. Diener and Emmons (1984) show that there is no compensation between positive and negative emotions, arguing that we can experience such compensation only in very short periods. If we consider a long period of time, we might remember having experienced opposite emotions, but we would not represent them as an average emotional level.

The third pillar of SWB is the *eudaimonic dimension*. It allows capturing aspects of SWB strongly based on inter-subjective relationships, such as shared values, social recognition and belonging. This dimension cannot be called subjective in the strict sense, because it is not based only on the evaluations, opinions or perceptions of those directly involved, but also on the observation of behaviors, attitudes, relationships and results. *Eudaimonia* corresponds to a state of good psychological function that goes beyond conscious evaluation or emotional perception and concerns the realization of one's individual potential (Waterman 2008 and 2011). Some authors emphasize the egocentric perception of the need for realization and self-determination (Deci and Ryan 2008), other ones (*e.g.* Nussbaum 2011) the aspect of sharing and recognition within a wider system of values. A common definition of the concept of *eudaimonia* – and the identification of indicators to measure it – is not yet available.

Thus, we have just seen how the three dimensions of SWB can be arranged on the basis of an increasing level of definitional complexity and operational criticality and a decreasing level of agreement in the scientific community. In any case, the debate on SWB and its indicators is still very active and involves further concepts, such as the flourishing (Diener et al. 2010; Seligman 2011; Huppert and So 2013), which we will not consider here. The three pillars of SWB are, in turn, defined on the basis

---

[3]Developed by a group of scholars at the New England Medical Center in Boston coordinated by J. H. Ware (1993).

of non-elementary concepts (OECD 2013). Moreover, the correlations between the three main dimensions are not so strong (Clark and Senik 2011; Huppert and So 2013) as to allow ignoring the contribution of one of them in the evaluation of SWB. This confirms the need to reveal all three main dimensions of SWB and to define interpretative tools that can safeguard the multidimensionality of the concept (OECD 2013).

Comparing different groups of respondents requires synthesizing the information gathered on very large databases. By doing this, surveys become accessible and usable to a wider public, supporting democratic participation and political decisions. However, the need to reduce the complexity can lead to the risk of excessive simplification of the concepts and to distortive interpretations of the information collected. In particular, the identification of the measurement model and the choice of the synthesis method are critically important steps, which have an impact on the results. An explicit or implicit conceptual model always guides the interpretation of the relationship between indicators. According to Maggino (2017), the main distinction is between reflective and formative models. The reflective model refers to a latent variable that exists irrespective of the units of measurement and the units of analysis. The indicators must intercept it, interpret it, and *measure it*. This means that the indicators are interchangeable and that the internal consistency is of fundamental importance: if two indicators are uncorrelated, they do not measure the same concept. The psychometric tools, usually adopted to measure SWB in statistical surveys, use a reflective model. The indicators' effectiveness consists in their capacity to approximate the measure of a characteristic (mental health, intelligence, ability). In the study of social phenomena, the most common approach is, instead, the formative (or constructivist) one. According to this model, the indicators do not depend on the latent variable, but they determine its nature and characteristics. The formative is a bottom-up explanatory approach. The internal consistency is of minimal importance: two uncorrelated indicators can both be useful and even indispensable for the knowledge of multidimensional phenomena. Therefore, omitting an indicator means omitting part of the construct. In this paper, we deal with a formative measurement model[4].

The correct definition of the conceptual model allows to correctly interpret the relationships between the indicators and to correctly identify the procedures for synthesizing them (Maggino 2017). Thus, it is important to use the correct method of synthesis, considering the nature of the data. As previously written, many studies process ordinal information as a continuous quantitative variable, using then synthesis methods traditionally adopted for this type of variable (e.g. the arithmetic mean). We consider these methods unsuitable to synthesize ordinal data and misleading for conclusions. Furthermore, in the analysis of complex

---

[4]It should be made clear that the choice of the measurement model does not depend on a free choice of the researcher, but exclusively on the nature of the latent variable measured (Alaimo and Maggino 2020).

phenomena, we must focus our attention on interrelations, rather than on causal relationships among variables.

In this work we apply our analysis to data deriving from the *ad-hoc* module on SWB of the *European Union Survey on Income and Living Conditions* (hereinafter: Eu-SILC), adopted in the 2013 edition. These data are analysed by Eurostat in two different reports. The first one (Eurostat 2015) examines data from this module, aggregated at country level. It studies life satisfaction (hereinafter: LS) according to some socio-economic dimensions (e.g. the working condition), and the meaning of life (hereinafter: MoL) in relation to sex, age class and LS. The values of the meaning of life are higher than those of LS, and the two variables (considering data aggregated at the national level) seem to have an almost linear relationship, with the Pearson[5] correlation coefficient (Eurostat 2013) of 0.56. Maybe, the homogeneity in the structure of the two questions and the relative answers, as well as the recoding of the two items into the same three classes, [6] could have influenced the homogeneity of the distribution of the aggregate values. Then it appears necessary to carry out a micro-data processing to better understand the relationships between the two variables. As concerning the emotional state (hereinafter: ES), the module adopted the MHI battery (from the SF-36 questionnaire) with questions relating to five affects[7]. The questions have a very different structure compared to those on LS and MoL, because they concern the frequency of each affect in a limited period of time (four weeks). Furthermore, the five modalities of response[8] are clearly ordinal. However, the Eurostat report analyzes only happiness, in relation to age group, family structure, work condition, LS and MoL. The second Eurostat report (Eurostat 2016) compares three multivariate regression models that consider overall LS as a dependent variable. The third model includes mental well-being as independent variable, considering the mental status[9] calculated on the MHI battery – as a factor influencing SWB, and not as one of its components. None of the two Eurostat reports considers the conjoint contribution of the three dimensions in expressing the SWB level, although this is one of the recommendations of the OECD guidelines. But above all, all the three variables expressing the dimensions of SWB are analysed using statistical methods typical of cardinal variables, thus not taking into account their ordinal nature.

In this paper, we present an application to the synthesis of SWB indicators taking into account their ordinal nature. In particular, we propose a two-step synthesis process: first, we address the synthesis of the indicators used to measure the ES

---

[5]The choice of Pearson's correlation coefficient reveals the questionable assumption that variables are quantitative.

[6]The two items are both revealed on a scale from 0 to 10. Recoding defined 3 modalities: low if the level is less than six, median if the level is between 6 and 8 and high for levels 9 and 10.

[7]Q: How much of the time, during the past four weeks have you been/felt: Very nervous; Down (in the dump); Calm and peaceful; Downhearted and depressed; Happy.

[8]A: All of the time; Most of the time; Some of the time; A little of the time; None of the time.

[9]Mental status is calculated as the average score of the answers to the five questions on emotional states, reporting the value on a scale from 0 to 100.

and then move on to the synthesis of the three dimensions of the SWB. Finally, we propose an analysis of the differences in the SWB level between different sub-populations of respondents.

We identified the sub-populations according to the self-defined labour status. The relationship between labour status and SWB is widely established in the litera-ture (Gallie et al. 2012; Diener et al. 2018).

Our analysis intends to safeguard the multidimensionality of the phenomenon, applying the Partially Ordered Set (hereinafter: poset) methodology to the micro-data analysis.

2 The Issue of Subjective Indicators Synthesis

Methods to synthesize subjective measures expressed with ordinal characters have become more and more sophisticated, while the possibility of making increasingly complex calculations with simple computers has made such methods more acces-sible to scholars. There are two main kinds of synthesis approaches: aggregative-compensatory or non-aggregative (Maggino 2017).

The synthesis of the indicators of ES that uses the arithmetic mean is an example of aggregative-compensatory approach (Ware et al. 1993; Eurostat 2016). This approach is based on a reflective measurement model and it assumes that the five questions measure one and only one latent variable. According to this model, it is admissible to aggregate information allowing a value of an indicator that is consistent with the concept to be compensated by a value of another indicator that appears to be reverse. The model supposes the existence of a high correlation between the indicators that concur to detect the latent variable, which can be interpreted as an expression of a high internal consistency. Thus, in a compensatory approach using, for instance, the arithmetic mean to synthesize different measures, a respondent expressing the [5, 3, 1] combination on a set of three ordinal-scale responses from 1 to 5 would be assimilated without distinction to another responding [3, 3, 3]. It is evident how the use of an aggregative method *flattens* the differences between the two different combinations, making equal two profiles that are actually different (Alaimo and Maggino 2020). Furthermore, the use of arithmetic mean for the synthesis is conceptually wrong if we consider each affect as a partially independent element of the emotional status, but primarily because the five variables expressing affects are ordinal and not cardinal. In brief, statistical tools for the synthesis of indicators based on an aggregative approach are inadequate for describing and dealing with multidimensional systems of ordinal data (Fattore et al. 2011). For this reason, the choice of a non-aggregative method for the synthesis of SWB variables is the most correct choice. In this work we adopted the poset methodology. It grounds on partial orders as an application of discrete mathematics and it is not compensatory. It identifies response profiles basing on possible combinations of modalities and defines sorting criteria for these profiles.

A quite complete presentation of the basic definitions and main elements of this methodology can be found in Fattore 2016 and 2017.

Using poset[10], it is possible to analyze the different levels of a multidimensional concept, making a synthesis by variables. It is also possible to compare different sub-populations of respondents on providing summary measures for each subgroup. This paper takes the basic elements of the methodology for granted.

3   Data

Eu-SILC is an European harmonized official statistical survey[11], carried out since 2004 and structured in a longitudinal and a transversal component. It periodically adopts *ad-hoc* modules that examine in depth topics of particular importance. The 2013 *ad-hoc* module measures the state of well-being from a subjective perspective, and consists of 22 questions, on overall LS, satisfaction with specific aspects of life (including interpersonal relationships), MoL, emotional states (including happiness), trust, physical security, availability of someone to talk to and someone to ask for help. The survey units are households and family members over 16-years old. The Italian survey collected 38,039 respondents in 2013. Within the subset of people who answered to the *ad-hoc* well-being module (25,500 respondents), we select those between 26 and 65 years old (15,354 records). We have chosen to represent the three dimensions of SWB through the LS the MoL and the five *affects*. We have also selected other variables that we consider important to study the differences in SWB levels in different subgroups: labour status and gender. In particular, concerning the labour status (employed, unemployed, retired, etc.) we used the one indicated by the respondent. In fact, we consider that the self-attributed status is more significant in defining a relationship with the perceived level of well-being. Naturally the variable status is categorical non-sortable.[12]

---

[10]There is a large literature on the treatment and synthesis of multidimensional systems of ordinal data using non-aggregative methods, allowing the construction of synthetic measures without the aggregation of the scores of basic indicators. Within this approach, poset has become a reference over the years, as demonstrated by many works in different fields of research (for instance, see: Annoni and Bruggemann 2009; Fattore et al. 2015; Carlsen and Bruggemann 2017; Arcagni et al. 2019). However, poset can also be suitable for quantitative data (see: Fattore 2018; Alaimo 2020; Alaimo et al. 2020a, b, c), allowing the overcoming of some limitations of the aggregative methods.

[11]EC Regulation n.1177/2003.

[12]Full-time employees (EFT), Part-time employees (EPT), Full-time self-employed (SEFT), Part-time self-employed (SEPT), Unemployed (UNE), Students (STU), Retired (RET), Unfit to work (UNF), Fulfilling domestic care (HOU), Other inactive (INA).

4 Exploratory Data Analysis

The first step in our analysis has been the exploratory analysis concerning the
bivariate relationships between the variables of well-being. In order to test the
existence and the strength of linear relationships among the seven variables of
SWB considered, we calculate the correlation coefficients (Kendall's Tau-b)[13]. A
high correlation coefficient would indicate a potential high internal consistency
between dimensions, which should allow us to reduce the set of items. As can be
seen from Fig. 1, reporting the correlation plot, the coefficients are not very high.
The higher values are between the affects *down* and *depressed* (0.63) and between
calm and *happy* (0.62). The correlation between MoL and LS is equal to 0.49. The
relationships between these variables and those expressing the ES are weak. The
results of the exploratory analysis seem to confirm our hypothesis that the different
variables considered do not fall on the same continuum and, therefore, constitute
elements of knowledge that cannot be neglected in the synthesis.

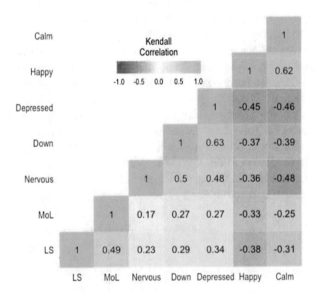

Fig. 1 Correlation plot between variables of subjective well-being: life satisfaction; meaning of
life; the five affects representing life satisfaction. N = 15,354

---

[13] We only measure linear relationship. It is therefore perfectly possible that while there is strong
non-linear relationship between the variables, r is close to 0.

5   Application of the Partial Ordering Methodology

The exploratory analysis confirms that the three main dimensions of SWB cannot be summarised in a single measure using aggregative compensatory approaches without losing significant information. This synthesis involves a very delicate two-step process. The first step defines the ES by synthesizing the five affects; in the second one, the ES becomes an input variable, together with LS and Mol, to define a SWB measure, which allows us to assess the level of SWB of the population of respondents. The poset methodology allows creating a non-aggregative and non-compensatory synthesis of the three dimensions of well-being, based on solid mathematical criteria (Fattore et al. 2015; Fattore and Bruggemann 2017). In particular, it seems the most suitable approach to synthesize the five variables (*affects*) expressing ES. In effect, these variables are weakly correlated, they are expressed in terms of frequency of subjective experience and their modalities represent the order and not absolute values (they do not enjoy the properties of a ratio scale and the different levels are not equidistant). We believe that the five *affects* do not lie in a continuum, which identifies a latent variable: we cannot say that a frequent happiness and an equally frequent depression give rise to an average level of ES (even if the temporal reference is limited to the last four weeks).

A first problem is the definition of the ES poset, due to the criticality linked to the number of profiles expected in a set of five variables each with five modalities (3125). In fact, each partially ordered set that we can draw is one of the possible sets (linear extensions) generated by the comparison of profiles. The number of possible linear extensions in a set of 3125 profiles is enormous and makes computation impossible. Thus, to overcome computational problems, we have recoded all the five variables in three modalities (Table 1). By doing this, the number of the possible profiles becomes 243. Figure 2 reports the distribution of the answers according to each affect. Only 1.2% of profiles are "homogeneous" (i.e. [1,1,1,1,1], [2,2,2,2,2], [3,3,3,3,3]). The 38% of respondents have a profile of this type (in particular, 26% of total population present a profile [3,3,3,3,3])[14].

A second problem is to know the specific ES value to be attributed to each respondent, starting from the values assumed by the 5 affects. According to Fattore

Table 1 Recoding of affect variables

Modalities	Variables				
	Nervous	Down	Calm	Depressed	Happy
Always or most of the time	1	1	3	1	3
Sometimes	2	2	2	2	2
A little or none of the time	3	3	1	3	1

---

[14]For an example of the distribution of homogeneous profiles within the total ones, see: Conigliaro (2018).

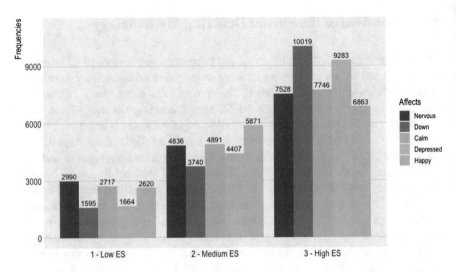

Fig. 2 Distribution of emotional status affects: *nervous*; *down*; *calm*; *depressed*; *happy*. N = 15,354

(2017), we can consider poset evaluation as a way to derive a complete order out of a partial order: once the evaluation scores have been assigned to the elements of a poset, they can be ordered in a linear order. Through a simple procedure (Fattore 2017), it is possible to assign to each element of a finite poset a score representing its position in a "low-high" axis. In this way, we obtain the *average rank*, i.e. the average position assumed by a profile in each of the considered linear extensions. We cannot use the average rank – representing itself a synthesis – as an expression of the level of ES, as it is a measure of (average) positioning of the profile in the general order. Furthermore, if we apply the quantile criteria to split the profile into groups, we could find in a single group profiles expressing very different levels of ES and profiles of the same level in contiguous groups. In this way, we risk committing errors in attributing an ES level to individual respondents. In addition to the position in the general order, we need to take into account the situation in terms of ES deprivation of each profile, by establishing a criterion to assign the response profiles to the "deprived" or "not-deprived" category. We can do this by defining a *threshold profile*. This choice allows characterizing the distribution according to a conceptual definition of the phenomenon (it may contemplate the meaning attributed to the item in a particular culture or other considerations). Starting from the partial order among profiles, the aim of our analysis is to identify the deprived profiles, with respect to the threshold, in each linear extension of the poset (Fattore 2016). For the ES, we have defined as threshold profile the combination [2,2,3,2,3][15]. It identifies

---

[15]Other thresholds could also be added to better characterize and distinguish profiles. However, in this case we have not identified another suitable point of discrimination at conceptual level.

those respondents who have experienced negative emotions for most of the time during the reporting period and positive emotions sometimes or never. We consider them as people in a low emotional state. After setting the threshold, an *identification function*[16] is defined, to assign deprivation membership scores in [0,1] to the poset elements (Fattore 2016)[17]. In particular, it expresses the quote of events in which the profile falls into the area of discomfort, considering the different linear extensions. The *severity* is the average of the graph distance of the profile from the first profile above all threshold elements, and it is equal to 0 for profiles above the threshold. Both the identification function and the severity function are meaningful to describe the deprivation in the poset: the first one describes the deprivation ambiguity, the second one its intensity. We cannot use only the identification function as an absolute value expressing the ES level, because not all profiles with the same level of it are equally deprived. Indeed, their deprivation can show very different levels of severity. Taking into account the two information, we can assess the (probable) deprivation condition of a profile and consequently of the subject expressing it.

In order to construct a synthetic indicator of SWB, we need to know the specific value of the ES of each respondent. In this way, we can treat it as an input variable for the second stage of synthesis. To do this, we decided to summarize the three functions explained above in a single index, so as to have a measure that takes into account the position of profiles in the general order and the intensity and ambiguity of their deprivation with respect to the chosen threshold. We could have calculated the different functions upon the theoretical set of profiles (i.e. 243 profiles) or upon the actual dataset (in which we observe 230 profiles of the 243 possible ones). We decided to consider the theoretical set, so as to obtain values of the functions not influenced by the respondents' population. To obtain a synthesis, we decided to use the *Mazziotta-Pareto Index*[18] (hereinafter: MPI), a composite index for summarizing a set of indicators that are assumed not fully substitutable (Mazziotta and Pareto 2016). Our composite (constructed by synthesizing the average rank, the identification function and the severity function[19]) is positive, i.e., increasing values of the index correspond to positive variations of the phenomenon. The composite assumes values between 87 (profile [1,1,1,1,1]) and 121 (profile [3,3,3,3,3]). Table

---

[16]For a complete definition of the identification function and its computation, please see: Fattore et al. 2012.

[17]All the operations were carried out using the R package PARSEC (Arcagni and Fattore 2018).

[18]MPI is a partially non-compensatory composite indicator based on a non-linear function which, starting from the arithmetic mean of the normalized indicators (indicators are standardized by means of a variant of *z-scores*, which transforms the indicators into distributions with mean 100 and standard deviation 10) introduces a penalty for the units with unbalanced values of the indicators. We chose this method because various analyses have shown that it is more robust than others are (for instance, see: Mazziotta and Pareto 2015; Alaimo 2020). For more information on the MPI, please see: Mazziotta and Pareto 2016 and 2017.

[19]In the normalization, it is necessary to define the polarity of the basic indicators, i.e. the sign of the relation between the indicator itself and the phenomenon to be measured. Therefore, the type of composite we want to construct defines polarity. In our case, the average rank has positive polarity, while the identification and the severity functions negative.

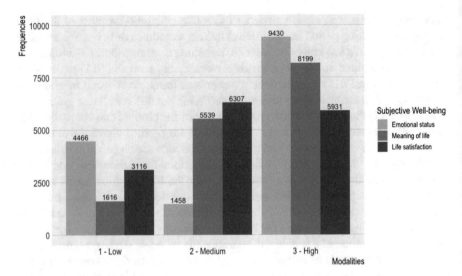

Fig. 3 Distribution of subjective well-being variables: emotional status; meaning of life; life satisfaction. N = 15,354

A1 in appendix reports the values of the MPI assigned to each profile. In order to use the new ES variable in the poset of subjective well-being with MoL and LS, we have discretized the composite index according to the following criteria:

- profiles with MPI ≤ 99 obtain a value of 1;
- profiles with $100 < MPI \leq 110$ obtain a value of 2;
- profiles with MPI ≥ 110 obtain a value of 3.

Then, we have attributed the values of the new discretized ES variable to the Eu-SILC dataset. We also have recoded LS and MoL into three modalities following the recoding adopted by the Italian National Institute of Statistics – Istat (2003) for LS[20]. Figure 3 shows the distribution of the frequencies of the three variables of SWB in the dataset.

The second step of our analysis consists in defining a synthetic "measure" of SWB, using the three variables representing its dimensions (MoL, LS and ES). Figure 4 shows on the left the Hasse diagram representing the 27 SWB profiles and on the right the distribution of respondents for each profile (the nodes have a size proportional to the frequency of respondents). To highlight different levels of SWB we defined three thresholds: [3,1,2], [3,2,1] and [2,2,2]. The analysis of the distribution highlighted that most of respondents declared high levels of MoL. We have established that a subjective profile, to be defined at a "good" SWB level, must have at least the dimension MoL at 3 (i.e. from 8 to 10 in the original scale) and one of the other two dimensions at 2. At the same time, we have also defined another

---

[20]Recoding method: (0–5) = 1 – "Low"; (6–7) = 2 – "Medium"; (8–10) = 3 – "High".

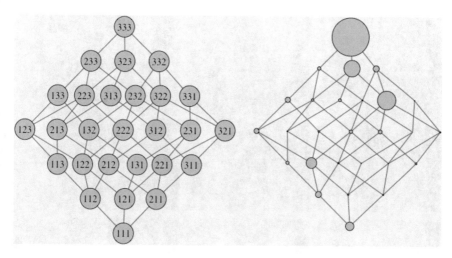

Fig. 4 Subjective well-being: Hasse diagram of the three dimensions; distribution of respondents in each profile

threshold, the "low" one, identifying it in a profile that presents a medium level in all three dimensions at the same time. This means that people who have at most 2 in one or two dimensions and less than 2 in the rest are definitely in poor SWB condition.

Considering that the SWB levels could be influenced by other variables (e.g. sex, working condition, etc.), we analyzed and compared it in function of some of them, taking into account different sub-populations. Let's consider, for example, the case of the labour status. Consistently with the aggregate level analysis (Eurostat 2015), the unemployed and other persons excluded from work (such as full and permanent unfitness) show lower levels for all subjective well-being measures. There are also differences in the distribution of levels for each of the three SWB dimensions, depending on the different conditions. Figure 5 shows an example of differences based on labour status (we consider three categories: unemployed, part-time employees and full-time employees) and gender. Considering full-time employees, we can see that MoL has a similar distribution in all gender categories, while there are some small differences in LS and ES levels. Part-time employees show some differences in the distribution of males and females, with the latter having higher levels of MoL and LS, while lower ones of ES. For the unemployed, SWB levels are lower for all respondents and for all dimensions, with women reporting higher levels in all items than males.

These are the result of the analyses at micro-data level and the adoption of a method that respects the multidimensional nature of the concept. They suggest more precise interpretative hypotheses of the relationships between labour status and SWB. Thus, a correct analysis of SWB levels in a population must be carried out taking into account other variables. On the basis of these findings, we decided to identify a series of sub-sets, generated from the initial dataset on the basis of the

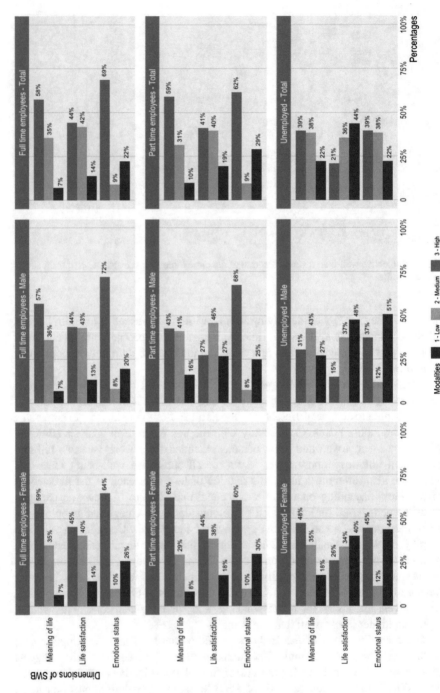

Fig. 5 Subjective well-being dimensions distribution according to gender and labour status. N = 15,354

Table 2 Synthetic measures of subjective well-being by labour status and gender: sub-populations; poverty gap; wealth gap; number of observations

Sub-populations	Poverty Gap	Wealth Gap	Observations
Total	0.361	0.671	15,354
Full time employees	0.292	0.699	6,076
Part time employees	0.361	0.689	1,206
Unemployed	0.511	0.551	1,400
Male	0.344	0.670	7,131
Female	0.375	0.672	8,223
Full time employees - female	0.313	0.702	2,590
Full time employees - male	0.274	0.695	3,486
Part time employees - female	0.359	0.706	1,009
Part time employees - male	0.298	0.360	197
Unemployed - female	0.482	0.589	732
Unemployed - male	0.536	0.502	668

values of two context variables, gender and labour status[21]. We studied the SWB levels in the sub-populations through two procedures.

The first procedure consists in comparing SWB levels in different sub-populations using two synthetic measures, called *poverty gap* and *wealth gap*[22]. The name of these measures is a reference to synthetic poverty measures calculated using the AF method (Alkire et al. 2015). These measures are the average values of relative severity and relative wealth[23] and can be used to compare different populations. In this work, higher values of poverty gap indicate a population at a lower level of ES and higher values of wealth gap reveal good ES. Table 2 shows the *poverty gap* and *wealth gap* values in the different sub-populations studied.

The results highlight that the unemployed have worse levels of poverty gap and wealth gap than the total population. The worst level of poverty gap belongs to unemployed men. Full-time employees have better levels in both measures then total population. These results were quite predictable. However, women with a part-time employment register the highest level of wealth. One possible interpretation of this result is that part-time work is often, but not always, a choice for many women allowing them to reconcile labour with private life. This is just a hypothesis, which should be supported by detailed analyses, for example, regarding the characteristics of work, working time and income.

---

[21] We consider three categories of labour status (unemployed, part-time employees and full-time employees), three categories of gender (female, male and total population) and their interactions.

[22] For a definition of the two functions and their computational procedures, please see: Arcagni and Fattore 2018.

[23] The relative wealth is the average graph distance from the maximum threshold element, over the sampled linear extensions divided by the maximum wealth.

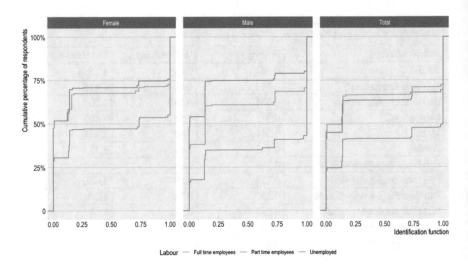

Fig. 6 Cumulative frequencies distribution of identification function according to labour status and gender

In the second procedure, we compare the distribution of cumulative frequencies of the identification (Fig. 6) and relative severity (Fig. 7) in the different sub-populations. Looking at the total population, in Fig. 6 we can see that a half of the part-time employees and more than 40% of full-time employees have zero or very low identification level (between 0 and 0.04 of the horizontal axis), while less than 30% have a high or absolute identification. The share of unemployed persons with a zero or low level of identification is low (just above 25%), while over half of them present a high level. Females perform better than total population compared to full-time and part-time employees, with over 50% presenting zero or very low identification level (about 25% have a high or absolute identification). Even the population of female unemployed tends to be better than the total unemployed (for instance, 30% have zero or very low identification level). The male population has a similar trend to female one for full-time employees. Considering part-time employees, we can see that about 30% of men have a low identification value, compared to 50% of women; similar considerations can also be made for the unemployed. The results highlight the differences in SWB status for women and men in different working conditions, confirming a generally better status for women – unemployed and part-time employees – than for men (same result as in the first procedure, see Table 2).

In the same way we can compare the distribution of the cumulated relative frequencies of the relative severity value (Fig. 7), i.e. the depth of the fragility of the profile among the deprived profiles, comparing it with the most deprived profile. The value is almost nil for almost 70% of part-time employees, more than 60% for the full-time and quite 40% for unemployed. It is very high for 14% of

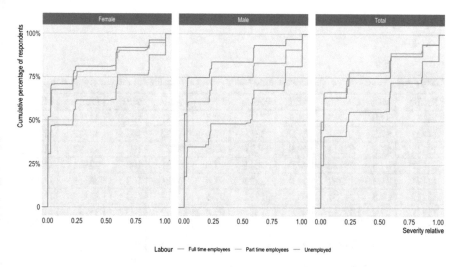

Fig. 7 Cumulative frequencies distribution of severity function according to labour status and gender

the unemployed and for just 5% of full-time employees. With respect to gender differences, the considerations made above can be confirmed.

6 Conclusions

The relationship between labour status and SWB is widely established in the literature. However, this notion is always evaluated at the general level comparing aggregated data or else at individual psychological level.

The present analyses confirm this relationship at individual level, in wide population of respondents. Applying the analyses to micro-data supplied by an official statistical survey represents a further step that allows us to compare a consistent number of really subjective conditions. It is in line with the recommendations of the Commission on the Measurement of Economic Performance and Social Progress and following international guidelines on measuring well-being in wide populations.

The adopted methodology of synthesis (poset) perfectly fits the framework of SWB providing a satisfying tool to analyse micro-data and synthesize indicators expressed in ordinal way. Moreover, the non-compensatory nature of posets make them the perfect approach to analyze SWB data. This research experiments also a multi-stage application of this methodology, identifying critical issues and proposing solutions.

A.1 Appendix

Table A.1 Values of the Mazziotta Pareto Index – MPI for profiles of the emotional status

Profile	MPI	Profile	MPI	Profile	MPI	Profile	MPI	Profile	MPI
11111	87.0	21113	91.3	22222	97.4	32231	102.8	22332	112.0
11211	87.2	13131	92.8	13321	97.4	32222	102.9	32322	112.0
12111	87.3	32211	92.8	33121	97.5	23132	103.0	32232	112.0
21111	87.3	23112	92.9	33211	97.6	33131	103.0	22233	112.0
11121	87.3	12231	93.0	13222	97.6	23321	103.1	23313	112.1
11112	87.5	31131	93.0	31231	97.6	33212	103.1	23322	112.1
11311	87.9	23121	93.0	21331	97.8	13232	103.3	33132	112.1
12211	88.0	21321	93.0	32221	97.9	23231	103.4	32223	112.2
31111	88.0	13221	93.0	23131	98.0	11323	103.4	33213	112.2
13111	88.1	31221	93.1	12331	98.1	12332	103.4	13233	112.3
11131	88.1	33111	93.1	31222	98.1	32321	103.4	32133	112.3
11221	88.1	21312	93.2	22231	98.2	33311	103.4	31332	112.4
21211	88.1	12321	93.3	31321	98.2	22331	103.5	33312	112.4
12121	88.2	22221	93.3	32122	98.2	23222	103.5	33123	112.5
22111	88.2	13311	93.4	32212	98.2	31331	103.7	33231	112.5
11212	88.2	13122	93.4	31132	98.2	13331	103.7	23133	112.6
12112	88.3	23211	93.5	22123	98.3	31232	103.8	13323	112.8
21112	88.3	22311	93.5	22312	98.3	13322	103.8	32313	112.8
11113	88.4	13212	93.5	23311	98.4	33221	103.9	31233	112.8
21121	88.6	32112	93.6	23122	98.4	23123	103.9	31323	112.9
11122	88.9	21222	93.6	33112	98.6	32312	104.0	21333	112.9
12311	89.3	21231	93.7	12223	98.7	22323	104.0	12333	113.2
13112	89.6	13113	93.7	31213	98.7	33113	104.1	33223	118.6
11231	89.7	31311	93.7	21322	98.8	31313	104.1	32332	118.6
23111	89.7	12132	93.7	31123	98.8	23312	104.1	33322	118.6
11312	89.8	11331	93.8	21232	98.8	32213	104.1	23332	118.7
11213	89.9	32121	93.8	32131	98.8	32123	104.2	33232	118.9
13211	90.0	22131	93.8	13132	98.8	31322	104.2	33133	118.9
12131	90.0	12312	94.0	21313	99.0	23213	104.2	33313	118.9
12113	90.0	11133	94.1	22132	99.2	13223	104.3	32233	119.0
22121	90.0	22122	94.1	31312	99.2	33122	104.4	33331	119.0
22211	90.1	12222	94.1	21223	99.3	32132	104.6	23233	119.0
13121	90.1	22212	94.2	11332	99.3	21332	104.7	31333	119.1
21221	90.1	12123	94.2	12232	99.3	22232	104.7	13333	119.1
31211	90.1	31122	94.3	13123	99.5	21133	104.7	23323	119.5

(continued)

Table A.1 (continued)

Profile	MPI	Profile	MPI	Profile	MPI	Profile	MPI	Profile	MPI
22112	90.2	31212	94.3	32113	99.5	13313	105.1	32323	119.5
32111	90.2	21213	94.3	12313	99.6	12233	105.2	22333	119.8
31121	90.2	11232	94.3	23113	99.6	21233	105.3	33323	119.8
21131	90.3	22113	94.5	13213	99.8	11333	105.4	33332	119.9
31112	90.3	21132	94.6	12133	99.8	13133	105.4	33233	120.0
21311	90.4	11322	94.6	11233	100.1	31223	105.4	32333	120.0
12122	90.4	31113	94.7	13312	101.0	31133	105.5	23333	120.0
21212	90.4	21123	94.8	22223	101.2	22133	105.5	33333	121.1
11321	90.4	12213	94.9	12322	101.3	33321	111.2		
12221	90.5	11313	95.0	22322	101.5	23331	111.4		
11222	90.5	11223	95.5	21323	101.7	23223	111.5		
21122	90.5	23221	96.8	23212	101.7	32331	111.5		
11132	90.5	22321	97.1	12323	101.8	23232	111.5		
11123	90.6	32311	97.2	22313	101.8	13332	111.8		
12212	90.6	13231	97.3	22213	102.6	33222	111.9		

References

Alaimo, L. S. (2020). *Complexity of social phenomena: Measurements, analysis, representations and synthesis*. Unpublished doctoral dissertation, University of Rome "La Sapienza", Rome, Italy.

Alaimo, L. S., & Maggino, F. (2020). Sustainable development goals indicators at territorial level: Conceptual and methodological issues—The Italian perspective. *Social Indicators Research, 147*, 383–341. https://doi.org/10.1007/s11205-019-02162-4.

Alaimo, L. S., Arcagni, A., Fattore, M., & Maggino, M. (2020a). Synthesis of multi-indicator system over time: A Poset-based approach. *Social Indicators Research*. https://doi.org/10.1007/s11205-020-02398-5.

Alaimo, L. S., Arcagni, A., Fattore, M., Maggino, M., & Quondamstefano, V. (2020b). Measuring equitable and sustainable well-being in Italian regions: The non-aggregative approach. *Social Indicatros Research*. https://doi.org/10.1007/s11205-020-02388-7.

Alaimo, L. S., Ciacci, A., & Ivaldi, E. (2020c). Measuring sustainable development by non-aggregative approach. *Social Indicators Research*. https://doi.org/10.1007/s11205-020-02357-0.

Alkire, S., Foster, J. E., Seth, S., Santos, M. E., Roche, J. M., & Ballon, P. (2015). *Multidimensional Poverty Measurement and Analysis: Chapter 5 – The Alkire-Foster Counting Methodology* (OPHI Working Paper N. 86). Oxford: Oxford Poverty & Human Development Initiative. Internet resource. https://ophi.org.uk/resources/ophi-working-papers.

Annoni, P., & Bruggemann, R. (2009). Exploring partial order of European countries. *Social Indicators Research, 92*(3), 471–487. https://doi.org/10.1007/s11205-008-9298-4

Arcagni, A., & Fattore, M. (2018). parsec: Partial orders in socio-economics. *R package version 1.2.1*. https://CRAN.R-project.org/package=parsec

Arcagni, A., Barbiano di Belgiojoso, E., Fattore, M., & Rimoldi, S. M. L. (2019). Multidimensional analysis of deprivation and fragility patterns of migrants in Lombardy, using partially ordered sets and self-organizing maps. *Social Indicators Research, 141*(2), 551–579. https://doi.org/10.1007/s11205-018-1856-9

Carlsen, L., & Bruggemann, R. (2017). Fragile state index: Trends and developments. A partial order data analysis. *Social Indicators Research, 133*(1), 1–14. https://doi.org/10.1007/s11205-016-1353-y

Clark, A. E., & Senik, C. (2011). *Is happiness different from flourishing? Cross-country evidence from the ESS*. PSE Working Papers n.2011–04. https://doi.org/10.3917/redp.211.0017

Conigliaro, P. (2018). *Labour status and subjective well-being. A micro-level analysis and a multidimensional approach to well-being*. Working papers of the Ph.D. Course in Applied Social Sciences, Department of Social Sciences and Economics, Sapienza University of Rome. WP n. 4/2018. https://web.uniroma1.it/disse/sites/default/files/WP4_Conigliaro_0.pdf

Deci, E. L., & Ryan, R. M. (2008). Self-determination theory: A macro-theory of human motivation, development and health. *Canadian Psychology, 49*(3), 182–185. https://doi.org/10.1037/a0012801

Diener, E., & Emmons, R. A. (1984). The Independence of positive and negative affects. *Journal of Personality and Social Psychology, 47*(5), 1105–1117. https://doi.org/10.1037//0022-3514.47.5.1105

Diener, E., Wirtz, D., Tov, W., Kim-Prieto, C., Choi, D., Oishi, S., & Biswas-Diener, R. (2010). New Well-being measures: Short scales to assess flourishing and positive and negative feelings. *Social Indicators Research, 97*, 143–156. https://doi.org/10.1007/s11205-009-9493-y

Diener, E., Oishi, S., & Tay, L. (2018). Advances in subjective Well-being research. *Nature Human Behaviour, 2*, 253–260. https://doi.org/10.1038/s41562-018-0307-6

Eurostat. (2013). *Eu-Silc module on wellbeing – Assessment of the implementation*. https://ec.europa.eu/eurostat/documents/1012329/1012401/2013+Module+assessment.pdf.

Eurostat. (2015). *Quality of life, fact and views*. Luxembourg: Publications Office of the European Union. https://doi.org/10.2785/59737

Eurostat. (2016). *Analytical report on subjective Well-being*. Luxembourg: Publications Office of the European Union. https://doi.org/10.2785/318297

Fattore, M. (2016). Partially ordered sets and the measurement of multidimensional ordinal deprivation. *Social Indicators Research, 128*, 835–858. https://dx.doi.org/10.1007/s11205-015-1059-6

Fattore, M. (2017). Synthesis of indicators: The non-aggregative approach. In F. Maggino (Ed.), *Complexity in society: From indicators construction to their synthesis* (pp. 192–212). Cham: Springer. https://doi-org-443.webvpn.fjmu.edu.cn/10.1007/978-3-319-60595-1_8

Fattore, M. (2018). Non-aggregated indicators of environmental sustainability. *Silesian Statistical Review/Slaski Przeglad Statystyczny, 16*(22), 7–22. https://doi.org/10.15611/sps.2018.16.01

Fattore, M., & Bruggemann, R. (2017). *Partial order concepts in applied sciences*. Cham: Springer. Internet resources. https://doi.org/10.1007/978-3-319-45421-4

Fattore, M., Bruggemann, R., & Owsiński, J. (2011). Using poset theory to compare fuzzy multi-dimensional material deprivation across regions. In S. Ingrassia, R. Rocci, & M. Vichi (Eds.), *New perspectives in statistical modeling and data analysis* (pp. 49–56). Berlin/Heidelberg: Springer. https://doi-org-443.webvpn.fjmu.edu.cn/10.1007/978-3-642-11363-5_6

Fattore, M., Maggino, F., & Colombo, E. (2012). From composite indicators to partial orders: Evaluating socio-economic phenomena through ordinal data. In F. Maggino & G. Nuvolati (Eds.), *Quality of life in Italy: Research and reflections* (pp. 41–68). Dordrecht: Springer. https://doi.org/10.1007/978-94-007-3898-0_4

Fattore, M., Maggino, F., & Arcagni, A. (2015). Exploiting ordinal data for subjective well-being evaluation. *Statistics in Transition: Journal of the Polish Statistical Association, 16*, 409–428. https://doi.org/10.21307/stattrans-2015-023

Gallie, D., Gosetti, G., & La Rosa, M. (2012). *Qualità del Lavoro e della Vita Lavorativa: Cosa èCambiato e Cosa sta Cambiando*. Milano: FrancoAngeli.

Huppert, F. A., & So, T. T. C. (2013). Flourishing across Europe: Application of a new conceptual framework for defining Well-being. *Social Indicators Research, 110*(3), 837–861. https://doi.org/10.1007/s11205-011-9966-7

Macri, E. (2017). Label scale and rating scale in subjective well-being measurement. In G. Brulé & F. Maggino (Eds.), *Metrics of subjective well-being: Limits and improvements* (pp. 185–200). Cham: Springer. https://doi.org/10.1007/978-3-319-61810-4_9

Maggino, F. (2017). Dealing with synthesis in a system of indicators. In F. Maggino (Ed.), *Complexity in society: From indicators construction to their synthesis* (pp. 115–137). Cham: Springer. https://doi.org/10.1007/978-3-319-60595-1_5

Mazziotta, M., & Pareto, A. (2015). Comparing two non-compensatory composite indices to measure changes over time: A case study. *Statistika-Statistics and Economy Journal, 95*(2), 44–53.

Mazziotta, M., & Pareto, A. (2016). On a generalized non-compensatory composite index for measuring socio-economic phenomena. *Social Indicators Research, 127*(3), 983–1003. https://doi.org/10.1007/s11205-015-0998-2

Mazziotta, M., & Pareto, A. (2017). Synthesis of indicators: The composite indicators approach. In F. Maggino (Ed.), *Complexity in society: From indicators construction to their synthesis* (pp. 161–191). Cham: Springer. https://doi.org/10.1007/978-3-319-60595-1_7

Michalos, A. C. (1985). Multiple discrepancy theory. *Social Indicators Research, 16*(4), 347–413. https://doi.org/10.1007/BF00333288

Nussbaum, M. C. (2011). *Creating capabilities. The human development approach.* Cambridge, MA: The Belknap Press of Harvard University Press. https://doi.org/10.4159/harvard.9780674061200

Organisation for Economic Cooperation and Development – OECD. (2013). *Guidelines on measuring subjective well-being.* Paris: OECD Publishing. https://doi.org/10.1787/9789264191655-en

Seligman, M. E. P. (2011). *Flourish: A visionary new understanding of happiness and Well-being.* New York: Free Press.

Seligman, M. E. P., & Csíkszentmihályi, M. (2000). Positive psychology. An introduction. *American Psychologist, 55*(1), 5–14. https://doi.org/10.1037/0003-066X.55.1.5

Stiglitz, J. E., Sen, A., & Fitoussi, J. P. (2009). *Report by the Commission on the measurement of economic performance and social progress.* Institut national de la statistique et des études économiques – INSEE. https://www.insee.fr/fr/publications-et-services/dossiers_web/stiglitz/doc-commission/RAPPORT_anglais.pdf

Ware, J. E., Snow, K. K., Kosinski, M., & Gandek, B. (1993). *SF-36 health survey – Manual and interpretation guide.* Boston: Nimrod Press.

Waterman, A. S. (2008). Reconsidering happiness: A Eudaimonist's perspective. *The Journal of Positive Psychology, 3*(4), 234–252. https://doi.org/10.1080/17439760802303002

Waterman, A. S. (2011). Eudaimonic identity theory: Identity as self-discovery. In S. J. Schwartz, K. Luyckx, & V. L. Vignoles (Eds.), *Handbook of identity theory and research* (Vol. 1). New York: Springer. https://doi.org/10.1007/978-1-4419-7988-9_16

Part IV
Software Aspects

Deep Ranking Analysis by Power Eigenvectors (DRAPE): A Study on the Human, Environmental and Economic Wellbeing of 154 Countries

Cecile Valsecchi and Roberto Todeschini

1 Introduction

Multi-criteria decision making (MCDM) is a decision process applicable in presence of multiple criteria which are often in contrast with each other (Ivlev et al. 2015; Kumar et al. 2017; Ho et al. 2010; Triantaphyllou 2000). A decision process consists of (1) definition and structuring of the problem, (2) model development to compare and rank the alternatives in a transparent way, (3) elaboration of an action plan. In general, the main aim of a decision process is to generate information and solutions in an effective way providing a good understanding of the problem. The simplest approaches in a MCDM belong to the so-called scoring methods, such as desirability/utility functions and simple average scoring (Pavan and Todeschini 2008a). Other methods comparing objects pairwise are called outranking methods, such as dominance functions (Pavan and Todeschini 2008b). Furthermore, partial order ranking methods, such as Hasse diagram technique, highlight the conflicting information by identifying incomparable objects (Pavan and Todeschini 2009).

Recently, a new approach based on a development of the Power-Weakness Ratio (PWR), called Deep Ranking Analysis by Power Eigenvectors (DRAPE) was proposed (Todeschini et al. 2015, 2019). Indeed, this method is based on the Power-Weakness Ratio proposed by Sir Kendall in (Kendall 1955) and later implemented by Ramanujacharyulu in (Ramanujacharyulu 1964) exploiting the ability of the eigenvalue/eigenvector technique in ranking the objects by taking into account the whole information present in the data.

C. Valsecchi · R. Todeschini (✉)
Milano Chemometrics and QSAR Research Group, Department of Earth and Environmental Sciences, University of Milano-Bicocca, Milan, Italy
e-mail: roberto.todeschini@unimib.it

The DRAPE approach offers a sequential ranking procedure and allows a simple interpretation of the obtained ranking in terms of the importance of the original criteria.

Here the DRAPE approach is applied to study the sustainability of 154 countries, expressed as a function of human, environmental and economic wellbeing indicators (i.e. criteria) with the aim to highlight the potentiality of this multivariate ranking approach.

This data set is only used as a case study to apply the new proposed ranking method; a deep discussion and sociological analysis about sustainability as well as the links between sustainability factors is out of the scope of this paper.

2   Theory

2.1   DRAPE

The Deep Ranking Analysis by Power Eigenvectors (DRAPE) approach (Todeschini et al. 2019; Valsecchi et al. 2020) allows to rank objects described by multiple criteria starting from the definition of tournament table and of Power-Weakness Ratio (Ramanujacharyulu 1964). In addition, this approach provides *a-posteriori* information about the contributions of the initial criteria to the ranking by means of a retro-regression analysis. The method will be briefly described; further details are available in (Todeschini et al. 2019).

2.1.1   The Tournament Table and Power-Weakness Ratio

Let $\mathbf{X}$ be a data matrix being comprised n objects described by p criteria. A basic tournament table $(\mathbf{T}^W)$ is a weighted count matrix of size $n \times n$ obtained by defining its elements (t_{ij}^W) as follows:

$$t_{ij}^W = \sum_{k=1}^{p} w_k \cdot \delta_{ij,k} \quad \text{where} \quad \delta_{ij,k} = \begin{cases} 1 & \text{if } x_{ik} \rhd x_{jk} \\ 0.5 & \text{if } x_{ik} \triangleq x_{jk} \\ 0 & \text{if } x_{ik} \lhd x_{jk} \end{cases} \quad \text{and} \quad \sum_{k=1}^{p} w_k = 1$$

$$(1)$$

where w_k is the weight given to the k-th variable and x_{ik} is the value of the k-th variable for the i-th object. The weights can be established by the user according to any prior knowledge or reflecting defined priorities; if no weighting scheme is adopted, all the weights are set by default to $1/p$ (i.e., all the criteria have equal weights).

The t_{ij}^W value can be easily interpreted as the proportion of cases where the object i dominates the object j.

In order to emphasize the extreme winner/loser objects, a threshold t^* is used to smooth the original tournament table as the following:

$$t_{ij}^{W} = \begin{cases} 0.5 & if \ 1 - t^* \leq t_{ij}^{W} \leq t^* \quad 0.5 \leq t^* < 1 \\ t_{ij}^{W} & otherwise \end{cases} \tag{2}$$

that is, the basic tournament table $\mathbf{T}^{W}$ is transformed by replacing the original entries fulfilling the threshold condition with the value of 0.5, which corresponds to a draw. This basically means to neglect the differences between two objects for some criteria.

The threshold selection is not user-defined but automatically performed based on the different actual values of the basic tournament table. Indeed, we extract from the basic tournament table the different entry values greater than 0.5 and use them one at a time as the threshold to generate a new smoothed tournament table. It can be noted that the threshold $t^* = 0.5$ corresponds to the original basic tournament matrix and it is always included in the set $\mathbf{t}^*$ of selected thresholds. Since different weighting schemes w can be adopted to modulate the relevance of the criteria in the decision process and for each of them a different set of thresholds $\mathbf{t}^*$ can be obtained, a family of tournament tables $\{\mathbf{T}^{W}[\mathbf{t}^*]\}$ is finally generated.

The tournament tables are asymmetrical, nonnegative and irreducible and, applying the eigenvalue/eigenvector technique, according to the Perron–Frobenius theorem (Keener 1993), they always give at least a positive eigenvalue to which corresponds a positive eigenvector $\mathbf{L}$. Thus, from the tournament table and its transpose matrix we can calculate two eigenvectors $\mathbf{L}$ and $\mathbf{L}^*$. The loadings of $\mathbf{L}$ rank the objects from the best to the worst one, while the loadings of $\mathbf{L}^*$ give a reverse ranking, that is, from the worst to the best object. For each object, it is then calculated the following score, called Power Weakness Ratio (PWR):

$$PWR_i = \frac{\alpha + L_i}{\alpha + L_i^*} \qquad \alpha = \frac{1}{n} \tag{3}$$

where α is a small empirical correction parameter, defined as the reciprocal of the number n of objects, which avoids singularities or spikes in the case of zero or very small eigenvector entries. This PWR function encodes both the power and the weakness of each object over the remaining $n - 1$ ones, resulting in a trade-off between how many times an object "defeats" the other objects and how many times an object is "defeated" by the others.

Once all the objects have been ordered by the corresponding PWR values, a PWR diagram can be built by defining the vertical axis in terms of the PWR values of the objects.

A reliable ranking is thus obtained by shrinking the PWR values, usually represented by several decimal digits, to one or two decimal digits, representing the degree of resolution.

To measure the degree of total ordering provided by PWR or, in other words the degree of not-distinguishable objects, we used the standardized Shannon entropy H:

$$H = \frac{-\sum_{k=1}^{q} p_k \cdot \log_2 (p_k)}{\log_2(n)} \qquad 0 \le H \le 1 \qquad (4)$$

where q is the number of equivalent ranking levels (*i.e.*, the number of the different PWR values, rounded to two decimal digits), and p_k is the fraction of objects in the k-th level. If all the objects have different PWR values, a total order without draws is obtained and the following relationships hold:

$$q = n \qquad p_k = 1/n \quad \forall k, \quad k = 1, 2, \dots, q \qquad H = 1$$

On the other side, if all the objects have the same PWR value, no ranking is obtained, and the following relationships hold:

$$q = 1 \qquad p_1 = n/n = 1 \qquad H = 0$$

In general, the standardized Shannon entropy is expected to decrease with the increase of the threshold value, since increasing the smoothing of the tournament table leads to have more objects placed at the same ranking level.

2.1.2 Deep Ranking Analysis

The set of families of tournament tables obtained considering different thresholds $t^*(\mathbf{w})$ is automatically pruned by eliminating those matrices whose ranking has a correlation higher than 0.995 (settled as default value) with those of the previous ones.

By exploiting the PWR rankings obtained with the different thresholds of the tournament table, a deep ranking analysis is performed by Principal Component Analysis (PCA) on the matrix having n rows (the objects) and K columns (the different PWR rankings). The first component is enough to explain almost the whole variability of the data. The scores of the first component represent the consensus ranking (CPWR) of the objects, while the loadings explain the role played by each threshold in determining the global ranking scores. Thus, the result is a deep ranking analysis able to summarize all the different PWR rankings.

2.1.3 Retro-regression Analysis

Given the CPWR or a PWR ranking of the objects, a regression can be carried out to evaluate how the original ordering criteria encoded in the $\mathbf{X}$ data matrix are related to the final ranking. Indeed, a regression method can be used to obtain the standardized regression coefficients explaining the relationships between $\mathbf{X}$ (i.e., the original criteria that are the independent variables) and CPWR or PWR, which is the dependent variable (i.e., the vector "response" $\mathbf{y}$). We used the ridge regression method, which enables the calculation of the regression coefficients even in the case of more criteria than objects and avoids spurious spikes of the coefficients. The ridge model equation can be defined both for the single PWR rankings (at different thresholds) and for the consensus ranking CPWR as:

$$\mathbf{b}_{RR} = \left(\mathbf{X}^{\mathrm{T}} \cdot \mathbf{X} + k \cdot \mathbf{I}\right)^{-1} \cdot \mathbf{X}^{\mathrm{T}} \cdot \mathbf{y} \tag{5}$$

where k is a scalar value and $\mathbf{I}$ is the identity matrix. The value of k was fixed to 0.01 and not optimized through some validation procedure, since we are interested in the interpretability of the model (*i.e.* in its fitting) rather than in its predictive capability.

The ridge regression coefficients $\mathbf{b}_{RR}$ are standardized by the usual expression:

$$b_j^* = b_j \cdot \frac{s_j}{s} \qquad j = 1, p \tag{6}$$

where s_j and s are the standard deviations of the j-th variable and the response $\mathbf{y}$ (*i.e.*, CPWR or PWR), respectively, and b_j is the ridge regression coefficient of j-th variable.

The analysis of the retro-regression coefficients allows the *a-posteriori* interpretation of the obtained rankings in terms of relevance of the original criteria and their evolution according to the applied degree of smoothing given by the threshold.

2.2 Comparison with Other Ranking Techniques

We compared the results obtained by the DRAPE approach with other well-known traditional ranking techniques (Pavan and Todeschini 2008a; Pavan and Todeschini 2009). The comparison was performed with the dominance functions (DOM) (Keller et al. 1991), utility (UTI) functions (Zionts and Wallenius 1976) and simple additive ranking (SAR) (Zimmermann and Gutsche 1991).

The considered ranking methods are mathematically represented by the following formulas, denoting n as the number of objects, p the number of variables (criteria), w_j the weight of the j-th variable, r_{ij} the rank of the i-th objects for the j-th variable; f_{ij} is the linear transform of the i-th object for the j-th variable into the interval [0, 1], W^+ is the sum of the variable weights when object i dominates object m, W^- is the sum of the variable weights when object i is dominated by object m:

$$\mathrm{UTI}_i = \sum_{j=1}^{p} w_j \cdot f_{ij} \tag{7}$$

$$SAR_i = \frac{\sum\limits_{j=1}^{p} w_j \cdot r_{ij}}{n} \tag{8}$$

$$DOM_i = \sum_{m=1}^{n} \frac{1 + W^+}{1 + W^-} \quad m \neq i \tag{9}$$

All the values obtained by the formulas 7, 8 and 9 were always scaled in the interval [0, 1].

3   Dataset

As a case study we considered the countries levels of sustainability used in the calculation of the Sustainable Society Index (SSI), developed by Sustainable Society Foundation (Van de Kerk and Manuel 2008) and freely available at http://www. ssfindex.com. The last available data coming from the sixth edition of the SSI (2016) were considered.

According to the Brundtland definition, a sustainable society is a society (1) that meets the needs of the present generation, (2) that does not compromise the ability of future generations to meet their own needs, (3) in which each human being has the opportunity to develop itself in freedom, within a well-balanced society and in harmony with its surroundings (WCDE 1987). Being based on this definition, for 154 countries the SSI collects 21 indicators, grouped into 7 categories belonging to 3 wellbeing dimensions: human (9 indicators), environmental (7 indicators) and economic (5 indicators) wellbeing (see Table 1).

The 21 indicators are based on different public data sources circulated by international organizations such as: Food and Agriculture Organization of the United Nations (FAO), United Nations Educational, Scientific and Cultural Orga- nization (UNESCO), World Health Organization (WHO), World Economic Forum (WEF) and International Monetary Fund (IMF). Further information about how the indicators are calculated, are provided on the SSI website (SSI n.d.). All the indicators are in range [0, 10] in the direction of sustainability, i.e. a value of 10 always indicates the maximal sustainability.

We refer to the countries using three letters following the ISO 3166-1 alpha-3 standard published by the International Organization for Standardization (ISO). The country codes are reported in Appendix. Moreover, we codified the 21 indicators as reported in the last column of Table 1. Therefore in this work we expand the considerations made in the previous paper (Todeschini et al. 2019) considering all the 21 wellbeing indicators.

Table 1 Description and coding of the 21 indicators of sustainability taken from (SSI n.d.)

Wellbeing dimension	Category	Indicator	Indicator's based on	Code
Human	Basic needs	Sufficient food	Number of undernourished people in % of total population	H1
		Sufficient to drink	Number of people in % of total population, with sustainable access to an improved water source	H2
		Safe sanitation	Number of people in % of total population, with sustainable access to improved sanitation	H3
	Personal Development & Health	Education	Gross enrolment ratio for primary, secondary & tertiary education (combined)	H4
		Healthy life	Life expectancy at birth in number of healthy life years	H5
		Gender equality	Gender gap index	H6
	Well-balanced society	Income distribution	Ratio of income of the richest 10% to the poorest 10% people in a country	H7
		Population growth	5-years change in total population size (% of total population)	H8
		Good governance	Sum of the six worldwide governance indicators	H9
Environment	Natural resources	Biodiversity	10 years change in forest area and Size of protected land area (in % total land area)	N1
		Renewable water resources	Annual water withdrawals (m3 per capita) as % of renewable water resources	N2
		Less consumption	Ecological footprint minus carbon footprint	N3
	Climate &energy	Less energy use	Energy use (tonnes of oil equivalent per capita)	N4
		Energy savings	Change in energy use over 4 years (%)	N5
		Less greenhouse gases	CO2 emissions per person per year	N6
		Renewable energy	Consumption of renewable energy as % of total energy consumption	N7
Economy	Transition	Organic farming	Area for organic farming in % of total agricultural area of a country	E1
		Genuine savings	Genuine savings (adjusted net savings) as % of gross National Income (GNI)	E2
	Economy	GDP	Gross domestic product per capita, PPP, current international $	E3
		Employment	Number of unemployed people in % of total labour force	E4
		Less public debt	The level of public debt of a country in % of GDP	E5

4 Software

All the algorithms and calculations were carried out in MATLAB (version 2019a) by means of routines written by the authors. The DRAPE toolbox v. 2 (new version) is available for download at http://www.michem.unimib.it/download/matlab-toolboxes/drape-toolbox-for-matlab/.

5 Results and Discussion

In order to analyse the effect of different weighting schemes, the following two approaches were adopted:

1. In the first case an equal weight ($w_k = 1/21$) was assigned to each indicator, i.e. the initial contribution of the indicators is equal to 0.0048. Therefore, henceforth this first approach will be called equal.
2. Then, we weighted equally the three wellbeing dimensions which have thus a contribution of one third each. In this case the sum of the weights of the indicators belonging to a dimension has to be equal to 0.33. Therefore, each indicator of human wellbeing has a lower weight ($w_H = 0.0367$) than those of environment wellbeing ($w_N = 0.0471$) and economy wellbeing ($w_E = 0.066$). From now, this second approach will be called proportional since the initialized weights are inversely proportional to the number of indicators per dimension.

Furthermore, we assessed the DRAPE ranking method with both equal and proportional weight initialization approaches considering: (1) all countries together and (2) countries grouped by continents (Europe, Africa, North, South and Central America and Asia-Oceania). The CPWR ranking was considered. For Europe alone we took into account only 18 out of 21 indicators because H1, H2 and H3 were almost constant. For sake of visualization a resolution of one decimal digit was used.

5.1 All Countries

The DRAPE method was applied on 154 countries described by the 21 wellbeing indicators as summarized in Fig. 1. A weight (equal or proportional) is assigned to each indicator prior to the construction of the basic tournament table and of the progressively smoothed ones. The different thresholds found in the basic tournament table after the correlation pruning are 10 (0.5, 0.55, 0.59, 0.6, 0.62, 0.67, 0.71, 0.74, 0.79 and 0.81) using the equal weighting scheme and 35 for the proportional one. It can be noticed that the higher the thresholds selected the more the ties in the consensus ranking, i.e. more objects (points) have the same PWR value. This is

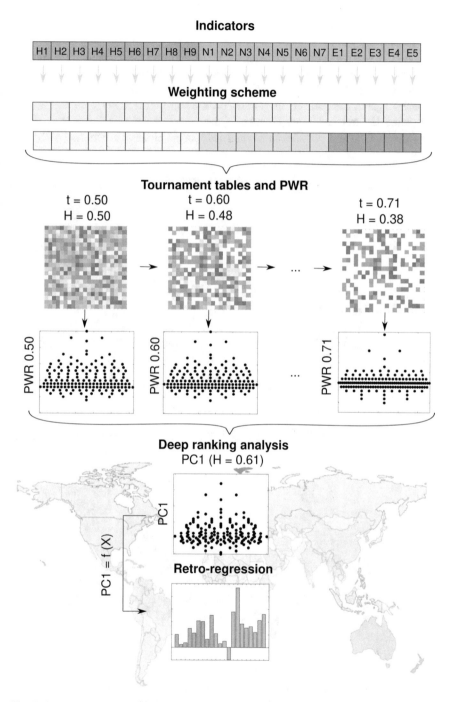

Fig. 1 Scheme of the DRAPE procedure applied to the data countries case study. Starting from the creation of the tournament tables applying a weighting scheme (equal or proportional) for the wellbeing indicators; the PWR and entropy are calculated and a deep ranking analysis through PCA is performed. Finally, the retro-regression analysis allows the interpretations of the indicator's importance over the ranking

due to the progressive smoothing of the tournament table (formula 3) which lead to smaller entropy values.

The PCA was then performed using the PWR rankings for the retained thresholds (i.e. thresholds leading to a ranking with a correlation lower than 0.995 with the others). The CPWR ranking plots are shown in Fig. 2. It is evident that the European countries (green circles) are in general on the top of the CPWR plots, together with some Central-American countries (black circles) such as Costa Rica (CRI) and Cuba (CUB) and two Oceania countries (purple circles): Australia (AUS) and

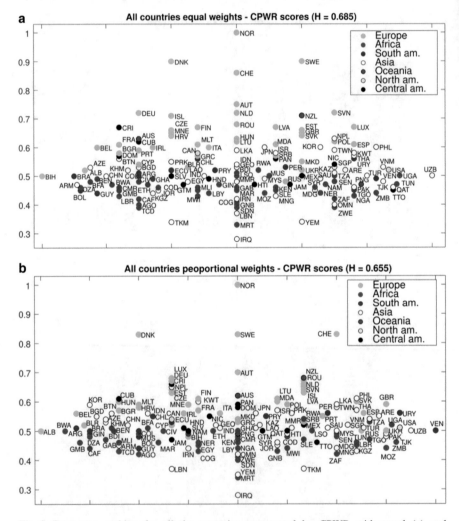

Fig. 2 Consensus ranking for all the countries represented by CPWR with equal (**a**) and proportional (**b**) initial weights for the indicators. Regression coefficients at threshold 0.5 for equal (**c**) and proportional (**d**) initial weights (the coefficient of determination R^2 is equal to 0.92 and 0.89 for the equal and proportional weight initializations, respectively)

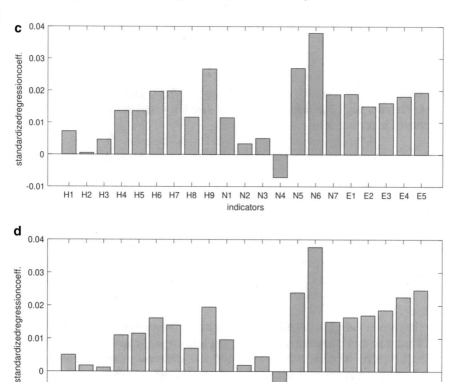

Fig. 2 (continued)

New Zealand (NZL). Asian countries (white circles) show high variability in their ranks, being some of them located in the middle positions and other in the bottom positions. The majority of the (blue, yellow and black circles) American countries (blue circles) are located in the middle-down positions, while the African countries (red circles) are in the middle-down and bottom of the rank.

The different initialization of the weights involves small variations in the ranking. For example, Switzerland (CHE) is the fourth most sustainable country considering the indicators with equal weights (Fig. 2a), but when the indicators are proportionally weighted (Fig. 2b) it becomes second in the rank order together with Sweden (SWE) and Denmark (DNK).

The equal initialization of weights is intrinsically unbalanced towards human well-being indicators, as they are more numerous (9) than environmental (7) and economic (5) indicators. Therefore, considering equally relevant the three dimensions of wellbeing (i.e. using a proportional weight initialization) Switzerland may be more sustainable than Denmark for environmental and/or economic reasons.

The CPWR ranking was then interpreted by carrying out the retro-regression in function of the original data matrix X and looking at the regression coefficients. Regarding the environmental indicators, for both weighting Schemes N4 (energy use) has a not significant coefficient as N2 (renewable water resources) and N3 (less consumption). On the other hand, N5 (energy savings) and N6 (less greenhouse gases) have a high coefficient suggesting that the sustainability ranking can be explained in particular by these indicators.

5.2 Continents

5.2.1 Europe

Slight differences between the two weighting schemes were noticed also considering the 40 European countries described by 18 wellbeing indicators, being H1 (sufficient food), H2 (sufficient to drink) and H3 (safe sanitation) almost constant. The countries in the two CPWR rankings (Fig. 3a, b) show small variations in the ranking.

The corresponding standardized regression coefficients show a similar behaviour, but those obtained using the equal weighting scheme are in general lower (Fig. 3a).

In particular, the most influent indicators are H5 (healthy life), N5 (energy savings) and E5 (less public debt). Moreover, N1 (biodiversity), N3 (less consumption), N4 (less energy use) and E3 (GDP) are not influent over the ranking.

5.2.2 America

Interestingly, the H5 indicator (healthy life) is relevant also for the 25 American countries ranking (Fig. 4), but with a high negative coefficient. This indicate that H5 is inversely related to the ranking suggesting that the most sustainable American countries are those with a less healthy lifestyle. In this case good governance (H9) and less greenhouse gases (N6) are the indicator with the most relevant positive coefficient for the equal and proportional weighting scheme, respectively. It can be noticed for example that Canada (CAN) and Jamaica (JAM) decrease while Panama (PAN) increases their CPWR scores considering the weights as proportional.

5.2.3 Africa

In the interpretation of the 44 African countries ranking, E3 (GDP) and N4 (less energy use) have a negative coefficient (Fig. 5); while H3 (safe sanitation), H9 (good governance) and N7 (renewable energy) have the greatest positive coefficients in the equal weights initialization case, while H5 (healthy life), N2 (renewable water resources), E1 (organic farming), E2 (genuine savings), E4 (employment) and E5 (less public debt) become relevant in the proportional weights initialization. In general, using the proportional weighting scheme compared to the equal one, economic wellbeing indicators have a higher coefficient, that is a higher influence on the African sustainability ranking. For this reason, countries like Tunisia (TUN) and Lesotho (LSO) decrease their ranking position switching from equal to proportional weights initialisation.

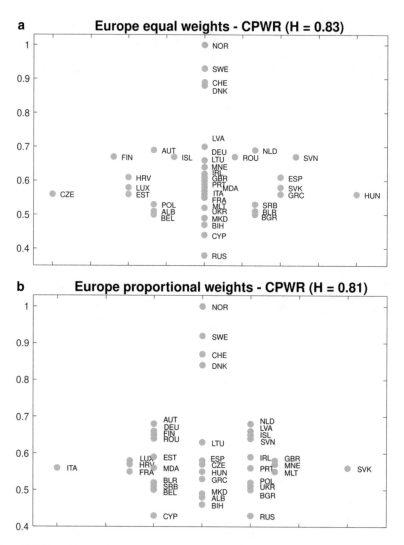

Fig. 3 CPWR ranking and standardized regression coefficients for equal (**a** and **c**) and proportional (**b** and **d**) weights initializations for Europe. CPWR values are represented on the vertical axis; H stands for the standardized entropy (the coefficient of determination R^2 is equal to 0.92 and 0.90 for the equal and proportional weight initializations, respectively)

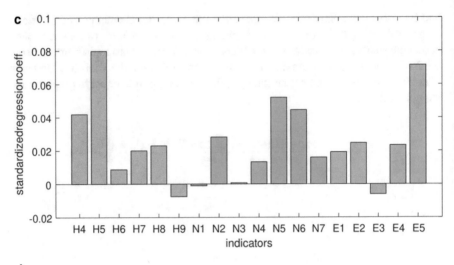

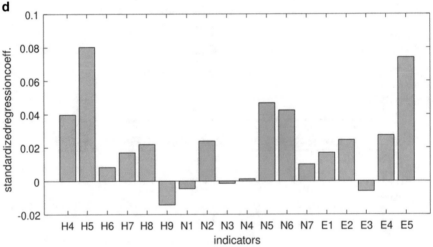

Fig. 3 (continued)

5.2.4 Asia-Oceania

Regarding Asia-Oceania, N4 (energy use) and N6 (less greenhouses gases) have the most influential coefficients using the equal weighting scheme (Fig. 6c), the first with a negative and the second with a positive value. On the other hand, applying a proportional weighing scheme reduces the negative influence of N4. According to the considered indicators, New Zealand (NZL) is the most sustainable country in Asia-Oceania. Although the coefficients seem not to vary significantly using the two initializations of weights (Fig. 6c, d), there are still some ranking differences; for example weighting equally all the 21 indicators, Australia (AUS) is in the fourth position after Sri Lanka (LKA) and South Korea (KOR), while it descends significantly by weighting equally the three dimensions (i.e. proportional approach, Fig. 6a, b). Indeed, even small variations in coefficients imply different rankings, especially for countries with similar values of indicators.

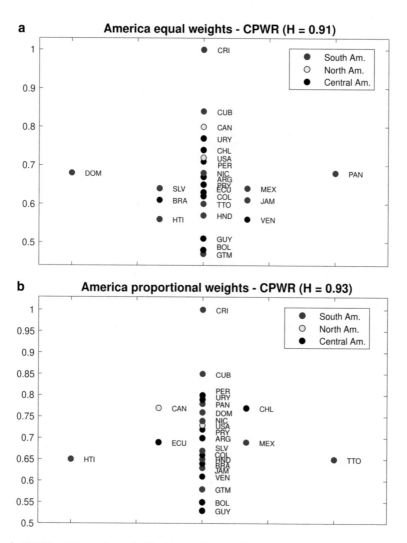

Fig. 4 CPWR ranking and standardized regression coefficients for equal (**a** and **c**) and proportional (**b** and **d**) weights initializations for America. CPWR values are represented on the vertical axis; H stands for the standardized entropy (the coefficient of determination R^2 is equal to 0.93 both for equal and proportional weighting schemes)

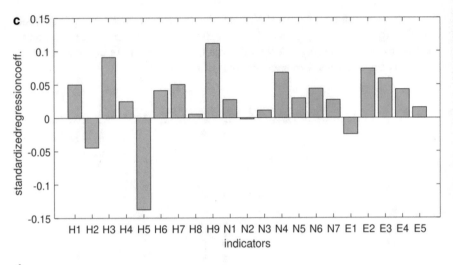

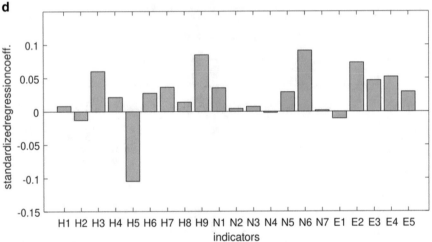

Fig. 4 (continued)

5.3   Comparison with Traditional Ranking Methods

In order to verify the similarity as well as the discrepancy between different ranking methods, the CPWR DRAPE rankings are compared to those obtained from the following traditional ranking techniques: dominance functions (DOM), simple average ranking (SAR) and utility functions (UTI).

As an example, the rankings for all the considered methods were applied to American countries and the comparisons are reported in Fig.7a when equal weights are given to all the criteria and in Fig.7b when proportional weights are given to the criteria.

All the methods agree to identify Costa Rica (CRI) as the most sustainable American country, while Guatemala (GTM), Haiti (HTI) and Bolivia (BOL) are in the last position for SAR, DOM, UTI and CPWR, respectively using equal weights for the indicators. With a proportional weighting scheme, all the methods locate Guyana (GUY) in the last position.

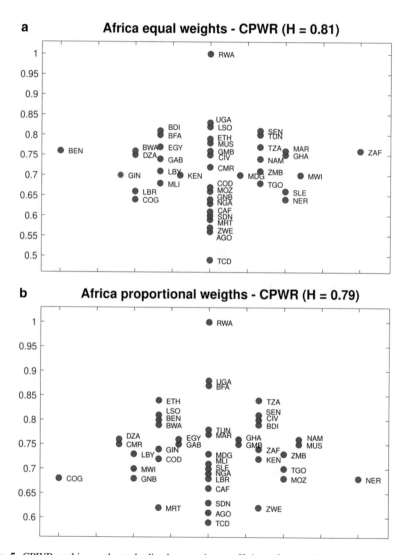

Fig. 5 CPWR ranking and standardized regression coefficients for equal (**a** and **c**) and proportional (**b** and **d**) weights initializations for Africa. CPWR values are represented on the vertical axis; H stands for the standardized entropy (the coefficient of determination R^2 is equal to 0.82 for the equal weights and 0.77 for the proportional ones)

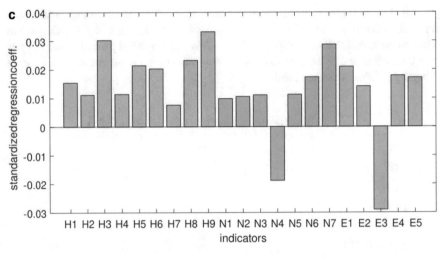

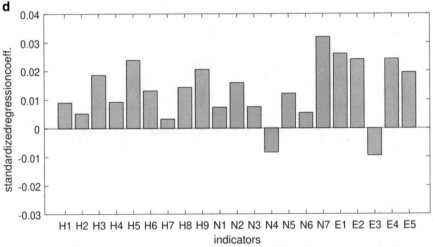

Fig. 5 (continued)

In general, it can be noted that, as expected, the different ranking methods show similar trends but not exactly coincide.

6 Conclusions

The Deep Ranking Analysis by Power Eigenvectors method was applied to 154 countries described by 21 indicators belonging to three wellbeing dimensions taken from the Sustainable Society Index (SSI) website. In order to study the effect of a different initialization of weights on the ranking we considered all the 21 indicators (1) with the equal weights and (2) with weights contribution proportional to the three wellbeing dimensions. Since the human wellbeing dimension has the greatest number of indicators, the equal approach is more oriented to human-based sustainability than the proportional approach which on the other hand offers a more balanced trade-off between human, environmental and economic

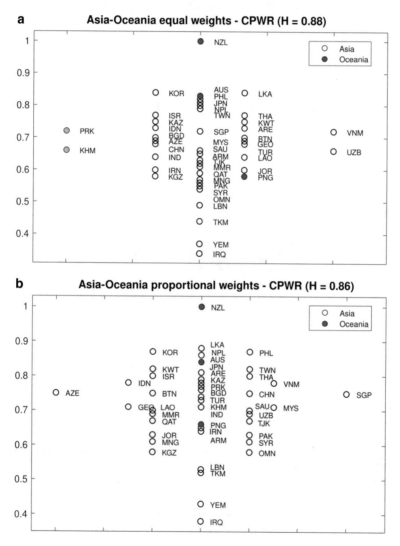

Fig. 6 CPWR ranking and standardized regression coefficients for equal (**a** and **c**) and proportional (**b** and **d**) weights initializations for Asia-Oceania. CPWR values are represented on the vertical axis; H stands for the standardized entropy (the coefficient of determination R^2 is equal to 0.95 for the equal weights and 0.94 for the proportional ones)

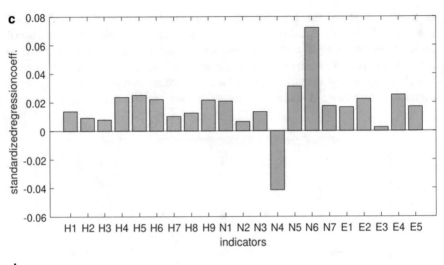

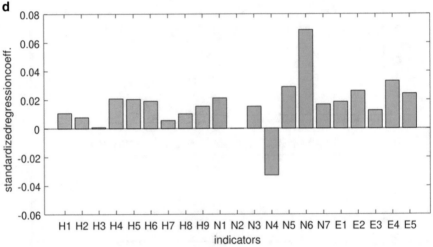

Fig. 6 (continued)

sustainability, accordingly to the considered indicators. The coefficients explaining the sustainability ranking provided by the *a-posteriori* retro-regression analysis take different values when single continents are considered. For instance, the healthy life indicator (H5) showed an opposite contribution to the ranking of the European and American countries. Furthermore, several indicators, which are almost irrelevant in some continents, become fundamental for the interpretation of the ranking in other continents.

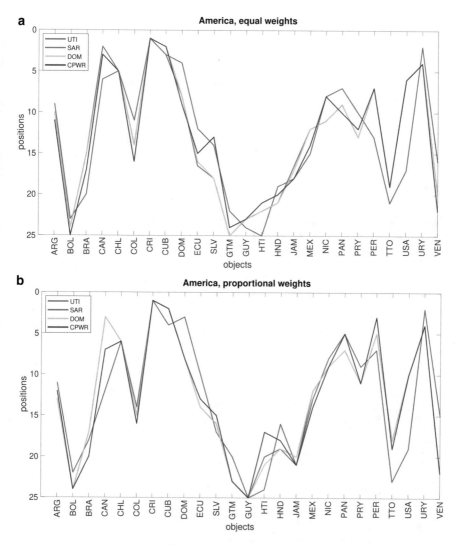

Fig. 7 Ranking positions according to dominance functions (DOM, yellow line), simple average ranking (SAR, orange line), utility functions (UTI, blue line) and DRAPE consensus PWR (CPWR, purple line), applied to American countries

In conclusion, this simple case study has shown the great ability of the DRAPE approach to combine several criteria in a single ranking that, being quantitative, allows very informative comparisons. In addition, the DRAPE approach has allowed to customize the ranking with the application of different weighting schemes and to interpret it by means of a retro-regression analysis.

A.1 Appendix 1

Country	ISO3	Country	ISO 3	Country	ISO 3	Country	ISO 3	Country	ISO 3
Albania	ALB	Congo, rep	COD	Iceland	ISL	Mali	MLI	Romania	ROU	United Kingdom	GBR
Algeria	DZA	Costa Rica	CRI	India	IND	Malta	MLT	Russian Federation	RUS	United States of America	USA
Angola	AGO	Côte d'Ivoire	CIV	Indonesia	IDN	Mauritania	MRT	Rwanda	RWA	Uruguay	URY
Argentina	ARG	Croatia	HRV	Iran	IRN	Mauritius	MUS	Saudi Arabia	SAU	Uzbekistan	UZB
Armenia	ARM	Cuba	CUB	Iraq	IRQ	Mexico	MEX	Senegal	SEN	Venezuela	VEN
Australia	AUS	Cyprus	CYP	Ireland	IRL	Moldova, republic of	MDA	Serbia	SRB	Vietnam	VNM
Austria	AUT	Czechia	CZE	Israel	ISR	Mongolia	MNG	Sierra Leone	SLE		
Azerbaijan	AZE	Denmark	DNK	Italy	ITA	Montenegro	MNE	Singapore	SGP		
Bangladesh	BGD	Dominican rep.	DOM	Jamaica	JAM	Morocco	MAR	Slovakia	SVK		
Belarus	BLR	Ecuador	ECU	Japan	JPN	Mozambique	MOZ	Slovenia	SVN		
Belgium	BEL	Egypt	EGY	Jordan	JOR	Myanmar	MMR	South Africa	ZAF		
Benin	BEN	El Salvador	SLV	Kazakhstan	KAZ	Namibia	NAM	Spain	ESP		
Bhutan	BTN	Estonia	EST	Kenya	KEN	Nepal	NPL	Sri Lanka	LKA		
Bolivia	BOL	Ethiopia	ETH	Korea north	PRK	Netherlands	NLD	Sudan	SDN		
Bosnia and Herzegovina	BIH	Finland	FIN	Korea south	KOR	New Zealand	NZL	Sweden	SWE		
Botswana	BWA	France	FRA	Kuwait	KWT	Nicaragua	NIC	Switzerland	CHE		
Brazil	BRA	Gabon	GAB	Kyrgyzstan	KGZ	Niger	NER	Syria	SYR		
Bulgaria	BGR	Gambia	GMB	Laos	LAO	Nigeria	NGA	Taiwan	TWN		
Burkina Faso	BFA	Georgia	GEO	Latvia	LVA	Norway	NOR	Tajikistan	TJK		
Burundi	BDI	Germany	DEU	Lebanon	LBN	Oman	OMN	Tanzania	TZA		
Cambodia	KHM	Ghana	GHA	Lesotho	LSO	Pakistan	PAK	Thailand	THA		
Cameroon	CMR	Greece	GRC	Liberia	LBR	Panama	PAN	Togo	TGO		
Canada	CAN	Guatemala	GTM	Libya	LBY	Papua New Guinea	PNG	Trinidad and Tobago	TTO		
Central African Republic	CAF	Guinea	GIN	Lithuania	LTU	Paraguay	PRY	Tunisia	TUN		
Chad	TCD	Guinea-Bissau	GNB	Luxembourg	LUX	Peru	PER	Turkey	TUR		
Chile	CHL	Guyana	GUY	North Macedonia	MKD	Philippines	PHL	Turkmenistan	TKM		
China	CHN	Haiti	HTI	Madagascar	MDG	Poland	POL	Uganda	UGA		
Colombia	COL	Honduras	HND	Malawi	MWI	Portugal	PRT	Ukraine	UKR		
Congo	COG	Hungary	HUN	Malaysia	MYS	Qatar	QAT	United Arab Emirates	ARE		

References

Ho, W., Xu, X., & Dey, P. K. (2010). Multi-criteria decision making approaches for supplier evaluation and selection: A literature review. *European Journal of Operational Research, 202,* 16–24. https://doi.org/10.1016/j.ejor.2009.05.009.

Ivlev, I., Vacek, J., & Kneppo, P. (2015). Multi-criteria decision analysis for supporting the selection of medical devices under uncertainty. *European Journal of Operational Research, 247,* 216–228. https://doi.org/10.1016/J.EJOR.2015.05.075.

Keener, J. P. (1993). Perron-Frobenius theorem and the ranking of football teams. *SIAM Review, 35,* 80–93. https://doi.org/10.1137/1035004.

Keller, H. R., Massart, D. L., & Brans, J. P. (1991). Multicriteria decision making: A case study. *Chemometrics and Intelligent Laboratory Systems, 11,* 175–189. https://doi.org/10.1016/0169-7439(91)80064-W.

Kendall, M. G. (1955). Further contributions to the theory of paired comparisons. *Biometrics, 11,* 43. https://doi.org/10.2307/3001479.

Kumar, A., Sah, B., Singh, A. R., et al. (2017). A review of multi criteria decision making (MCDM) towards sustainable renewable energy development. *Renewable and Sustainable Energy Reviews, 69,* 596–609. https://doi.org/10.1016/J.RSER.2016.11.191.

Pavan, M., & Todeschini, R. (2008a). Total-order ranking methods. In *Scientific data ranking methods: Theory and applications* (pp. 51–73). Elsevier Science.

Pavan, M., & Todeschini, R. (2008b). *Scientific data ranking methods : Theory and applications.* Elsevier Science.

Pavan, M., & Todeschini, R. (2009). Multicriteria decision-making methods. In *Comprehensive chemometrics* (pp. 591–629). Elsevier.

Ramanujacharyulu, C. (1964). Analysis of preferential experiments. *Psychometrika, 29,* 257–261. https://doi.org/10.1007/BF02289722.

SSI. (n.d.) SSI website. http://www.ssfindex.com. Accessed 10 Nov 2020.

Todeschini, R., Grisoni, F., & Nembri, S. (2015). Weighted power-weakness ratio for multi-criteria decision making. *Chemometrics and Intelligent Laboratory Systems, 146,* 329–336. https://doi.org/10.1016/j.chemolab.2015.06.005.

Todeschini, R., Grisoni, F., & Ballabio, D. (2019). Deep Ranking Analysis by Power Eigenvectors (DRAPE): A wizard for ranking and multi-criteria decision making. *Chemometrics and Intelligent Laboratory Systems, 191,* 129–137. https://doi.org/10.1016/j.chemolab.2019.06.005.

Triantaphyllou, E. (2000). Multi-criteria decision making. *Methods,* 5–21.

Valsecchi, C., Ballabio, D., Consonni, V., Todeschini, R. (2020). Deep Ranking Analysis by Power Eigenvectors (DRAPE): A polypharmacology case study. Chemometrics and Intelligent Laboratory Systems, 203, 104001. https://doi.org/10.1016/j.chemolab.2020.104001.

Van de Kerk, G., & Manuel, A. R. (2008). A comprehensive index for a sustainable society: The SSI – the sustainable society index. *Ecological Economics, 66,* 228–242. https://doi.org/10.1016/j.ecolecon.2008.01.029.

WCDE. (1987). WCED our common future, chair: Gro Harlem Brundtland. World Commission on Environment and Development.

Zimmermann, H.-J., & Gutsche, L. (1991). *Multi-criteria-analysis* (pp. 21–33).

Zionts, S., & Wallenius, K. (1976). Interactive programming method for solving the multiple criteria problem. *Management Science, 22,* 652–663. https://doi.org/10.1287/mnsc.22.6.652.

PyHasse, a Software Package for Applicational Studies of Partial Orderings

Rainer Bruggemann, Adalbert Kerber, Peter Koppatz, and Valentin Pratz

1 Introduction

Often a decision is to be found for a set of objects (synonym: options, alternatives) which are highly complex. Their complexity prevents that a single criterion, i.e., property is sufficiently characterizing these objects in a manner that a decision can be built on. Therefore, objects are to be characterized by a set of properties, which indicate as to how far the objects are matching with the criterion, relevant for the decision. In Bruggemann and Patil (2011) the term "Multi-indicator system was introduced, just to express that the complexity of objects has its counterpart in a whole set of indicators.

Once a set of indicators is found and a method established, as to how far these indicators can be quantified, the next problem is, how to distill a decision. This kind of question lead to many sophisticated Multi-criteria decision aid (MCDA) systems, as an example PROMETHEE may be mentioned (Brans and Vincke 1985). The field of MCDA is still increasing and a good overview can be found in Figueira et al. (2005).

R. Bruggemann (✉)
Leibniz-Institute of Freshwater Ecology and Inland Fisheries, Berlin, Germany
e-mail: brg_home@web.die

A. Kerber
Department of Mathematics, University of Bayreuth, Bayreuth, Germany
e-mail: Kerber@uni-bayreuth.de

P. Koppatz
TH Wildau Technical University of Applied Sciences Wildau, Wildau, Germany
e-mail: peter.koppatz@th-wildau.de

V. Pratz
University of Heidelberg, Gundelfingen, Germany
e-mail: pyhasse@valentinpratz.de

© The Author(s), under exclusive license to Springer Nature Switzerland AG 2021
R. Bruggemann et al. (eds.), *Measuring and Understanding Complex Phenomena*,
https://doi.org/10.1007/978-3-030-59683-5_18

Within the field of MCDA two problems arise:

(i) How to quantify the many parameters needed beyond the data matrix, and how to understand their roles.

(ii) The final quantity on which a decision should be based is often a mathematical complex combination of entries of the data matrix and other (supporting) parameters. Therefore it is difficult to trace back how a decision was found and what was the role of the different inputs of the MCDA.

Partial order theory can be a helpful tool. Although partial order is often seen as a MCDA too, its very idea is that only the entries of the data matrix should be considered. Often partial order leads to graphical representations, the so-called Hasse diagrams (see e.g. Bruggemann and Patil 2011). The examination of the Hasse diagrams lead to evaluation (finally to a decision) and to an exploration (finally a trace back identifying the role of the single indicators and of their values). Although both, evaluation and exploration are simply with respect to the mathematics need, the manual management can be very tedious and error-prone. This fact caused the development of software. Beside others, the software PyHasse was developed. This contribution is thought of as a brief introduction into PyHasse and its further development.

2   The Mathematical Basis of PyHasse

The idea behind the software package PyHasse was to support the interested researcher in the evaluation of a data matrix, consisting of several rows, (several objects) which are characterized by several indicators, the columns of the matrix. In that sense an object x is characterized by a set of indicator values, which is ordered and considered as a data profile for each object. Partial order comes into play, by a simultaneous analysis of the data profiles of the objects, i.e., by evaluating the central equation (of the Hasse diagram technique (HDT)):

$$x \leq y : \iff q(i, x) \leq q(i, y) \text{ for all indicators, } q(i), \text{ (syn.attributes)} \qquad (1)$$

The indicators characterize the objects x, y with respect to the criterion under which the decision is to be performed. We call X the set of objects. (For details about HDT, see e.g. Bruggemann et al. 2001; Bruggemann and Patil 2010, 2011; Newlin and Patil 2010; Patil and Taillie 2004).

If for any two objects x, y Eq. (1) does not hold, then it is said: x is incomparable with y, denoted in most mathematical papers as

$$x \parallel y.$$

Sets equipped with the order relation (1) are called partially ordered sets (abbr. posets), denoted as $(X, \leq)$. The graphical presentation of posets can be done by Hasse diagrams and is extremely useful, as long the graphs are not too complex.

The simple Eq. (1) has many facets, for example the notion of conflicts, or of co-monotony, or of separability etc. This fact requires algorithms, which are (as mentioned above) most often pretty simple, but awkward to perform manually. Correspondingly PyHasse includes

(a) graph-theoretical,
(b) order-theoretical,
(c) combinatorial and
(d) statistical aspects.

Due to the school of F. Wille (see Ganter and Wille 1996) one also can associate

(e) artificial intelligence with partial order theory, see Ganter and Wille 1996 and a more general approach, based on the concept of t-norms, (Bruggemann et al. 2011; Kerber 2017a, b; Bruggemann and Kerber 2018).

An example of (a) is given by disjoint subsets X1, X2 of X, for which is true:

$$x \in X1, y \in X2 \Rightarrow x \parallel y \qquad (2)$$

Subsets with the property (2) are called separated subsets. Usually objects belonging to different separated subsets are not connected in directed graphs (for details, see Bruggemann and Voigt 2011). The identification of such separated subsets is very useful in the sense of exploring the role of indicators and their values.

An example of (b) and (c) is the construction of a linear order, based on the linear extensions of a poset $(X, \leq)$. Linear extensions of a poset are linear orders which preserve the order found in the poset $(X, \leq)$. As an example, consider the object set $X = \{a, b, c\}$ for which Eq. 1 is analyzed with the following results: $a < b$, $a < c$. Nothing is said about the relation between b and c, because for b and c the Eq. 1 does not hold, then the sequences $(a < b < c)$ and $(a < c < b)$ represent the original poset whose order relations are preserved. The set of linear extensions can be further analyzed leading to many useful concepts (for details see Bruggemann and Carlsen 2011).

An example for (d) is once again given by the set of linear extensions, which can be evaluated by statistical tools; another example is the combination of cluster analysis with partial order, as shown in (Bruggemann and Carlsen 2014). Other examples are concerned with the problem of data noise (Bruggemann and Carlsen 2016) and those attempting to combine statistical proximity with a proximity concept, based on partial order theory (Bruggemann et al. 2014a).

3 Development of PyHasse

The software PyHasse is clearly not the only one, which may be useful in application of partial order. An important software package, PARSEC, is based on the statistical software R, see for details Arcagni 2017. The early development of software for the analysis of partially ordered setspartial order, for example of WHASSE was described by Bruggemann and Halfon 1995, Bücherl et al. 1995 and Halfon 2006.

The development of PyHasse began 2007. The programming language is Python. Python is an interpreter language and convinced the first author due to its clear programming style and powerful libraries, see for instance Weigend 2003, 2006, Von Löwis and Fischbeck, 2001. When the programming of PyHasse started, Python 2.6 was used. The name" PyHasse" arises from Python and from the usual output, a Hasse diagram. An overview is available, describing the state of 2014, see Bruggemann et al. 2014b,

The graphical user interface is based on Tkinter, a language derived from the TCl/Tk-system. Since around 2013 a PyHasse version was made available, which is accessible via Internet, a web-browser based system (Koppatz and Bruggemann 2017). In its very end it implemented a subset of PyHasse modules, accessible in a browser based application and should be made generally available for the mathematicians. This aim, however is coupled with a series of rules, which do not have directly to do with algorithmic aspects, but with the general readability and with conformity in the structure with other Python-open software. A predefined installation of all PyHasse modules is also available as virtual machines (VM). A virtual machine is a computer within a computer. A predefined sandbox is used to emulate hardware and deploying all software for a specific use case. The use of a virtual machine frees the user from extensive and complex installation steps, except for the installation of the virtual machine itself. Nowadays installing a virtual machine is an easy step.

The extra requirements for PyHasse running in the browser caused a very slow development, so actually only seven modules are available. In contrast, the "old" PyHasse version ("PyHasse conventional") is in a fast development, because this software system is not only thought of as useful for other interested scientists, but is also the main working tool for its developer, R. Bruggemann. Actually PyHasse conventional contains more than 140 programs (called "modules") which cover a wide range from general tasks to very special ones, according to the actual scientific research fields of its developer. An overview can be found in [Bruggemann et al. 2014b, where also PyHasse – conventional is structured in different classes of applications. Due to the difficulties indicated above in the development of PyHasse-online, an interim solution is planned (see below).

4   An Example of Application of PyHasse

In the following sections the application of both PyHasse software packages is demonstrated. We apply both variants of PyHasse to the generation of a Hasse diagram of eight Rhine regions, which are characterized by 4 indicators, describing the pollution of the herb layer by lead, cadmium, zinc and sulfur (in mg/kg dry mass). These indicators are denoted by the chemical symbols, i.e. Pb, Cd, Zn and S. The idea behind this monitoring study by the environmental protection agency of Baden-Württemberg [...] (in the following often abbreviated as "bawue") is to identify transport mechanisms.

4.1   Data Matrix

The basic assumption is a complete data matrix. Here the eight regions are characterized by the pollution of the herb layer by four indicators, therefore the data matrix has 8 rows (for the objects) and 4 columns (for the indicators). For application in PyHasse-conventional it is urgently required that the (0,0)-position is filled with a (dummy) text. Therefore the typical data matrix, suitable for both packages looks, as shown in Fig. 1.

The text "bawueks" is located in the (0,0)-position of the matrix. It is a good idea to use the "dummy" text for a brief information about the data matrix.

bawueks	Pb	Cd	Zn	S
01	1	0,04	21	1540
10	1	0,03	29	1780
24	1,7	0,18	39	1740
31	1,1	0,15	28	1740
19	0,8	0,01	18	4030
43	0,5	0,11	39	4030
52	2	0,23	36	4030
56	1	0,11	34	1970

Fig. 1 data matrix, see text, ks stands for German "Krautschicht" ("herblayer")

Here "bawueks" indicates that regions in the south-west of Germany, in Baden-Wuerttemberg are considered, and that the herb layer is the target of pollution. The columns are separated by tabs. The regions defining the rows are coded by 01, 10, 24, etc., the indicators are abbreviated by Pb, Cd, Zn and S for lead, cadmium, zinc and sulfur, resp.

There should be no gap, i.e. if m indicators and n objects are supposed, then all n*m entries must be filled by numbers, preferentially is the decimal separation the dot, however a comma as in Fig. 1 is accepted too.

4.2 PyHasse-Online

4.2.1 Overview

Here, the online version is considered. After selection of the web-site pyhasse.org the user sees, what is shown in Fig. 2.

In Fig. 2 Main parts of the website:

1. About PyHasse
2. Programs & Docs
3. References
4. Interview
5. Jupyter Notebooks
6. News

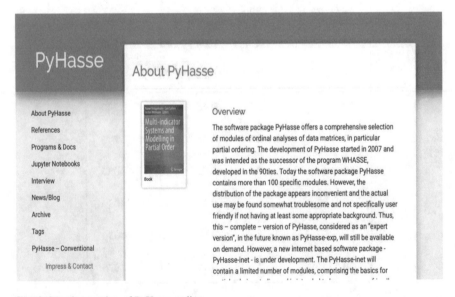

Fig. 2 First impression of PyHasse-online

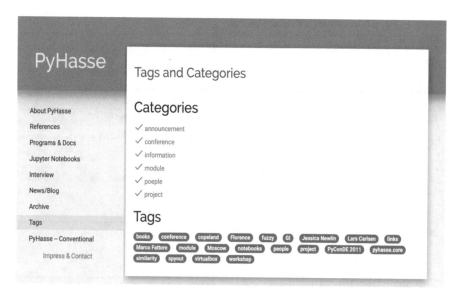

Fig. 3 Tags and Categories linking to all posts

7. Archive
8. Tags
9. PyHasse – Conventional

Ad 1.: "Programs & Docs": This web site informs technically about Python, virtual machines and the installation procedures. Furthermore, access to the seven modules, actually available in online. A table contains also links to background information and examples with basic introductions as Jupyter Notebooks. This is an additional step to document how to use PyHasse modules.

Ad 2: "About PyHasse and Interviews": presents some background information.

Ad 3: References: Some publications that reference to partial order theory and its applications (updated from time to time).

Ad 4: News, Archive and Tags: Adresses news about people, conferences, modules and other activities around PyHasse. All posts are tagged and accessable within a tag page also grouped by categories (Fig. 3).

In Table 1 the seven modules are briefly characterized (available in the last column of the table following the Link "Programs & Docs". Note, however, all seven modules available online have a uniform main menu, shown below:

- Home: More information, settings, etc. For example properties for HDT's (colors for nodes and labels, fonts,...) .
- Sets: Access to data matrices and their administration.
- General Info: Beside others, the Hasse diagram.

Table 1 The seven modules in PyHasse-online

Name	Main task	Remarks
Spyout	Generation of a Hasse diagram.	Basic module. Recommended for the beginner, and applied in this section
Antichain	Analysis of conflicts	Antichain: a subset of the set of objects, mutually incomparable
Chain	Generation of chains and their characterization	Chain: a subset of the set of objects, mutually comparable
Copeland	The well-known decision support system based on a concept of Copeland	
LPOM	Approximative generation of a weak order for the objects of the selected data matrix	
Fuzzy	Concept of De Walle et al. (1998), where the relations between two objects are evaluated by fuzzy techniques.	
Similarity	Two data matrices with the same objects, but with different indicator sets are compared	Currently in restructuring

- Module specific: Calculations depending specifically on the selected module. For example in the module chains, one has access to the set of objects, which are mutually comparable.
- Export: A still not fully implemented possibility to generate results, embeddable in other programs.

4.2.2 Application of Spyout on the Data Matrix, Shown in Fig. 1

Selection of the button "Spyout" leads to a user interface, shown in Fig. 4.

In Fig. 5 a part of Fig. 4 is shown.

First of all a set is to be selected, for which a partial order analysis is intended. This set should contain a data matrix (once again: objects are row, attributes, column defining). Therefore the button "SETS" is important. If no set is selected, a simple one is used. Here the user has s three possibilities to upload his own matrix. The most important one is to select a file from one of the user folders. After selection of the file and "submit" and activating the uploaded set, the user can get several information. Here the main purpose is to obtain the Hasse diagram, see Fig. 6.

There is a pretty good graphical editor, by which the user can modify the graph, shown in Fig. 6, however under preservation of the order relations. In Fig. 6 a good example for separated subsets can be found: $X1:=\{24, 31\}$ and $X2 = \{19, 43\}$, for no pair x,y with $x \in X1$ and $y \in X2$ a order relation, due to Eq. 1 can be found. This kind of separatedness indicates often special data structures. The knowledge of

Fig. 4 After selection of "Spyout"

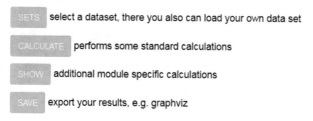

Fig. 5 Section out of Fig. 4

such data structures is helpful in tracing back, why an object has a certain position in comparison to others within a Hasse diagram.

4.3 Application of PyHasse-Conventional

4.3.1 Some Remarks to the First Steps

Actually the software package PyHasse-conventional, often and in the text below just called PyHasse, is now downloadable from the website pyhasse.org (following

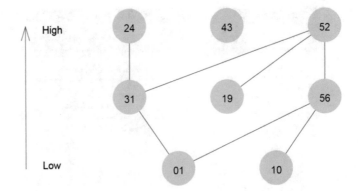

Fig. 6 Hasse diagram of the 8 regions along river Rhine. The partial order is defined by the values of the four indicators lead, cadmium, zinc and sulfur

the instructions in README.txt). The user has to install Python, version 2.6 or 2.7, then to store the five libraries

- rmod2: basic algorithms,
- raioop2: mainly user interactions, graphics.
- polib: mainly around drawing Hasse diagrams,
- pstat and stats: taken from the Internet and programmed by G. Strangman http://old.lwn.net/1998/1210/a/stats.py.html, in order to make standard statistical procedures accessible,

within the folder of Python, and finally he should define a folder for all the modules, help- and about-texts (on the CD under "RainerHasse"). Usually by the implementation of Python an icon "Idle" will be located at the desktop, from where the user can start his activity.

4.3.2 Application of hdsimpl

Some modules in PyHasse are so complicated that the developer (the first author) decided to program simpler versions, where only the major tasks are available. The module "hdsimpl" (actually version 04_1) is thought of as a beginners program, just to generate the Hasse diagram. Its user interface is shown in Fig. 7.

In most of the PyHasse-modules the sequence of activities is roughly given by the vertical arrangement of the buttons. The module hdsimpl follows this rule.

The first two buttons "about" and "help" obviously inform more closely about hdsimpl: "about" informs briefly about the module, whereas "help" is a tutorial and includes often references for supporting publications; most often with a more theoretical background.

Then a file with the wanted data matrix is to be selected "Select filename", the main button is "Draw HD (Hasse Diagram)", whereas the next ones are most

Fig. 7 User interface of
PyHasse, module
hdsimpl04_1.py

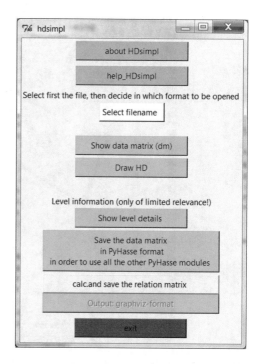

often not needed. However the user should close his calculations by using the "exit" button.

After pressing "Select filename" a browser is opened, and the user can select the data matrix he wants to be analyzed. In the case, considered here, it is the file with the name 0_BawueRheinks.txt.

After the selection of the filename, a window pops up, by which the format can be seleted, corresponding to the format, the data were stored. Usually the txt-file is to be selected, however other formats are tolerated too.

The next step is in most cases the main step: Draw HD. The result is shown in Fig. 8.

Clearly the Hasse diagram in Fig. 8 is the same as that of Fig. 6.

Beside the directed, triangle free graph of the Hasse diagram, an additional information is given in the header of the window:

$$\text{Utotal} = 18.$$

This informs about the content of the set $U = \{(x,y) \in X^2, \text{ with } x||y\}$. X is the set of objects, here of the 8 Rhine-regions. There are 18 pairs of objects, for which the HDT-equation is not fulfilled, i.e. some indicator values of for example region 31 have larger values than those of region 19, whereas there are some other indicators, whose values of 31 are less than those of 19.

Concerning the buttons, "cover relation" and "show equivalence class" the reader should visit the standard literature, for example Bruggemann and Patil 2011.

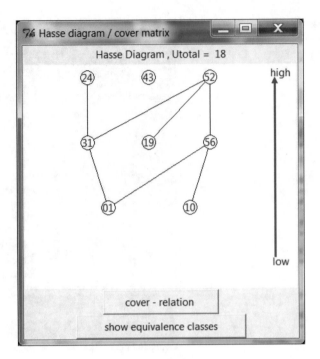

Fig. 8 Hasse diagram, obtained by applying hdsimpl, after selection of the data matrix, with 8 Rhine-regions and the four chemical pollution indicators

5   A Newer Example

In two papers (Grisoni et al. 2015; Todeschini et al. 2015) (see also a contribution in this book) a method is proposed and discussed in the International Conference on Partial Orders in Applied Sciences: "Towards an Understanding of Complex Phenomenon: Applying Partial Order Theory to Multi-Indicator Systems.", which is based on tournaments. The entries of the tournament matrix $t(i1,i2)$ ($i1$, $i2$ indicate the objects) are defined as follows:

$$t(i1, i2) = \Sigma\, w(j) * \delta(j, i1, i2)\,;\ w(j)\ \text{are suitable selected weights, with}\ \Sigma$$
$$w(j) = 1\ \text{and}\ w(j) \in [0, 1] \tag{3}$$

and

$$\delta(j, i1, i2) = \begin{cases} 1 \text{ if } q(i1, j) > q(i2, j) \\ 0.5 \text{ if } q(i1, j) = q(i2, j) \\ -1 \text{ if } q(i1, j) < q(i2, j) \end{cases} \tag{4}$$

Fig. 9 Graphical
user-interface of method 1 of
the group of Todeschini, see
Grisoni et al. 2015,
Todeschini et al. 2015

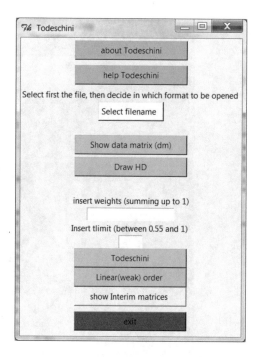

The entries t(I,j) in turn can be related to the zeta–matrix, i.e. to the adjacency matrix of the directed graph, describing the cover – relations of the partial order. Figure 9 shows the graphical interface.

The data of Rhine pollution are once again selected ("Select filename"), the Hasse diagram based on the original data matrix (see Fig. 1) ("Draw HD") is already shown (see Fig. 8). The specific part starts with the entry field ("inserts weights"), where (0.25, 0.25, 0.25, 0.25) is selected. By tlimit too large and too low entries of the tournament-matrix are filtered out. Here, for demonstration tlimit = 0.67 is selected, following the recommendation, that tlimit should be in the range [0.55,1]. Pressing the button "Todeschini" a Hasse diagram is obtained, resulting from

- weighting and
- cutting by tlimit.

There is no more an isolated element, and the number of incomparabilities is heavily reduced. In fact, the poset, presented in Fig. 10 is an enriched one of that, shown in Fig. 8. By the buttons "Linear (weak) order" a linear or weak order is obtained, following an idea originally proposed by Copeland and recently discussed by Al-Sharrah (2010), and by "show interim matrices" different matrices are shown, which may be useful for a closer inspection.

The new Pyhasse program has standard routines, taken from the libraries and only the specific part, related to the work of Todeschini's group had to be newly programmed.

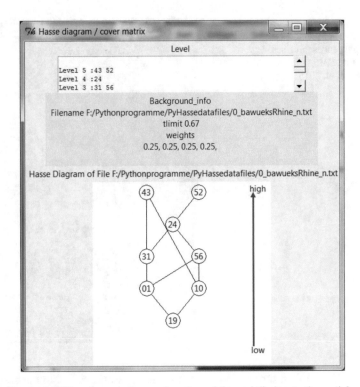

Fig. 10 Result of additional parameters, such as the weights and tlimit onto the partial order

6 Discussion

6.1 Three Pillars of PyHasse

Nowadays the PyHasse development has three pillars:

1. PyHasse – conventional: This pillar has the largest set of modules and is changing continuously. If there is a new and interesting idea, then it will be first programmed and made available for this pillar. A map of modules is hopeless, as in the meantime around 150 modules are available.
2. PyHasse-Internet: There are some few modules available. However following the pretty strict rules of the Python community there are to obey many rules, which are not only influencing the algorithmic structure but also the plain layout of a Python program. These rules are for sure necessary to guarantee a good documentation and some standards, however they slow down somewhat the development.
3. PyHasse via notebook. As there is still a gap between PyHasse-conentional and PyHasse-Online. Therefore the last years were characterized by the search for suitable software solutions and a standardized development and distribution of new modules. A solution is provided by the notebook concept.

6.2 Notebook

The notebook concept led to a new approach to software development and the use of all modules:

1. the use of a widespread software solution for scientific research – "Anaconda".
2. "Anaconda" is a Python-based software that provides good support for mathematical calculations, including the presentation of results.
3. Jupyter Notebooks are part of the software package "Anaconda", but can also installed without "Anaconda" in a plain virtual environment.

Jupyter Notebooks are an additional software variant to start investigating your own data. With a standard way to install python packages in a virtual environment "pip" or "conda" the investigation can be started. The website presents some sample data for each module to start fast. Later on it is possible to import other datasets and look for insights of this data.

This combination of documentation and application will contribute to a better understanding of both the calculation procedure and the handling of the software modules.

We plan the automatic generation of Jupyter Notebooks from the existing tests that are part of each PyHasse module. Later on also all graphical representations for results (present at the online versions) will be back.

With every Jupyter Notebook, all necessary calculation steps and comments are displayed in cells Fig. 11.

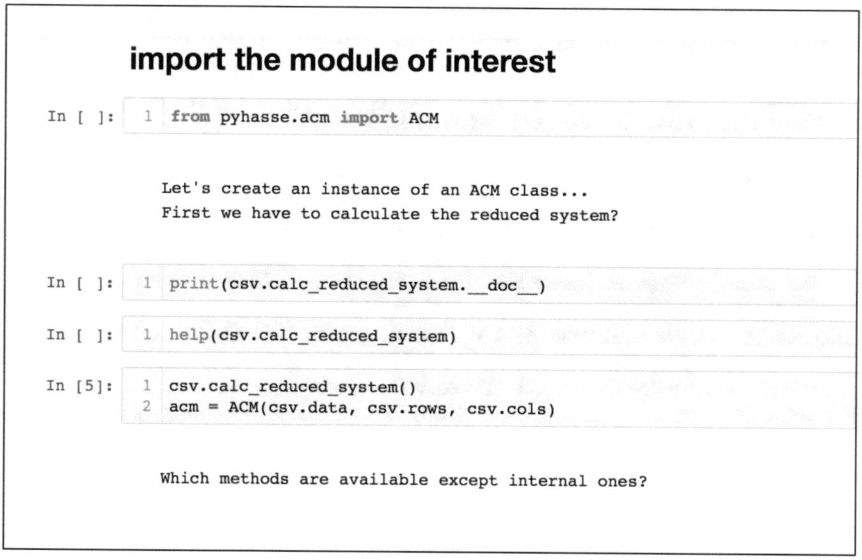

import the module of interest

```
In [ ]:    1  from pyhasse.acm import ACM


           Let's create an instance of an ACM class...
           First we have to calculate the reduced system?


In [ ]:    1  print(csv.calc_reduced_system.__doc__)

In [ ]:    1  help(csv.calc_reduced_system)

In [5]:    1  csv.calc_reduced_system()
           2  acm = ACM(csv.data, csv.rows, csv.cols)


           Which methods are available except internal ones?
```

Fig. 11 A screen shot of Jupyter

This allows us to document the usage of our own modules and the steps of a calculation from the import of the data to the calculation results.

References

Al-Sharrah, G. (2010). Ranking using the Copeland score: A comparsion with the Hasse diagram. *Journal of Chemical Information and Modeling, 50*, 785–791.

Arcagni, A. (2017). PARSEC: An R package for partial orders in socio-economics. In R. B. Marco Fattore (Ed.), *Partial order concepts in applied sciences* (pp. 275–289). Cham: Springer.

Brans, J. P., & Vincke, P. H. (1985). A preference ranking organisation method (the PROMETHEE method for multiple criteria decision – Making). *Management Science, 31*, 647–656.

Bruggemann, R., & Carlsen, L. (2011). An improved estimation of averaged ranks of partially orders. *MATCH Communications in Mathematical and in Computer Chemistry, 65*, 383–414.

Bruggemann, R., & Carlsen, L. (2014). Incomparable: What now II? Absorption of incomparabilities by a cluster method. *Quality & Quantity, 49*, 1633–1645.

Bruggemann, R., & Carlsen, L. (2016). An attempt to understand Noisy Posets. *MATCH Communications in Mathematical and in Computer Chemistry, 75*, 485–510.

Bruggemann, R., & Halfon, E. (1995). *Theoretical Base of the Program "Hasse"*. GSF-Bericht 20/95, München-Neuherberg.

Bruggemann, R., & Kerber, A. (2018). Fuzzy logic and partial order; first attempts with the new PyHasse-program L_eval. *MATCH Communications in Mathematical and in Computer Chemistry, 80*, 745–768.

Bruggemann, R., & Patil, G. P. (2010). Multicriteria prioritization and partial order in environmental sciences. *Environmental and Ecological Statistics, 17*, 383–410.

Bruggemann, R., & Patil, G. P. (2011). *Ranking and prioritization for multi-indicator systems – Introduction to partial order applications*. New York: Springer.

Bruggemann, R., & Voigt, K. (2011). A new tool to analyze partially ordered sets – Application: Ranking of polychlorinated biphenyls and alkanes/alkenes in river main, Germany. *MATCH Communications in Mathematical and in Computer Chemistry, 66*, 231–251.

Bruggemann, R., Halfon, E., Welzl, G., Voigt, K., & Steinberg, C. (2001). Applying the concept of partially ordered sets on the ranking of near-shore sediments by a battery of tests. *Journal of Chemical Information and Computer Sciences, 41*, 918–925.

Bruggemann, R., Kerber, A., & Restrepo, G. (2011). Ranking objects using fuzzy orders, with an application to refrigerants. *MATCH Communications in Mathematical and in Computer Chemistry, 66*, 581–603.

Bruggemann, R., Scherb, H., Schramm, K.-W., Cok, I., & Voigt, K. (2014a). CombiSimilarity, an innovative method to compare environmental and health data sets with different attribute sizes example: Eighteen Organochlorine pesticides in soil and human breast milk samples. *Ecotoxicology and Environmental Safety, 105*, 29–35.

Bruggemann, R., Carlsen, L., Voigt, K., & Wieland, R. (2014b). PyHasse software for partial order analysis. In R. Bruggemann, L. Carlsen, & J. Wittmann (Eds.), *Multi-indicator systems and modelling in partial order* (pp. 389–423). New York: Springer.

Bücherl, C., Bruggemann, R., & Halfon, E. (1995). *Hasse-Ein Programm zur Analyse von Hasse-Diagrammen*. GSF-Bericht 19/95, München-Neuherberg.

Figueira, J., Greco, S., & Ehrgott, M. (2005). *Multiple criteria decision analysis, state of the art surveys*. Boston: Springer.

Ganter, B., & Wille, R. (1996). *Formale Begriffsanalyse Mathematische Grundlagen*. Berlin: Springer.

Gary Strangman.: http://old.lwn.net/1998/1210/a/stats.py.html. Assessed 13 Oct 2018.

Grisoni, F., Consonni, V., Nembri, S., & Todeschini, R. (2015). How to weight Hasse matrices and reduce incomparabilities. *Chemometrics and Intelligent Laboratory Systems, 147*, 95–104.

Halfon, E. (2006). Hasse diagrams and software development. In R. Bruggemann & L. Carlsen (Eds.), *Partial order in environmental sciences and chemistry* (pp. 385–392). Berlin: Springer.

Kerber, A. (2017a). Evaluation and exploration, a problem-oriented approach. *Toxicological & Environmental Chemistry, 99*, 1270–1282.

Kerber, A. (2017b). Evaluation, considered as problem orientable mathematics over lattices. In R. B. Marco Fattore (Ed.), *Partial order concepts in applied sciences* (pp. 87–103). Cham: Springer.

Koppatz, P., & Bruggemann, R. (2017). PyHasse and cloud computing. In M. Fattore & R. Bruggemann (Eds.), *Partial order concepts in applied sciences* (pp. 291–300). Cham: Springer.

Newlin, J., & Patil, G. P. (2010). Application of partial order to stream channel assessment at bridge infrastructure for mitigation management. *Environmental and Ecological Statistics, 17*, 437–454.

Patil, G. P., & Taillie, C. (2004). Multiple indicators, partially ordered sets, and linear extensions: Multi-criterion ranking and prioritization. *Environmental and Ecological Statistics, 11*, 199–228.

Todeschini, R., Grisoni, F., & Nembri, S. (2015). Weighted power-weakness ratio for multicriteria decision making. *Chemometrics and Intelligent Laboratory Systems, 146*, 329–336.

Von Löwis, M., & Fischbeck, N. (2001). *Python 2 – Einführung und Referenz der objektorientierten Skriptsprache*. München: Addison-Wesley.

Van de Walle, B., De Baets, B., & Kerre, E. (1998). Characterizable fuzzy preference structures. *Annals of Operations Research, 80*, 105–136

Weigend, M. (2003). *Python – Ge-packt*. Bonn: mitp-Verlag.

Weigend, M. (2006). *Objektorientierte Programmierung mit Python*. Bonn: mitp-Verlag.

Index

Printed in the United States
by Baker & Taylor Publisher Services